PRINCIPLES AND TECHNIQUES

OF

APPLIED MATHEMATICS

PRINCIPLES AND TECHNIQUES OF APPLIED MATHEMATICS

BERNARD FRIEDMAN

New York University

DOVER PUBLICATIONS, INC.

NEW YORK

Published in Canada by General Publishing Company, Ltd., 30 Lesmill Road, Don Mills, Toronto, Ontario.

Published in the United Kingdom by Constable and Company, Ltd., 3 The Lanchesters, 162–164 Fulham Palace Road, London W6 9ER.

This Dover edition, first published in 1990, is an unabridged and unaltered republication of the work first published by John Wiley & Sons, Inc., New York, in 1956.

Manufactured in the United States of America
Dover Publications, Inc., 31 East 2nd Street, Mineola, N.Y. 11501

Library of Congress Cataloging-in-Publication Data

Friedman, Bernard, 1915–1966.
Principles and techniques of applied mathematics / Bernard Friedman.
p. cm.
Reprint. Originally published: New York : Wiley, 1956.
Includes bibliographical references and index.
ISBN 0-486-66444-9
1. Mathematics. I. Title
QA37.2.F75 1990
510—dc20 90-43685
CIP

PREFACE

For many years the gap between pure mathematics and applied mathematics has steadily widened. On the one hand, the pure mathematicians are considering structures and systems which are becoming ever more abstract and general; on the other hand, the applied mathematicians are studying concrete and specific problems. It is well known, of course, that this gap between the two groups is really illusory, that the study of abstract systems can help in the solution of concrete problems, and that the study of specific problems may suggest interesting generalizations for the pure mathematicians.

This book was written in an attempt to show how the powerful methods developed by the abstract studies can be used to systematize the methods and techniques for solving problems in applied mathematics. Such a systematic treatment requires a great deal of preparation by the student; consequently, more than half of the book is devoted to a study of abstract linear spaces and of operators defined on such spaces. However, in this treatment, the emphasis is not on the abstract theory but on the techniques which can be derived from this theory to solve specific problems. For example, Chapter 3 presents the elements of Laurent Schwartz's "Theory of Distributions" in a form which should make it more accessible to the people who would use it—the applied mathematicians, the physicists, and the engineers.

An introductory book on applied mathematics, such as this one, can by its very nature contain little that is new or original. However, it is believed that some of the techniques presented here, such as those for solving integral equations, for finding the Green's function for ordinary or partial differential equations, and for finding the spectral representation of ordinary differential operators, may be relatively unfamiliar to the general reader. The development and exposition of these techniques are the main purpose of this book.

As far as possible, I have attempted to present the subject so as to lay stress upon the ideas and not upon the minutiae of the proofs; consequently, many details, illustrations, and extensions of the text have been put into problems and appendices. It is recommended that the reader study the problems as well as the text in order to get a more complete knowledge of the subject.

A few words of explanation for the changes in notation and nomenclature should be given. The scalar product of two vectors x and y is denoted by $\langle x, y\rangle$ instead of the more conventional (x, y). This notation, which is a slight modification of that used by Dirac, has the advantage of not overworking the parenthesis. The term *Green's function* customarily is used to represent the kernel of the integral operator which inverts an ordinary or partial differential operator but only when the domain of the operator consists of functions which are zero on the boundary. Here the term is used also where the domain consists of functions which satisfy any linear homogeneous, not necessarily zero, boundary conditions. This usage, which is common among physicists, has many advantages to recommend it.

The subject matter of this book has been presented for several years as a one-year course in the Graduate School of New York University. The prerequisites for the course are a knowledge of linear algebra and complex integration.

Part of the work for this book was done on a research contract with the Air Force Cambridge Research Center. I wish to thank them for their support.

I wish to acknowledge with thanks the help and encouragement I received from colleagues and students at New York University. The following should be particularly mentioned: Professor R. Courant, Professor K. O. Friedrichs, Professor M. Kline, Professor W. Magnus, Professor N. Marcuvitz, and Mr. B. Levy. A special word of thanks is also due my wife for her help in editing, proof-reading, and indexing the manuscript.

BERNARD FRIEDMAN

March 1956

CONTENTS

1 LINEAR SPACES

2 SPECTRAL THEORY OF OPERATORS

1

LINEAR SPACES

Introduction

Many of the ideas and techniques used in applied mathematics to solve linear problems are generalizations of the ideas and techniques used in algebra to solve simultaneous linear equations. Such generalizations arise from the study of linear spaces. The theory of linear spaces is an extension of the theory of three-dimensional vector analysis. Subsequent sections will show that the theory of linear spaces includes as special cases n-dimensional Euclidean space, E_n ($n = 1, 2, 3, \cdots$), infinite-dimensional Euclidean space, E_∞, function spaces, etc. As an introduction to the study of linear spaces, let us consider the following example of three simultaneous linear equations.

Find the values of x_1, x_2, x_3 such that

$$\begin{aligned} a_{11}x_1 + a_{12}x_2 + a_{13}x_3 &= b_1, \\ a_{21}x_1 + a_{22}x_2 + a_{23}x_3 &= b_2, \\ a_{31}x_1 + a_{32}x_2 + a_{33}x_3 &= b_3. \end{aligned} \tag{1.1}$$

It is well known that the set of equations (1.1) will, in general, not have a solution if there exists a non-zero solution of the corresponding set of homogeneous equations; that is, the set of equations (1.1) when $b_1 = b_2 = b_3 = 0$. Suppose that $x_1 = x_1'$, $x_2 = x_2'$, $x_3 = x_3'$ is one solution of these homogeneous equations and that $x_1 = x_1''$, $x_2 = x_2''$, $x_3 = x_3''$ is another solution; then

$$\begin{aligned} x_1 &= \alpha x_1' + \beta x_1'', \\ x_2 &= \alpha x_2' + \beta x_2'', \\ x_3 &= \alpha x_3' + \beta x_3'', \end{aligned} \tag{1.2}$$

where α and β are arbitrary constants, is also a solution of the homogeneous equation.

We may express this fact in an interesting geometrical way. Let x' denote a vector with components x_1', x_2', x_3' and let x'' denote a vector with components x_1'', x_2'', x_3''; then the vector whose components are given by (1.2) is a vector in the plane determined by the vectors x' and x''. We see that if x' and x'' are vectors whose components are the solutions of the homogeneous equations corresponding to (1.1), the components of any vector in the plane determined by x' and x'' will also be a solution of the same homogeneous equations.

This geometric language, which is so intuitive and suggestive, can be extended to discuss more complicated linear problems. For example, consider the problem of solving n linear homogeneous equations in n unknowns $x_1, x_2, \cdots, x_n$. Again, if the set of numbers $x'_1, x'_2, \cdots, x'_n$ is one solution of the equations and the set $x''_1, x''_2, \cdots, x''_n$ is a second solution, then the set

$$\begin{aligned} x_1 &= \alpha x'_1 + \beta x''_1, \\ x_2 &= \alpha x'_2 + \beta x''_2, \\ &\cdots\cdots \\ x_n &= \alpha x'_n + \beta x''_n, \end{aligned}$$

is also a solution. If a vector x is now defined as a set of n numbers $x_1, x_2, \cdots, x_n$, we have the following geometrical result.

Let x' and x'' be vectors whose components are a solution of the homogeneous equations; then the components of any vector in the plane determined by x' and x'' will also be a solution of the same homogeneous equations.

The fact that we obtain the same result in different cases indicates the usefulness of this geometric approach. In this chapter we shall present this geometric viewpoint in a general form by means of axioms. The results we thus derive will be applicable to all situations for which the axioms hold.

Linear Vector Spaces

In the preceding discussion we defined vectors by means of its components. This approach is restrictive since it necessitates the introduction of a definite coordinate system to which the components refer. We know, however, that in three dimensions, at least, vectors are geometric objects existing independently of any coordinate system.

We shall now set up a framework which will be sufficiently general to contain the possible extensions of the concept of vector. Consider a collection $\mathcal{S}$ of elements which we denote by small Latin latters x, y, z, a, b, $\cdots$. Suppose that an operation, which we shall call addition and denote by $+$, is defined on the objects of $\mathcal{S}$. This operation should have the following properties:

(1) Any two elements x and y in $\mathcal{S}$ may be added, and the result is an element z in $\mathcal{S}$. We write $x + y = z$.

(2) The operation is commutative and associative, that is,

$$x + y = y + x,$$
$$(x + y) + w = x + (y + w).$$

(3) $\mathcal{S}$ contains a unique element 0, called the null or zero element, such that for x in $\mathcal{S}$,

$$x + 0 = x.$$

(4) For any x in $\mathcal{S}$, there exists an element, which we denote by $-x$, such that

$$x + (-x) = 0.$$

These concepts may be understood by considering $\mathcal{S}$ as the set of vectors in n-dimensional Euclidean space E_n. Any vector x of this space is a set of n real numbers,

$$x = (\xi_1, \xi_2, \cdots, \xi_n).$$

The operation of addition considered above corresponds to the customary addition of vectors which is defined as follows:

Let

$$y = (\eta_1, \eta_2, \cdots, \eta_n);$$

then

$$x + y = (\xi_1 + \eta_1, \xi_2 + \eta_2, \cdots, \xi_n + \eta_n).$$

If we define

$$0 = (0, 0, \cdots, 0)$$

and

$$-x = (-\xi_1, -\xi_2, \cdots, -\xi_n),$$

it is easy to see that these definitions satisfy properties 1 to 4 specified above.

The set of vectors in E_n, however, has some other properties. For example, if x is a vector in E_n, then $2x$, that is, the vector whose components are twice as large as the components of the vector x, is also in E_n. We shall now express this property in abstract terms.

Let the Greek letters $\alpha, \beta \cdots$ denote the numbers of some field.† We shall call these numbers *scalars*. We assume that to any scalar α and any vector x there is defined a vector written αx which we call the product of the vector x by the scalar α. The definition of the product should have the following properties:

$$\begin{aligned} \alpha(\beta x) &= (\alpha\beta)x, \\ (\alpha + \beta)x &= \alpha x + \beta x, \\ \alpha(x + y) &= \alpha x + \alpha y \\ 1 \cdot x &= x. \end{aligned} \tag{1.3}$$

† A field is a collection of numbers which contains the sum, difference, product, and quotient of any two numbers in the field. Of course, division by zero is excluded. The set of all rational numbers, the set of real numbers, and the set of complex numbers are examples of fields.

Here α and β are arbitrary scalars, and x and y are arbitrary vectors in $\mathcal{S}$.

In E_n the result of multiplying a vector by a scalar may be defined as the vector obtained by multiplying each of its components by the scalar. This definition will clearly have the properties (1.3).

Any space $\mathcal{S}$ which is closed under the operations of addition and of multiplication by a scalar is called a *linear vector space*, and its elements are called vectors. For example, E_n $(n = 1, 2, \cdots)$ is a linear vector space.

The concepts of an n-dimensional vector space may be extended to give the concept of an infinite-dimensional vector space E_∞. In E_n, a vector x was a set of n real numbers:

$$x = (\xi_1, \xi_2, \cdots, \xi_n);$$

in E_∞, a vector x is a countably infinite set of real numbers:

$$x = (\xi_1, \xi_2, \cdots).$$

The rule for addition in E_∞ is the expected one, namely, if x is defined as above and if

$$y = (\eta_1, \eta_2, \cdots),$$

then

$$x + y = (\xi_1 + \eta_1, \xi_2 + \eta_2, \cdots).$$

The product of x by a scalar α is the vector

$$\alpha x = (\alpha\xi_1, \alpha\xi_2, \cdots).$$

With these definitions it is clear that E_∞ is a linear vector space.

Besides these simple examples of vector spaces, there are many others. For example, all functions $f(t)$ continuous on the interval $0 \leq t \leq 1$ form a linear vector space with the function $f(t)$ considered as a vector. The components of the vector would be the values of the function at different points of the interval. The sum of two vectors $f(t)$ and $g(t)$ is the function $h(t)$, whose values are the sum of the values of $f(t)$ and $g(t)$. The result of multiplying a vector $f(t)$ by a scalar α is the function whose values are α times the values of $f(t)$.

Another example of a vector space which will be important in the study of differential and integral equations is the space of all real-valued functions $g(t)$ such that $g(t)^2$ is Lebesgue integrable over the interval (0, 1). We shall denote this space by $\mathcal{L}_2$. Addition of vectors and multiplication by a scalar are defined in this space in the same way as in the space of continuous functions.

Scalar Product in E_n and E_∞

There is one important concept in vector analysis which has not been used so far, that is the concept of the scalar product of two vectors. If x and y are vectors in three-dimensional space, their scalar product, which we shall write $\langle x, y\rangle$, is the sum of the products of corresponding components of the two vectors. Using the concept of the scalar product, we may define the length of x as the positive square root of the scalar product of x with itself. Also, two vectors are mutually perpendicular or *orthogonal* if and only if their scalar product is zero.

These ideas may be readily extended to the spaces E_n and E_∞. In E_n the scalar product of the vectors x and y is defined as follows:

$$\langle x, y\rangle = \xi_1\eta_1 + \xi_2\eta_2 + \cdots + \xi_n\eta_n, \tag{1.4}$$

whereas in E_∞ the scalar product is defined by the following infinite series:

$$\langle x, y\rangle = \xi_1\eta_1 + \xi_2\eta_2 + \cdots. \tag{1.5}$$

This definition applies only when the infinite series converges.

Just as in three-dimensional space, we may define the length of a vector in E_n or E_∞ as the positive square root of the scalar product of x with itself. If we write the length of x as $|x|$, we have in E_n

$$|x| = (\xi_1^2 + \xi_2^2 + \cdots + \xi_n^2)^{1/2},$$

whereas in E_∞

$$|x| = (\xi_1^2 + \xi_2^2 + \cdots)^{1/2}. \tag{1.6}$$

Again, this last definition applies only when the infinite series converges. Note that we shall also use $|\xi_1|$ to mean the absolute value of the scalar ξ_1.

If the infinite series in (1.6) converges, we shall say that the vector x has finite length; otherwise, the vector x has infinite length. Henceforth, we shall restrict E_∞ to be the space of all vectors with finite length, that is, E_∞ contains only those vectors

$$x = (\xi_1, \xi_2, \cdots)$$

such that the infinite series $\xi_1^2 + \xi_2^2 + \cdots$ converges. Of course, now it is no longer obvious that E_∞ is a linear vector space since, if both x and y have finite length, we cannot be sure that $\alpha x + \beta y$ also has finite length. The fact that E_∞ is a linear vector space will be proved in Problem 1.3. We note that Problem 1.2 proves that, if x and y have finite length, the definition (1.5) applies; consequently, the scalar product is defined for any two vectors in E_∞.

PROBLEMS

1.1. Prove that in E_n and E_∞ (assuming all series converge):

$$\langle \alpha x, y\rangle = \langle x, \alpha y\rangle = \alpha\langle x, y\rangle,$$

$$\langle x_1 + x_2, y\rangle = \langle x_1, y\rangle + \langle x_2, y\rangle,$$

$$|\alpha x + \beta y|^2 = \langle \alpha x + \beta y, \alpha x + \beta y\rangle = \alpha^2\langle x, x\rangle + 2\alpha\beta\langle x, y\rangle + \beta^2\langle y, y\rangle.$$

1.2. Prove that, if x and y have finite length in E_∞, the scalar product of x and y has a meaning and we have

$$|\langle x,y\rangle| \leq |x| \cdot |y|.$$

This result is known as the Cauchy-Schwarz inequality. (*Hint.* The square of the length of the vector $\alpha x_n + \beta y$ is non-negative for all values of α and β. Put

$$x_n = (\xi_1, \xi_2, \cdots, \xi_n, 0, 0, \cdots),$$
$$\alpha = |y|^2, \qquad \beta = \langle x_n, y\rangle$$

and use the last result of Problem 1.1 to show that

$$|\langle x_n, y\rangle| \leq |x_n| \cdot |y|.)$$

1.3. Prove that E_∞ is a linear vector space. (*Hint.* Use Problem 1.2 to show that $\alpha x + \beta y$ has finite length if x and y have finite length.)

1.4. Consider an n-dimensional complex space, $\bar{E}_n$, in which a vector x is a set of n complex numbers and for which the scalar product of x and y is defined by the formula

$$\langle x,y\rangle = \bar{\xi}_1\eta_1 + \bar{\xi}_2\eta_2 + \cdots + \bar{\xi}_n\eta_n.$$

Here, the bar denotes the complex conjugate. Show that

$$\langle x, y\rangle = \overline{\langle y, x\rangle},$$
$$\langle \alpha x, y\rangle = \bar{\alpha}\langle x, y\rangle,$$
$$|\langle x, y\rangle| \leq |x| \cdot |y|.$$

1.5. Consider an ∞-dimensional complex space, $\bar{E}_\infty$, in which a vector x is a set of countable infinite complex numbers and for which the scalar product of x and y is

$$\langle x, y\rangle = \bar{\xi}_1\eta_1 + \bar{\xi}_2\eta_2 + \cdots.$$

Show that if x and y have finite length, the sum defining (x, y) converges and

$$|\langle x, y\rangle| \leq |x| \cdot |y|.$$

Scalar Product in Abstract Spaces

In an abstract space $\mathcal{S}$, a *scalar product* is a scalar-valued function of two vectors x and y, written $\langle x, y\rangle$, such that

$$\langle x, y\rangle = \langle y, x\rangle, \tag{1.7}$$
$$\langle \alpha_1 x_1 + \alpha_2 x_2, y\rangle = \alpha_1\langle x_1, y\rangle + \alpha_2\langle x_2, y\rangle,$$

and such that

$$\langle x, x\rangle > 0 \tag{1.8}$$

if x is not the zero vector.

Equations (1.4) and (1.5) are definitions of the scalar product in E_n and E_∞, respectively. In $\mathcal{L}_2$ the scalar product of two vectors f and g is defined to be

$$\langle f, g\rangle = \int_0^1 f(t)g(t)\,dt.$$

By means of the scalar product, the *length* of a vector is defined as follows:

$$|x| = +\sqrt{\langle x, x\rangle}.$$

Two vectors, x and y, are said to be *orthogonal* or perpendicular if

$$\langle x, y\rangle = 0.$$

This definition agrees with that used in three dimensions.

The scalar product that has been characterized by properties (1.7) and (1.8) is the appropriate one for a vector space having the field of real numbers as the scalar multipliers. We shall find it convenient, sometimes, to use complex numbers as scalar multipliers, and we shall assume that the properties (1.3) of scalar multiplication are still valid. The space $\mathcal{S}$ will be extended to a space $\bar{\mathcal{S}}$ which contains vectors whose components are complex numbers, and properties (1.7) of the scalar product will be assumed to hold. However, the extended scalar product will no longer be real, and the length may be zero for a non-zero vector. For example, in E_2 the vector with components $(1, i)$ has zero length.

This difficulty of non-zero vectors having zero length may be removed by defining in $\bar{\mathcal{S}}$ a complex-type scalar product which has the following properties:

$$\begin{aligned} &\langle x, y\rangle = \overline{\langle y, x\rangle}, \\ (1.9) \qquad &\langle \alpha_1 x_1 + \alpha_2 x_2, y\rangle = \bar{\alpha}_1\langle x_1, y\rangle + \bar{\alpha}_2\langle x_2, y\rangle, \\ &\langle x, x\rangle > 0, \text{ if } x \neq 0. \end{aligned}$$

In Problems 1.4 and 1.5, we have defined for the spaces $\bar{E}_n$ and $\bar{E}_\infty$ scalar products which satisfy (1.9). In $\bar{\mathcal{L}}_2$, the space of all complex-valued functions such that

$$\int_0^1 |f(t)|^2\, dt < \infty,$$

the scalar product of $f(t)$ and $g(t)$ is defined by

$$\langle f, g\rangle = \int_0^1 \overline{f(t)}g(t)\, dt.$$

This type of scalar product is the one used in quantum mechanics and the theory of Hilbert spaces.†

Since in most of our work the results will be real numbers, and because it is tedious to use a complex notation which in the end is not necessary, we shall hereafter use the real-type scalar product, unless otherwise specified. However, we must permit all complex numbers to be used as scalar multipliers in order that every scalar polynomial equation have a root. Consequently, some non-zero vectors will have zero length; therefore, the term "length" will be used only for vectors over the real field. With this restriction the results obtained with the two types of scalar product do not differ much. Most of the results obtained with one can

† See Stone, *Linear Transformations in Hilbert Space and Their Applications to Analysis*, American Mathematical Society, New York, 1932.

be suitably modified to cover the other. In the problems we shall occasionally discuss some of the necessary modifications.

To summarize: We shall use a linear vector space with the field of complex numbers as scalar multipliers, but we shall use a real-type scalar product.

PROBLEMS

1.6. Prove the results of Problem 1.1 using the abstract definition of the real-type scalar product. Also, for a vector space over the field of real numbers prove the *Schwarz inequality*:

$$|\langle x, y\rangle| \leq |x| \cdot |y|.$$

(*Hint.* In the third result of Problem 1.1 put $\alpha = -\langle x, y\rangle$, $\beta = \langle x, x\rangle$.) Work out the corresponding results for a complex-type scalar product.

1.7. Find the value of α which makes $|x - \alpha y|$ a minimum. Show that for this value of α, the vector $x - \alpha y$ is orthogonal to y and that

$$|x - \alpha y|^2 + |\alpha y|^2 = |x|^2.$$

The vector αy is called the *projection* of x on y. Draw a diagram in E_2 to see the reason for this name. Consider both types of scalar product.

1.8. Prove that

(*a*) $\qquad |x + y|^2 + |x - y|^2 = 2|x|^2 + 2|y|^2,$

(*b*) $\qquad |x + y| \leq |x| + |y|,$

(*c*) $\qquad |x + y|^2 = |x|^2 + |y|^2$ if and only if $\langle x, y\rangle = 0$.

For the real-type scalar product the triangle inequality (*b*) has meaning only if x and y are vectors over the real field.

1.9. Consider the linear vector space of real continuous functions with continuous first derivatives in the closed interval (0, 1). Which of the following defines a scalar product?

$$\langle f, g\rangle = \int_0^1 f'(t)g'(t)\,dt + f(0)g(0) \quad \text{or} \quad \langle f, g\rangle = \int_0^1 f'(t)g'(t)\,dt.$$

Convergence and Complete Spaces

A sequence of vectors $x_1, x_2, \cdots$ in $\mathcal{S}$ is said to *converge* to a vector x in $\mathcal{S}$ if, given $\varepsilon > 0$, there exists an integer $N = N(\varepsilon)$ such that

$$|x - x_n| < \varepsilon \tag{1.10}$$

for all $n > N$. The vector x is called the *limit* of the sequence, and we write

$$x = \lim x_n.$$

An infinite sum of vectors

$$y_1 + y_2 + \cdots$$

will be said to *converge to a sum* x if the partial sums

$$x_n = y_1 + y_2 + \cdots + y_n$$

converge to a limit x. In a finite-dimensional Euclidean space E_m, (1.10) implies that each component of the sequence of vectors x_n converges in the ordinary sense to the corresponding component of the limit vector x.

By means of the triangle inequality (Problem 1.8*b*), it is easily proved that if x_n converges to x, then, given $\varepsilon > 0$, there exists an integer N_1 such that

$$|x_n - x_m| < \varepsilon \tag{1.11}$$

for all integers n and m greater than N_1. A sequence of vectors for which (1.11) holds is said to *converge in the Cauchy sense.*

It is well known that if a sequence of vectors in E_n converges in the Cauchy sense, it will converge in the sense of (1.10). However, in a general vector space this result need no longer be true since the limit vector to which the sequence seems to be converging need not belong to the space. For example, consider the linear vector space of all functions $f(t)$ continuous on the closed interval (0, 1). We use the scalar product

$$\langle f, g\rangle = \int_0^1 f(t)g(t)\, dt.$$

Now consider the following sequence of continuous functions:

$$\begin{aligned} f_n(t) &= 0, & 0 \le t \le \frac{1}{2} + \frac{1}{2n} \\ &= n\left(t - \frac{1}{2}\right) + \frac{1}{2}, & \frac{1}{2} - \frac{1}{2n} \le t \le \frac{1}{2} + \frac{1}{2n} \\ &= 1, & \frac{1}{2} + \frac{1}{2n} \le t \le 1 \end{aligned}$$

for $n = 1, 2, \cdots$. It is easy to show that these functions converge in the Cauchy sense; that is, given $\varepsilon > 0$, there exists an integer N_1 such that

$$\int_0^1 (f_n - f_m)^2\, dt < \varepsilon$$

for all n and m greater than N_1. However, the limit of this sequence is the function

$$\begin{aligned} f(t) &= 0, \; 0 \le x \le \tfrac{1}{2} \\ &= 1, \; \tfrac{1}{2} \le x \le 1, \end{aligned}$$

which is not continuous and therefore does not belong to the space considered.

The natural way to remove this difficulty is to extend the space of continuous functions so that it will contain the limit function. If this extension is done for all Cauchy sequences, we shall obtain $\mathcal{L}_2$, the space of functions whose square is integrable in the sense of Lebesgue.† Suppose that we start with $\mathcal{L}_2$ and consider sequences that converge in the Cauchy sense; we may ask whether we will have to extend the space still further to take care of limits of such sequences. The answer is no. The space $\mathcal{L}_2$ is *complete*; that is, whenever a sequence $f_n(t)$ in $\mathcal{L}_2$ converges in the Cauchy sense, there exists a function $f(t)$ in $\mathcal{L}_2$ such that $f_n(t)$ converges to $f(t)$.

We shall henceforth assume that our linear vector spaces are complete. If we start with a space that is not complete, it may always be extended to a complete space by a process similar to that which Cantor used to extend the set of rational numbers to the set of all real numbers, irrational as well as rational.‡

A complete linear vector space with a complex-type scalar product is called a *Hilbert space*.

Linear Manifolds and Subspaces

In the introductory section we showed that any vector in the plane determined by two independent vector solutions of a set of linear homogeneous equations was also a solution of those equations. This result will be true in the general linear vector spaces also if we define in such a space the concepts of independence of vectors and of the plane determined by two vectors.

A set of vectors $x_1, x_2, \cdots, x_m$ is called *linearly dependent* if there exist scalars $\alpha_1, \alpha_2, \cdots, \alpha_m$, not all zero, such that

$$\alpha_1 x_1 + \cdots + \alpha_m x_m = 0. \tag{1.12}$$

Note that if the set $x_1, \cdots, x_m$ is linearly dependent, any larger set $x_1, \cdots x_m, x_{m+1}, \cdots, x_{m+p}$ will also be linearly dependent since we may take

$$\alpha_{m+1} = \alpha_{m+2} = \cdots = \alpha_{m+p} = 0,$$

and then we shall have

$$\alpha_1 x_1 + \cdots + \alpha_m x_m + \cdots + \alpha_{m+p} x_{m+p} = 0.$$

If, whenever (1.12) holds, it follows that

$$\alpha_1 = \alpha_2 = \cdots = \alpha_m = 0,$$

then we say that the set $x_1, \cdots, x_m$ is *linearly independent*.

† The reader unfamiliar with the theory of Lebesgue integration may assume that all integrals are Riemann integrals. For practical purposes there is no difference between the two theories. The Lebesgue theory is more useful for theoretical purposes because certain theorems hold for Lebesgue integrals but not for Riemann integrals. For a fuller discussion, see Titchmarsh, *Theory of Functions*, Chapter XI, Clarendon Press, Oxford, 1939.

‡ See C. C. MacDuffee, *Introduction to Abstract Algebra*, John Wiley and Sons, New York, 1940, Chapter VI.

Instead of defining the concept of a plane determined by two vectors, we shall define a more general concept which includes that of line, plane, or any higher-dimensional space. Such a concept is that of a *linear manifold*, defined as follows:

If a collection $\mathcal{M}$ of vectors in $\mathcal{S}$ is such that for all scalars α and β it contains the vectors $\alpha x + \beta y$ whenever it contains the vectors x and y, then $\mathcal{M}$ is a linear manifold.

The following are examples of linear manifolds:

(*a*) In E_3 any plane through the origin. Note that a linear manifold must always contain the zero vector for it contains $x - x$, where x is any vector in the manifold.

(*b*) In E_n any E_{n-j}, where $j = 0, 1, 2, \cdots, n$.

(*c*) In $\mathcal{L}_2$ the set of functions such that

$$\int_0^1 f(t) \sin t \, dt = 0.$$

(*d*) If $b_1 = b_2 = b_3 = 0$, the solutions of (1.1) form a linear manifold in E_3.

A set of vectors $x_1, x_2, \cdots, x_k$ in $\mathcal{M}$ is said to *span* or to determine $\mathcal{M}$ if every vector y in $\mathcal{M}$ can be represented as a linear combination of $x_1, x_2, \cdots, x_k$; that is, for any vector y in $\mathcal{M}$, there exist scalars $\alpha_1, \alpha_2, \cdots, \alpha_k$, depending on y, such that

$$y = \alpha_1 x_1 + \cdots + \alpha_k x_k. \tag{1.13}$$

The vectors $x_1, \cdots, x_k$ form a *basis* for $\mathcal{M}$ if they span $\mathcal{M}$ and also are linearly independent. In this case, the representation (1.13) is unique. For if the representation were not unique, there would exist another set of scalars such that

$$y = \beta_1 x_1 + \cdots + \beta_k x_k;$$

then, by subtraction,

$$0 = (\alpha_1 - \beta_1)x_1 + \cdots + (\alpha_k - \beta_k)x_k.$$

Since the basis vectors are independent, this implies

$$\alpha_1 - \beta_1 = \alpha_2 - \beta_2 = \cdots = \alpha_k - \beta_k = 0;$$

therefore the representation (1.13) must be unique.

A linear manifold $\mathcal{M}$ is said to be of *dimension* k if the basis consists of k vectors. For example, E_n is an n-dimensional space since the vectors $x_1 = (1, 0, \cdots, 0), \cdots, x_n = (0, 0, \cdots, 0, 1)$ form a basis for it. If no finite set of vectors spans the manifold, the dimension of the manifold is said to be infinite. For example, the linear manifold of continuous functions on the interval (0, 1) is infinite-dimensional. One of our most important tasks will be to determine an appropriate basis for infinite-dimensional function spaces.

It is an immediate consequence of the definition of dimension that any k-dimensional manifold contains k-independent vectors (the basis). We shall now prove:

Lemma. *Any set of $k + 1$ vectors in a k-dimensional linear manifold is linearly dependent.*

The proof is by induction. Suppose that $k = 1$ and let x_1 be a basis for the manifold. If y_1 and y_2 are any vectors in the manifold, then $y_1 = \alpha_1 x_1$, $y_2 = \alpha_2 x_1$ and we find that $\alpha_2 y_1 - \alpha_1 y_2 = 0$. This proves that y_1 and y_2 are linearly dependent. Now, suppose that every set of k vectors in any $(k - 1)$-dimensional manifold is linearly dependent. Let $y_1, \cdots, y_{k+1}$ be any set of $k + 1$ vectors in the k-dimensional manifold $\mathcal{M}$, and let $x_1, \cdots, x_k$ be a basis for $\mathcal{M}$. By the definition of a basis, each y_j $(1 \leq j \leq k + 1)$ can be expressed as a linear combination of the basis vectors. We may therefore write

$$y_j = \alpha_{1j} x_1 + \cdots + \alpha_{kj} x_k, \quad (j = 1, 2, \cdots, k + 1).$$

Not all the scalars α_{1j} are zero for, if they were, all the vectors y_j would be contained in the $(k - 1)$-dimensional subspace $\mathcal{M}_k$ spanned by the basis vectors $x_2, \cdots, x_k$; and then, by the induction hypothesis, any k and *a fortiori* any $k + 1$ of all the y_j vectors would be dependent. Suppose that $\alpha_{11} \neq 0$. Consider the k vectors $\alpha_{11} y_j - \alpha_{1j} y_1$, $(j = 2, 3, \cdots, k + 1)$. They belong to the $(k - 1)$-dimensional manifold $\mathcal{M}_k$; and by the induction hypothesis scalars $\beta_2, \cdots, \beta_{k+1}$, not all zero, exist such that

$$\beta_2(\alpha_{11} y_2 - \alpha_{12} y_1) + \cdots + \beta_{k+1}(\alpha_{11} y_{k+1} - \alpha_{1,k+1} y_1) = 0$$

or

$$\beta_2 \alpha_{11} y_2 + \cdots + \beta_{k+1} \alpha_{11} y_{k+1} - (\beta_2 \alpha_{12} + \cdots + \beta_{k+1} \alpha_{1,k+1}) y_1 = 0.$$

This shows that $y_1, y_2, \cdots y_{k+1}$ are dependent and completes the proof by induction.

Another way of stating the above lemma is:

In a k-dimensional manifold, the maximum number of linearly independent vectors is k.

Note that this implies that the dimension of a manifold is the same whatever basis is used.

In infinite-dimensional manifolds, the situation is more complicated. The definitions given previously must be modified as follows:

An infinite set of vectors $x_1, x_2, \cdots$ is said to *span* $\mathcal{M}$ if every vector y

in $\mathcal{M}$ can be expressed either as a linear combination of a finite number of the vectors $x_1, x_2, \cdots$ or as the limit of such linear combinations. An infinite set $x_1, x_2, \cdots$ is said to be *linearly independent* if

$$\alpha_1 x_1 + \alpha_2 x_2 + \cdots = 0$$

implies $\alpha_1 = \alpha_2 = \cdots = 0$. An infinite set of vectors $x_1, x_2, \cdots$ is a *basis* for $\mathcal{M}$ if every vector y in $\mathcal{M}$ can be expressed in a unique manner as a linear combination of a possibly infinite number of the vectors $x_1, x_2, \cdots$.

In contrast to the case for finite-dimensional manifolds, an infinite set of vectors which are linearly independent and which span $\mathcal{M}$ need not form a basis for $\mathcal{M}$. For example, let $\mathcal{M}$ be E_∞ and consider the following set of vectors:

$$f_j = (1, 0, 0, \cdots, 0, 1, 0, \cdots), \quad (j = 2, 3, \cdots).$$

Here the second unity is in the jth place. Obviously, the vectors f_j are linearly independent.

In Problem 1.16 we shall prove that the vectors $f_2, f_3, \cdots$ span E_∞. As an illustration of this statement, the sequence of sums

$$\frac{1}{n}(f_2 + f_3 + \cdots + f_{n+1})$$

converges, as n approaches infinity, to the vector e which has unity as its first component and has zero for all the other components.

However, the set of vectors f_j is not a basis for E_∞ because it is impossible to find scalars $\alpha_2, \alpha_3, \cdots$ such that

$$e = \alpha_2 f_2 + \alpha_3 f_3 + \cdots.$$

By comparing corresponding components of both sides of this equation, we see that, since $\alpha_2 = \alpha_3 = \cdots = 0$, the first components are unequal. Consequently, the vector e cannot be expressed in terms of the vectors f_j, and therefore they are not a basis for E_∞.

A linear manifold $\mathcal{M}$ is said to be *closed* if, whenever a sequence of vectors $x_1, x_2, \cdots$ in $\mathcal{M}$ converges to a limit, the limit of the sequence belongs to $\mathcal{M}$. A closed linear manifold is called a *linear subspace*. The distinction between manifold and subspace is illustrated by the following example.

In E_∞ consider the collection $\mathcal{F}$ of all vectors that have only a finite number of components different from zero. If x_1 and x_2 belong to $\mathcal{F}$, that is, if each has only a finite number of components different from zero, the vector $\alpha x_1 + \beta x_2$ also can have only a finite number of components different from zero, and so it too belongs to $\mathcal{F}$, thus making $\mathcal{F}$ a linear manifold. However, $\mathcal{F}$ is not a subspace as we see if we consider the following convergent sequence of vectors in $\mathcal{F}$:

$$x_1 = (1, 0, 0, 0, 0, \cdots),$$
$$x_2 = (1, \tfrac{1}{2}, 0, 0, 0, \cdots),$$
$$x_3 = (1, \tfrac{1}{2}, \tfrac{1}{3}, 0, 0, \cdots),$$

and so on, with x_n having as its first n components the numbers $1, \frac{1}{2}, \frac{1}{3}, \cdots, 1/n$, whereas all the rest of its components are zero. This sequence converges to the following vector x:

$$x = (1, \tfrac{1}{2}, \tfrac{1}{3}, \cdots);$$

this vector has no zero components and so is not in $\mathcal{F}$. $\mathcal{F}$, therefore, is an example of a linear manifold which is not a subspace.

The examples of linear manifolds given previously are also examples of linear subspaces.

The space $\mathcal{N}$ is said to be the *sum* of the two subspaces $\mathcal{M}_1$ and $\mathcal{M}_2$ written

$$\mathcal{N} = \mathcal{M}_1 + \mathcal{M}_2$$

if every vector in $\mathcal{N}$ can be written as the sum of a vector in $\mathcal{M}_1$ and a vector in $\mathcal{M}_2$. If every vector in $\mathcal{N}$ can be written in only one way as the sum of a vector in $\mathcal{M}_1$ and of a vector in $\mathcal{M}_2$, then $\mathcal{N}$ is called the *direct sum* of $\mathcal{M}_1$ and $\mathcal{M}_2$ and is written as follows:

$$\mathcal{N} = \mathcal{M}_1 \oplus \mathcal{M}_2.$$

For example, the space E_3 is the direct sum of any plane through the origin and any line not in the plane but meeting it at the origin, because any vector in E_3 can be written in a unique way as the sum of its projection on the plane and its projection on the line.

PROBLEMS

1.10. Prove that the set of vectors x orthogonal to a given vector y forms a linear subspace. (*Hint.* Use the Schwarz inequality in Problem 1.6.)

1.11. Prove that $\mathcal{L}_2$ is the direct sum of the subspace $\mathcal{M}$ spanned by the functions t and t^4 and the subspace $\mathcal{N}$ of all functions $g(t)$ such that

$$\int_0^1 g(t)t\,dt = \int_0^1 g(t)t^4\,dt = 0.$$

(*Hint.* If $f(t)$ is any function in $\mathcal{L}_2$, show that the linear equations

$$\int_0^1 f(t)t\,dt = \int_0^1 (\alpha t + \beta t^4)t\,dt$$
$$\int_0^1 f(t)t^4\,dt = \int_0^1 (\alpha t + \beta t^4)t^4\,dt$$

have a unique solution α, β; then show that $f(t) - \alpha t - \beta t^4$ belongs to $\mathcal{N}$.)

1.12. Prove that if $\mathcal{N}$ is the direct sum of $\mathcal{M}_1$ and $\mathcal{M}_2$, the zero vector is the only vector which is in both $\mathcal{M}_1$ and $\mathcal{M}_2$. Prove the converse, that if every vector in $\mathcal{N}$ can be written as the sum of a vector in $\mathcal{M}_1$ and a vector in $\mathcal{M}_2$, and if the zero vector is the only vector common to $\mathcal{M}_1$ and $\mathcal{M}_2$, then $\mathcal{N}$ is the direct sum of $\mathcal{M}_1$ and $\mathcal{M}_2$.

1.13. If $\mathcal{M}$ and $\mathcal{N}$ are linear subspaces, prove that the *intersection* of $\mathcal{M}$ and $\mathcal{N}$, that is, the set of all vectors which belong to both $\mathcal{M}$ and $\mathcal{N}$, is also a linear subspace.

1.14. Prove that every one-dimensional linear manifold is a subspace. (*Hint.* Suppose that x_1 is a basis for the subspace and suppose that $\alpha_n x_1$ is a Cauchy sequence. Then α_n converges to a limit α and $\alpha_n x_1$ converges to a limit αx_1.)

1.15. If $x_1, \cdots, x_k$ are linearly independent, prove that the greatest lower bound of $|\alpha_1 x_1 + \cdots + \alpha_k x_k|$ is greater than zero for all values of $\alpha_1, \cdots, \alpha_k$ such that $\alpha_1^2 + \cdots + \alpha_k^2 = 1$. Call this greatest lower bound γ, then we have

$$|\alpha_1 x_1 + \cdots + \alpha_k x_k| > \gamma(\alpha_1^2 + \cdots + \alpha_k^2)^{1/2}$$

for all values of $\alpha_1, \cdots, \alpha_k$. (*Hint.* The value of $|\alpha_1 x_1 + \cdots + \alpha_k x_k|$ is a continuous function of $\alpha_1, \cdots, \alpha_k$. Since the sphere $\alpha_1^2 + \cdots + \alpha_k^2 = 1$ is a closed set, the greatest lower bound is attained. If the bound were zero, this would contradict the independence.)

1.16. Prove the vectors $f_2, f_3, \cdots$ defined on page 13 span E_∞. (*Hint.* Let $x = (\xi_1, \xi_2, \cdots)$. Put $\sigma_n = \xi_1 - \xi_2 - \cdots - \xi_n$ and let τ_n be an integer larger than $n|\sigma_n|^2$; then the vectors $y_n = \xi_2 f_2 + \cdots \xi_n f_n + \dfrac{\sigma_n}{\tau_n}(f_{n+1} + \cdots + f_p)$ where $p = n + \tau_n$, converge to x.)

Representation of Linear Vector Spaces

Given a linear vector space $\mathcal{S}$, we may construct a basis for it in the following manner:

Choose an arbitrary vector x_1 to be one of the basis vectors. Either the vector x_1 spans $\mathcal{S}$, or there is in $\mathcal{S}$ a vector x_2 which is linearly independent of x_1 and which may be taken as another basis vector. Either the vectors x_1 and x_2 span $\mathcal{S}$, or there is a vector x_3 which is linearly independent of both x_1 and x_2. Add the vector x_3 to the set of basis vectors. If we continue this process, either it will end after a finite number of steps or it will continue indefinitely. If it ends after n steps, the vectors $x_1, \cdots, x_n$ will form a basis for $\mathcal{S}$. If the process continues indefinitely, the vectors $x_1, x_2, \cdots$ will form a basis for $\mathcal{S}$.†

In the case where the vectors $x_1, \cdots, x_n$ form a basis for $\mathcal{S}$, we shall show that the elements of $\mathcal{S}$ may be represented as sets of n numbers and that consequently $\mathcal{S}$ is the same as E_n. To prove this, consider any two elements x and y in $\mathcal{S}$. From the definition of a basis, we may write

$$x = \alpha_1 x_1 + \cdots + \alpha_n x_n,$$
$$y = \beta_1 x_1 + \cdots + \beta_n x_n.$$

† We assume that $\mathcal{S}$ possesses a countable basis.

Now, from (1.3) and the properties of addition, we see that

$$\alpha x = \alpha\alpha_1 x_1 + \cdots + \alpha\alpha_n x_n$$

and

$$x + y = (\alpha_1 + \beta_1)x_1 + \cdots + (\alpha_n + \beta_n)x_n.$$

This shows that if x is represented by the set of n numbers $(\alpha_1, \cdots, \alpha_n)$ and y by the set of n numbers $(\beta_1, \cdots, \beta_n)$, then $\mathcal{S}$ is the same as E_n as regards addition of vectors and multiplication of a vector by a scalar.

The two spaces may differ, however, in the definition of scalar product. In E_n, the scalar product of the vector $(\alpha_1, \cdots, \alpha_n)$ and the vector $(\beta_1, \cdots, \beta_n)$ is $\alpha_1\beta_1 + \cdots + \alpha_n\beta_n$. In $\mathcal{S}$ the scalar product of the vectors x and y is

$$\sum_1^n \sum_1^n \alpha_j \beta_k \langle x_j, x_k \rangle.$$

This will be the same as the usual scalar product defined in E_n if, and only if,

$$\begin{aligned} \langle x_j, x_k \rangle &= 0, \quad j \neq k \\ &= 1, \quad j = k, (j, k = 1, 2, \cdots, n); \end{aligned} \tag{1.14}$$

that is, if, and only if, the vectors $x_1, \cdots, x_n$ are mutually orthogonal and their lengths are unity.

A set of vectors is said to be *orthonormal*, or O.N., if they satisfy (1.14). From the above discussion we see that if $\mathcal{S}$ possesses an O.N. basis, the representation of elements in $\mathcal{S}$ by sets of n numbers will give correct results for scalar products also; consequently, $\mathcal{S}$ may be completely identified with E_n.

If $\mathcal{S}$ has an infinite O.N. basis, it may similarly be identified with E_∞. The next section will discuss a method of obtaining an O.N. basis for $\mathcal{S}$.

Orthogonalization

By a method known as the *Schmidt orthogonalization process* a set of mutually orthogonal vectors may be constructed from any set of linearly independent vectors $x_1, x_2, \cdots$. The construction is as follows:

Put

$$\begin{aligned} y_1 &= x_1 \\ y_2 &= x_2 - \frac{\langle y_1, x_2 \rangle}{\langle y_1, y_1 \rangle} y_1, \end{aligned} \tag{1.15}$$

so that y_2 is x_2 minus its projection on y_1. Note that

$$\langle y_2, y_1 \rangle = 0. \tag{1.16}$$

Similarly, we put

$$y_3 = x_3 - \frac{\langle y_1, x_3 \rangle}{\langle y_1, y_1 \rangle} y_1 - \frac{\langle y_2, x_3 \rangle}{\langle y_2, y_2 \rangle} y_2, \tag{1.17}$$

so that y_3 is x_3 minus its projection on the plane of y_1 and y_2. Note that, because of (1.16),

$$\langle y_3, y_1\rangle = \langle y_3, y_2\rangle = 0.$$

Suppose the vectors $y_1, y_2, \cdots, y_{j-1}$ $(j < k)$ have been defined; then we put y_j equal to x_j minus the sum of its projections on the vectors $y_1, y_2, \cdots, y_{j-1}$. Clearly, y_j is orthogonal to $y_1, \cdots, y_{j-1}$. Also, y_j is not the zero vector for, if it were, x_j would be linearly dependent on $y_1, y_2, \cdots, y_{j-1}$ and therefore dependent on $y_1, y_2, \cdots, y_{j-2}, x_{j-1}$ because of the definition of y_{j-1}. It follows that eventually x_j would be linearly dependent on $x_1, x_2, \cdots, x_{j-1}$, but this contradicts our original assumption that the vectors $x_1, x_2, \cdots, x_k$ are linearly independent. Consequently, the process generates a set of non-zero vectors that are mutually orthogonal.

The vectors y_j $(j = 1, 2, \cdots, k)$ can be normalized, that is, multiplied by appropriate constants so that their lengths are unity. Thus, put

$$z_j = y_j/|y_j|;$$

then

$$\langle z_j, z_j\rangle = 1.$$

Since the vectors y_j are mutually orthogonal, the vectors z_j are also mutually orthogonal. We write

$$\langle z_i, z_j\rangle = \delta_{ij}, \tag{1.18}$$

where δ_{ij}, the *Kronecker delta*, is defined as follows:

$$\begin{aligned} \delta_{ij} &= 0, \quad i \neq j, \\ &= 1, \quad i = j. \end{aligned} \tag{1.19}$$

A set of vectors satisfying (1.18) is said to form an *orthonormal* (*O.N.*) *set*. The Schmidt orthogonalization process shows that any given set of vectors may be transformed into an O.N. set; consequently, we may assume that every finite-dimensional subspace has an O.N. basis.

By the use of Problem 1.17, we may show that if an O.N. set of vectors spans an infinite-dimensional space, the O.N. set is a basis for the space. Suppose that $z_1, z_2, \cdots$ is such an O.N. set and let x be any vector. Since the O.N. set spans the space, then for any $\varepsilon > 0$ there exists a linear combination

$$x_k = \alpha_1 z_1 + \cdots + \alpha_k z_k$$

such that

$$|x - x_k| < \varepsilon.$$

However, Problem 1.17 shows that

$$|x - z_1\langle z_1, x\rangle - \cdots - z_k\langle z_k, x\rangle| < |x - x_k| < \varepsilon;$$

consequently, the sequence of sums

$$z_1\langle z_1, x\rangle + \cdots + z_k\langle z_k, x\rangle, \quad k = 1, 2, \cdots$$

converges to the limit x, and we may write

$$x = z_1\langle z_1, x\rangle + z_2\langle z_2, x\rangle + \cdots.$$

In a later section we shall discuss the problem of representing x in terms of a basis which is not O.N.

PROBLEMS

1.17. Suppose that $z_1, z_2, \cdots$ is an O.N. set. Find that linear combination $\alpha_1 z_1 + \cdots + \alpha_k z_k$ which best approximates a given vector x in the sense that

$$|x - \alpha_1 z_1 - \cdots - \alpha_k z_k|^2$$

is a minimum; then prove *Bessel's inequality*, namely,

$$\langle x, x\rangle \geq \langle z_1, x\rangle^2 + \cdots + \langle z_k, x\rangle^2.$$

(*Hint.* Write the square of the desired minimum length as a scalar product and show that

$$\alpha_j = \langle z, x\rangle, \quad j = 1, 2, \cdots, k.)$$

1.18. In the space of $\mathcal{L}_2$ functions over the interval $(-1, 1)$ with the scalar product

$$(f, g) = \int_{-1}^{1} f(t)g(t)\, dt$$

construct an O.N. set from the functions $1, t, t^2, t^3$. The polynomials so obtained are a few of the well-known Legendre polynomials.

Projection Theorem and Linear Functionals

Consider a plane through the origin in E_3 and a vector y which does not lie in the plane. Suppose that the vector y is projected on the plane; then the difference between y and its projection will be a vector perpendicular to the plane, that is, a vector perpendicular to all vectors x lying in the plane. This well-known geometric fact may be extended to abstract linear vector spaces. The result of this extension will have very useful applications in later sections.

Let $\mathcal{M}$ be a linear subspace in $\mathcal{S}$. If $\mathcal{M}$ is not the whole space $\mathcal{S}$, there exists at least one vector y in $\mathcal{S}$ such that y does not belong to $\mathcal{M}$. We state without proof the following

Projection Theorem: *If $\mathcal{M}$ is a subspace and y is not in $\mathcal{M}$, then there exists in $\mathcal{M}$ a vector w, called the projection of y on $\mathcal{M}$, such that $y - w$ is orthogonal to $\mathcal{M}$; that is,*

$$\langle y - w, x\rangle = 0$$

whenever x belongs to $\mathcal{M}$.

A proof of this theorem may be found in Appendix I.

To illustrate these concepts, assume $\mathcal{S}$ is E_3 and $\mathcal{M}$ is the linear subspace of all vectors whose components are equal, that is, x is in $\mathcal{M}$ if

$x = (\alpha_1, \alpha_2, \alpha_3)$, where $\alpha_1 = \alpha_2 = \alpha_3$. If y is the vector (1, 0, 0), then w will be the vector $(\frac{1}{3}, \frac{1}{3}, \frac{1}{3})$ and $y - w$, the vector $(\frac{2}{3}, -\frac{1}{3}, -\frac{1}{3})$. We have

$$\langle y - w, x\rangle = \tfrac{2}{3}\alpha - \tfrac{1}{3}\alpha - \tfrac{1}{3}\alpha = 0$$

exactly as stated in the Projection Theorem.

It is important to note that the theorem as stated requires $\mathcal{M}$ to be a linear *subspace*. The theorem would not be true if $\mathcal{M}$ were a linear manifold only. For example, let $\mathcal{S}$ be E_∞ and let $\mathcal{M}$ be the previously considered linear manifold of vectors which have only a finite number of components different from zero. Let

$$y = (1, \tfrac{1}{2}, \tfrac{1}{3}, \cdots, \frac{1}{n}, \cdots).$$

The vector y is not in $\mathcal{M}$, and yet we shall show that a projection w of y onto $\mathcal{M}$ does not exist. The proof is by contradiction. Suppose that w did exist in $\mathcal{M}$; then $y - w$ would be orthogonal to every vector in $\mathcal{M}$ and therefore orthogonal to the vector, all of whose components, except the nth component, are zero. This orthogonality implies that the nth component of $y - w$ is zero for any value of n; consequently, $y = w$, which is an obvious contradiction.

The Projection Theorem can be used to obtain a fundamental fact about the mapping of the linear vector space $\mathcal{S}$ into a one-dimensional space. Suppose that to every vector x in $\mathcal{S}$ a scalar, which we denote by $f(x)$, is assigned. We say that $f(x)$ is a *functional* defined on $\mathcal{S}$. If the functional $f(x)$ is such that

$$f(\alpha x + \beta y) = \alpha f(x) + \beta f(y), \tag{1.20}$$

it is called *linear*. For example, the length of a vector in E_n is a functional but not linear. The value of any one component of x in E_n is an example of a linear functional.

A functional $f(x)$ is *continuous* if

$$\lim x_n = x$$

implies that

$$\lim f(x_n) = f(x).$$

The functional is *bounded* if there exists a constant μ such that

$$|f(x)| < \mu|x|$$

for all x in $\mathcal{S}$. In Problem 1.23 it is shown that a bounded linear functional is continuous, and also the converse, a continuous linear functional must be bounded.

The following are examples of continuous linear functionals.

In E_3, let $x = (\alpha_1, \alpha_2, \alpha_3)$; then we may take $f(x) = \alpha_1 + \alpha_2 + \alpha_3$.

In the space $\mathcal{L}_2$ of square integrable functions $u(t)$ over (0, 1) we may take

$$f(u) = \int_0^1 tu(t)\, dt.$$

Note that in the first example $f(x)$ is equal to the scalar product $\langle z, x \rangle$, where z is the vector with components 1, 1, 1, whereas in the second example $f(u)$ is equal to the scalar product $\langle t, u(t) \rangle$. Given any linear vector space $\mathcal{S}$, it is apparent that every scalar product $\langle z, x \rangle$, where z is a fixed vector, is an example of a continuous linear functional. We shall prove the converse: every continuous linear functional is a scalar product. This is the content of

Theorem 1.1. *If $f(x)$ is a continuous linear functional, there exists a vector z in $\mathcal{S}$ such that*

$$f(x) = \langle z, x \rangle.$$

Since $f(x)$ assigns a real number to every vector in $\mathcal{S}$, we may consider the functional as a mapping of the vector space into a one-dimensional space, the real-number line. This consideration suggests that the functional is characterized essentially by one vector; we shall show this to be the case.

Let $\mathcal{M}$ be the set of all vectors y such that $f(y) = 0$. $\mathcal{M}$ is a linear manifold because $f(y_1) = f(y_2) = 0$ implies that $f(\alpha_1 y_1 + \alpha_2 y_2) = 0$. If $y_1, y_2, \cdots$ is a sequence of vectors in $\mathcal{M}$ which converges to a vector x then, because $f(x)$ is continuous,

$$f(x) = \lim f(y_n) = 0;$$

therefore, $\mathcal{M}$ is a subspace.

Clearly, the desired vector z must be orthogonal to $\mathcal{M}$. If $\mathcal{M}$ is the whole space, then for all x in $\mathcal{S}$, x is in $\mathcal{M}$; hence

$$f(x) = 0 = \langle 0, x \rangle$$

and we may take $z = 0$. If $\mathcal{M}$ is not the whole space, there exists in $\mathcal{S}$ a vector y_0, not in $\mathcal{M}$, such that $f(y_0) \neq 0$. From the Projection Theorem we know that y_0 has a projection w in $\mathcal{M}$ such that the non-zero vector $y = y_0 - w$ is orthogonal to $\mathcal{M}$. Note that

$$f(y) = f(y_0) \neq 0.$$

We put $z = \alpha y$ where the constant α is so chosen that

$$f(z) = \langle z, z \rangle. \tag{1.21}$$

Solving this equation, we have

$$\alpha = \frac{f(y)}{\langle y, y \rangle} \neq 0.$$

Since y was orthogonal to $\mathcal{M}$, we have

$$\langle x', z \rangle = 0$$

for any vector x' in $\mathcal{M}$.

We shall show that this vector z is the desired vector. Consider any vector x in $\mathcal{S}$. If

$$\beta = \frac{f(x)}{f(z)}$$

then

$$f(x - \beta z) = f(x) - \beta f(z) = 0; \tag{1.22}$$

consequently, the vector $x - \beta z$ is in $\mathcal{M}$ and therefore

$$\langle x - \beta z, z \rangle = 0.$$

Since, from (1.21) and (1.22), we get

$$\langle x, z \rangle = \langle x - \beta z + \beta z, z \rangle = \beta \langle z, z \rangle = \beta f(z) = f(x),$$

we have proved Theorem 1.1.

By the use of Theorem 1.1 we can now discuss the problem of representing a vector x in terms of an arbitrary basis. Suppose that $\mathcal{S}$ is a finite-dimensional space E_n, and let $x_1, x_2, \cdots, x_n$ be a basis of the space. We know that any vector x in E_n may be written as follows:

$$x = \alpha_1 x_1 + \alpha_2 x_2 + \cdots + \alpha_n x_n.$$

We shall show how the coefficients $\alpha_1, \alpha_2, \cdots, \alpha_n$ may be determined.

Consider the coefficient α_1. Its value depends upon the vector x, and therefore α_1 is a functional of x. Obviously, it is a linear functional. If we can show that α_1 is a bounded functional also, Theorem 1.1 will apply and, therefore, there will exist a vector z_1 such that we may write

$$\alpha_1 = \langle z_1, x \rangle.$$

Similarly, there will exist vectors $z_2, \cdots, z_k$ such that

$$\alpha_j = \langle z_j, x \rangle, \quad j = 1, 2, \cdots, k.$$

The proof that α_1 is a bounded functional follows from Problem 1.15. There it was shown that

$$|x| > \gamma(\alpha_1^2 + \alpha_2^2 + \cdots + \alpha_n^2)^{1/2}.$$

This implies that

$$|\alpha_1| < |x|/\gamma;$$

consequently, α_1 is a bounded functional.

We may then write

$$x = x_1\langle z_1, x \rangle + x_2\langle z_2, x \rangle + \cdots + x_n\langle z_n, x \rangle. \tag{1.23}$$

Suppose that we put $x = x_i$ $(i = 1, 2, \cdots, n)$ in (1.23); then we have

$$x_i = \sum_1^n x_j \langle z_j, x_i \rangle.$$

Since the vectors $x_1, \cdots, x_n$ are linearly independent, this equation implies that

$$(1.24) \qquad \langle z_j, x_i \rangle = \delta_{ij}, \ (i, j = 1, 2, \cdots, n).$$

We shall say that the vectors $z_1, \cdots, z_n$ are a *reciprocal basis* to the basis vectors $x_1, \cdots, x_n$. Our discussion has shown that, *given an arbitrary basis, there always exists a reciprocal basis satisfying* (1.24).† We shall use this result later.

PROBLEMS

1.19. If x is any vector in a linear subspace $\mathcal{M}$, y is any vector not in $\mathcal{M}$, and w is the projection of y on $\mathcal{M}$, show that

$$|y - w| \leq |y - x|.$$

(*Hint.* We have $|y - x|^2 = |y - w + w - x|^2 = |y - w|^2 + |w - x|^2$.)

1.20. In E_3 find the projection of the vector (1, 1, 1) on the plane $x_1 + 2x_2 + 3x_3 = 0$. Show that the length of the vector difference between the vector and its projection is equal to the distance of the vector from the plane.

1.21. Find all vectors in E_4 which are orthogonal to the subspace defined by the equations $x_1 - x_2 = x_3 - x_4 = 0$. The set of all such vectors is called the *orthogonal complement* of the given subspace.

1.22. Given $f(t)$ in $\mathcal{L}_2$ over $(0, \pi)$, let $g(t)$ be that linear combination of $\sin t$, $\sin 2t$, $\sin 3t$ which makes

$$\int_0^\pi [f(t) - g(t)]^2 \, dt$$

a minimum. Show that $g(t)$ is the projection of $f(t)$ on the subspace spanned by $\sin t$, $\sin 2t$, $\sin 3t$, that is, show that $f(t) - g(t)$ is orthogonal to any linear combination of $\sin t$, $\sin 2t$, $\sin 3t$.

1.23. Prove that a bounded linear functional is continuous. Prove that a continuous linear functional is bounded. (*Hint.* Suppose there exists a sequence of vectors x_n such that $|x_n| < 1$ and such that $|f(x_n)| < n$; then the sequence x_n/n converges to the zero vector and by continuity $f(x_n/n)$ converges to zero; but $|f(x_n/n)| > 1$.)

Linear Operators

An important concept in the theory of linear vector spaces is that of a mapping of the space onto itself. In E_3 a rotation of the entire space about a fixed axis, or a uniform expansion around a point, or a translation of the space are examples of operations that map the space onto itself. It is

† The distinction between the basis and the reciprocal basis is essentially that between covariant and contravariant vectors.

easy to see that the operation of rotation or of expansion can be represented by a matrix with elements α_{ij}, $(i = 1, 2, 3; j = 1, 2, 3)$, so that if ξ_1, ξ_2, ξ_3 are the components of an arbitrary vector in E_3 before the transformation and $\zeta_1, \zeta_2, \zeta_3$ the components of the transformed vector, then

$$\begin{aligned} \zeta_1 &= \alpha_{11}\xi_1 + \alpha_{12}\xi_2 + \alpha_{13}\xi_3, \\ \zeta_2 &= \alpha_{21}\xi_1 + \alpha_{22}\xi_2 + \alpha_{23}\xi_3, \\ \zeta_3 &= \alpha_{31}\xi_1 + \alpha_{32}\xi_2 + \alpha_{33}\xi_3. \end{aligned} \tag{1.25}$$

We write this in matrix form as

$$z = Ax, \tag{1.26}$$

where A denotes the matrix whose elements are α_{ij}.

The operation represented by the matrix A has the property that if the vector z_1 corresponds to the vector x_1, and the vector z_2 to the vector x_2, then the vector $\alpha z_1 + \beta z_2$ corresponds to the vector $\alpha x_1 + \beta x_2$. Such an operation is said to be *linear*.

The equations (1.25) determine the vector z if the vector x is given. It is also important to solve the inverse problem: given z, determine x. This is essentially the problem of solving the set of simultaneous linear equations (1.25). We know that the existence and uniqueness of the solution depend on whether there exists a non-zero solution of the homogeneous equations

$$Ax = 0.$$

We shall find that similar results hold for more general operators.

We wish to give an abstract treatment of the theory of linear operators which will cover many of the cases that are met with in applied mathematics. Some of the operators to be considered are the following:

(1) n-rowed square matrices in n-dimensional space in which case (1.26) represents n simultaneous linear equations.

(2) Integral operators in $\mathcal{L}_2$ in which case (1.26) would symbolize the integral equation

$$g(t) = \int_\alpha^\beta K(t, s)u(s)\, ds, \tag{1.27}$$

where the kernel $K(t, s)$ is any continuous function of t and s.

(3) Translation operators in $\mathcal{L}_2$ in which case (1.26) might represent the following difference equation:

$$g(t) = u(t + 2h) - 2tu(t + h) + f(t)u(t).$$

(4) Differential operators in $\mathcal{L}_2$ in which case (1.26) might be a differential equation

$$g(t) = \left(\frac{d^2}{dt^2} + t^2\right)u(t). \tag{1.28}$$

This last case illustrates some of the difficulties that the abstract treatment faces. The differential operator cannot be applied to every function $f(t)$ in $\mathcal{L}_2$ for the following two reasons: first, not every function in $\mathcal{L}_2$ has a first derivative, let alone a second derivative; and, second, even if $f(t)$ has a second derivative, it need not belong to $\mathcal{L}_2$, and hence the result of the operation does not necessarily belong to $\mathcal{L}_2$. Consequently, the differential operator can be applied only to such functions in $\mathcal{L}_2$ which have second derivatives that belong to $\mathcal{L}_2$.

Abstractly, an *operator* L is a mapping that assigns to a vector x in a linear vector space $\mathcal{S}$ another vector in $\mathcal{S}$ which we denote by Lx. The set of vectors for which the mapping is defined is called the *domain* of L. The set of vectors y which are equal to Lx for some x in the domain is called the *range* of the operator. An operator is *linear* if the mapping is such that for any vectors x_1, x_2 in the domain of L and for arbitrary scalars α_1, α_2, the vector $\alpha_1 x_1 + \alpha_2 x_2$ is in the domain of L and

$$L(\alpha_1 x_1 + \alpha_2 x_2) = \alpha_1 L x_1 + \alpha_2 L x_2.$$

A linear operator is *bounded* if its domain is the entire space $\mathcal{S}$ and if there exists a single constant C such that

$$|Lx| < C|x|$$

for all x in $\mathcal{S}$. It is easy to show that a linear bounded operator is *continuous*; that is, if a sequence of vectors x_n converges to x, then $L(x_n)$ converges to $L(x)$. In this chapter we shall consider *linear bounded operators only*. Consequently, our theorems will not apply immediately to differential operators since these are always unbounded. However, we shall see in a later chapter that most of the theorems can be extended so that they will be valid for differential operators.

PROBLEMS

1.24. Consider the following operators in E_∞:

If $x = (\xi_1, \xi_2, \cdots)$, then

$$(a) \quad Cx = (0, \xi_1, \xi_2, \cdots),$$
$$(b) \quad Dx = (\xi_2, \xi_3, \xi_4, \cdots),$$
$$(c) \quad Ex = (\xi_1, \tfrac{1}{2}\xi_2, \tfrac{1}{3}\xi_3, \cdots),$$
$$(d) \quad Fx = (\xi_1, 2\xi_2, 3\xi_3, \cdots).$$

Which are bounded?

1.25. Let L be a linear operator. Prove that if L is bounded, it is continuous, and, conversely, if L is continuous, it is bounded. Prove also that, if L is bounded,

$$L(x_1 + x_2 + \cdots) = Lx_1 + Lx_2 + \cdots.$$

(*Hint.* Compare Problem 1.23.)

1.26. Put γ equal to the least upper bound of $|Lx|$ for $|x| = 1$. Show that $|Lx| \leq \gamma|x|$ for all x. Put δ equal to the least upper bound of $|\langle Lx, y\rangle|$ for

$|x| = |y| = 1$. Show that $|\langle Lx, y\rangle| \le \delta|x|\cdot|y|$ for all x and y. Prove that $\gamma = \delta$. (*Hint.* Put $y = Lx$ in $\langle Lx, y\rangle$ and prove $\gamma \le \delta$. Use the Schwarz inequality in Problem 1.6 to prove $|\langle Lx, y\rangle| \le \gamma|x|\cdot|y|$ and thus $\delta \le \gamma$.)

Representation of Operators

Every bounded linear operator may be represented by a matrix. The matrix will have a finite or an infinite number of rows according as the dimension of $\mathcal{S}$ is infinite or finite. To show this, let $e_1, e_2, \cdots$ be an O.N. basis in $\mathcal{S}$; then every x in $\mathcal{S}$ may be written in the form

$$x = \alpha_1 e_1 + \alpha_2 e_2 + \cdots.$$

Since Lx is also in $\mathcal{S}$, we may write

$$Lx = \beta_1 e_1 + \beta_2 e_2 + \cdots;$$

but L is a bounded linear operator, therefore by Problem 1.25

$$Lx = \alpha_1 Le_1 + \alpha_2 Le_2 + \cdots.$$

If we now put

$$Le_j = \Sigma\gamma_{ij}e_i, \quad j = 1, 2, \cdots,$$

we find that

$$\beta_i = \Sigma\gamma_{ij}\alpha_j, \quad i = 1, 2, \cdots.$$

Clearly, the right-hand side of this equation is the result of multiplying the matrix whose elements are γ_{ij} by the column vector whose elements are α_j. Consequently, in terms of the vectors $e_1, e_2, \cdots$ as a basis, L is represented by the matrix whose elements are γ_{ij} where

$$\gamma_{ij} = \langle e_i, Le_j\rangle.$$

Note that a matrix representing L can be found by using any basis and not necessarily an O.N. one. Of course, a change in the basis changes the matrix representing L. In Chapter 2 we shall study the different matrices which may represent a given operator L.

If the bounded operator L is such that its range is finite-dimensional, that is, if for every x in $\mathcal{S}$, Lx belongs to some k-dimensional manifold $\mathcal{M}$, we may obtain another useful representation for L as follows:

Let $x_1, \cdots, x_k$ be a basis for $\mathcal{M}$; then

$$Lx = c_1x_1 + \cdots + c_kx_k,$$

where the coefficients $c_1, \cdots c_k$ are scalars whose value depends on x. To indicate this dependence, we shall write $c_1 = c_1(x)$, etc. From the equations

$$L(\alpha x) = \alpha c_1(x)x_1 + \cdots + \alpha c_k(x)x_k$$

and

$$L(x + y) = [c_1(x) + c_1(y)]x_1 + \cdots + [c_k(x) + c_k(y)]x_k$$

it follows that the equivalent of (1.20), namely,

$$c_1(\alpha x + \beta y) = \alpha c_1(x) + \beta c_1(y)$$

is valid; consequently, c_1 is a linear functional of x. Similarly, the other coefficients $c_2, \cdots, c_k$ are also linear functionals.

Just as in the proof given on page 21 for the existence of a reciprocal basis, we may use Problem 1.15 to prove that these functionals are bounded. Then by Theorem 1.1 there exist vectors $y_1, \cdots, y_k$ such that

$$c_1 = \langle y_1, x\rangle, \cdots, c_k = \langle y_k, x\rangle.$$

This shows that

$$Lx = x_1\langle y_1, x\rangle + \cdots + x_k\langle y_k, x\rangle. \tag{1.29}$$

We shall now explain a useful notation which was introduced by Dirac in his book on quantum mechanics. We have shown that if a space has finite dimension k, that space is equivalent to E_k and every element in it could be represented by a set of k scalars. These k scalars could be considered arranged in a column (we call this a column vector) or in a row (row vector). Up to now, this distinction between row and column vectors was unimportant.† However, when we introduce the concept of mappings, the distinction becomes important. We shall use the symbol $x\rangle$ to indicate that x is represented as a column vector, whereas $\langle x$ will indicate that it is represented as a row vector.

This notation is consistent with the notation for scalar product since if we put

$$x\rangle = \begin{pmatrix} \xi_1 \\ \xi_2 \\ \cdot \\ \cdot \\ \cdot \\ \xi_k \end{pmatrix}$$

and

$$\langle y = (\eta_1, \eta_2, \cdots, \eta_k),$$

then

$$\langle y, x\rangle = \xi_1\eta_1 + \xi_2\eta_2 + \cdots + \xi_k\eta_k$$

is the ordinary matrix product of the row vector for y by the column vector for x. Note that the matrix product of the vector $x\rangle$ by the vector $\langle y$ in that order is the following matrix:

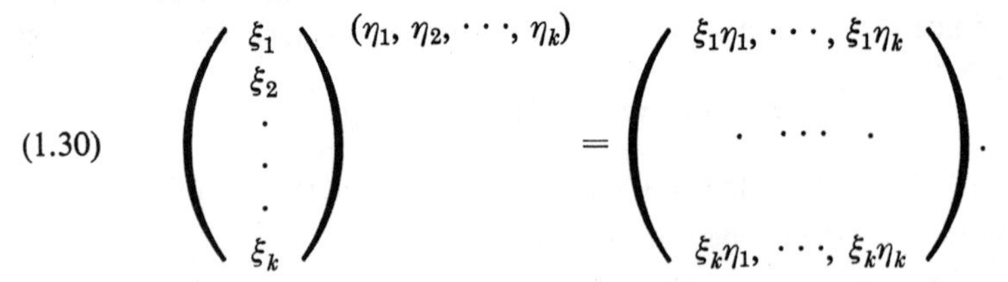

$$\begin{pmatrix} \xi_1 \\ \xi_2 \\ \cdot \\ \cdot \\ \cdot \\ \xi_k \end{pmatrix} (\eta_1, \eta_2, \cdots, \eta_k) = \begin{pmatrix} \xi_1\eta_1, \cdots, \xi_1\eta_k \\ \cdot \quad \cdots \quad \cdot \\ \xi_k\eta_1, \cdots, \xi_k\eta_k \end{pmatrix}. \tag{1.30}$$

† The distinction is essentially that between a covariant and a contravariant vector.

We shall use the symbol $x\rangle\langle y$ to denote the matrix in (1.30) and a sum of such symbols

$$x_1\rangle\langle y_1 + x_2\rangle\langle y_2 + \cdots + x_k\rangle\langle y_k$$

to denote the matrix sum of the corresponding matrices.

Now suppose that henceforth we represent x and the basis vectors $x_1, x_2, \cdots, x_k$ by column vectors; then (1.29) may be written as

$$Lx\rangle = x_1\rangle\langle y_1, x\rangle + \cdots + x_k\rangle\langle y_k, x\rangle.$$

The form of this equation suggests that we formally cancel the "common factor" $x\rangle$ on both sides. In this way, we get

$$L = x_1\rangle\langle y_1 + \cdots + x_k\rangle\langle y_k. \tag{1.31}$$

This formal equation is actually correct, for after introducing a basis it can be shown that L is represented by the matrix on the right-hand side of (1.31).

For another illustration of this notation, suppose that $\mathcal{S}$ is k-dimensional and that the vectors $z_1, z_2, \cdots, z_k$ form the basis reciprocal to $x_1, \cdots, x_k$. If y is any vector in $\mathcal{S}$, from (1.23) we have

$$y\rangle = x_1\rangle\langle z_1, y\rangle + \cdots + x_k\rangle\langle z_k, y\rangle;$$

consequently, the identity operator I in $\mathcal{S}$ may be written as follows:

$$I = x_1\rangle\langle z_1 + \cdots + x_k\rangle\langle z_k. \tag{1.32}$$

An operator that can be represented in the form (1.31) is called a *dyad* † or, if we wish to be more explicit, a *k-term dyad.* If a dyad is an integral operator, its kernel is said to be *degenerate.*

We shall also write

$$L = a_1\rangle\langle b_1 + a_2\rangle\langle b_2 + \cdots$$

if the infinite sum

$$a_1\langle b_1, x\rangle + a_2\langle b_2, x\rangle + \cdots$$

converges to Lx for all x.

The sum and the product of operators are defined in a similar way to the sum and product of matrices. If L_1 and L_2 are bounded linear operators, the sum $L_1 + L_2$ is defined by the equation

$$(L_1 + L_2)x = L_1x + L_2x,$$

and the product L_1L_2 is defined by the equation

$$(L_1L_2)x = L_1(L_2x).$$

Of course, just as with matrices, multiplication is not commutative; consequently, L_1L_2 need not equal L_2L_1.

† The name is taken from the corresponding concept in vector analysis.

However, the associative law for multiplication does hold. We may therefore form powers of operators, for example, L, L^2, etc. If we define multiplication of L with a scalar α by the equation

$$(\alpha L)x = \alpha(Lx),$$

we can define polynomial functions of operators. It is easy to show that the usual rules of algebra, except those which depend on the commutative law, will hold for such polynomials

From another viewpoint, the set of linear operators on $\mathcal{S}$ may be considered a linear vector space. In Problem 1.27, we shall show how a scalar product of two operators L_1 and L_2 in this space may be defined.

PROBLEMS

1.27. Define the scalar product of a dyad $x_1\rangle\langle y_1$ by another dyad $u_1\rangle\langle v_1$ as follows: The scalar product is

$$[x_1\rangle\langle y_1, u_1\rangle\langle v_1] = (\langle y_1, u_1\rangle)(\langle v_1, x_1\rangle).$$

Extend this definition to the scalar product of two k-term dyads by using the linearity of the scalar product. Show that this definition satisfies (1.7) and (1.8).

1.28. If A is a $k \times k$ matrix, show that it may be written as a k-term dyad. If I is the identity $k \times k$ matrix, prove that the scalar product, as defined in Problem 1.27, of A by I is the trace of A. (*Hint.* Use the vectors e_i, $(i = 1, 2, \cdots, k)$, which have their ith component equal to one and the other components zero, as a basis for the space.)

1.29. If L is a k-term dyad, show that for every x in $\mathcal{S}$ there exists a polynomial $p(t)$ of degree $k + 1$, at most, such that $p(L)x = 0$. (*Hint.* The $k + 1$ vectors $Lx, L^2x, \cdots, L^{k+1}x$ must be linearly dependent.)

Inversion of Operators

The fundamental problem in the theory of operators is the inversion of the operator, that is, given a vector a, to find a vector x such that

$$Lx = a. \tag{1.33}$$

We shall discuss how this may be done in some simple but nevertheless important cases.

Case 1. *L is a dyad.* Assume that L is a one-term dyad. If

$$L = a_1\rangle\langle b_1,$$

then (1.33) becomes

$$a_1\langle b_1, x\rangle = a. \tag{1.34}$$

Here the vectors a_1, b_1, and a are known and the vector x is to be determined.

Since the left-hand side of (1.34) is some scalar multiple of the vector a_1, the right-hand side also must be a multiple of a_1; consequently, (1.34) has a solution if and only if the vector a is a scalar multiple of a_1. Assume that $a = \alpha a_1$; then (1.34) becomes

$$\langle b_1, x\rangle = \alpha.$$

One solution of this equation is

$$x = \frac{\alpha}{|b_1|^2} b_1,$$

but if c is any vector orthogonal to b, we see that

$$x = \frac{\alpha}{|b_1|^2} b_1 + c \tag{1.35}$$

is also a solution. We leave it to the reader to prove that every solution must be of the form (1.35).

Suppose now that L is a k-term dyad. We may write

$$L = a_1\rangle\langle b_1 + a_2\rangle\langle b_2 + \cdots + a_k\rangle\langle b_k,$$

and then (1.33) becomes

$$a_1\langle b_1, x\rangle + a_2\langle b_2, x\rangle + \cdots + a_k\langle b_k, x\rangle = a. \tag{1.36}$$

Just as before, we see that, in general, a solution of this equation exists if and only if a is a vector in the linear manifold spanned by the vectors a_1, $a_2, \cdots, a_k$. Suppose that the vectors $a_1, \cdots, a_k$ are linearly independent and suppose that

$$a = \alpha_1 a_1 + \alpha_2 a_2 + \cdots + \alpha_k a_k;$$

then (1.36) reduces to

$$\langle b_1, x\rangle = \alpha_i, \quad i = 1, 2, \cdots, k.$$

To find x, we assume that

$$x = \xi_1 b_1 + \xi_2 b_2 + \cdots + \xi_k b_k,$$

and we get the following system of k linear equations in k unknowns:

$$\langle b_1, x\rangle = \xi_1\langle b_1, b_1\rangle + \xi_2\langle b_1, b_2\rangle + \cdots + \xi_k\langle b_1, b_k\rangle = \alpha_1,$$

$$\cdot \quad \cdot \quad \cdot \quad \cdot \quad \cdot \quad \cdot \quad \cdot \quad \cdot \quad \cdot \quad \cdot$$

$$\langle b_k, x\rangle = \xi_1\langle b_k, b_1\rangle + \xi_2\langle b_k, b_2\rangle + \cdots + \xi_k\langle b_k, b_k\rangle = \alpha_k.$$

If these linear equations do not have a solution, (1.36) also does not have a solution. If the above linear equations do have a solution, every solution of (1.36) will be of the form

$$x = \xi_1 b_1 + \cdots + \xi_k b_k + p,$$

where p is a vector orthogonal to all the vectors $b_1, b_2, \cdots, b_k$.

Case 2. *L is the identity plus a dyad D.* Again assume first that D is one term $a_1\rangle\langle b_1$. The equation $Lx = a$ becomes

$$Ix + Dx = x + a_1\langle b_1, x\rangle = a. \tag{1.37}$$

The method for solving this is to determine the value of $\langle b_1, x\rangle$ as follows.

Take the scalar product of (1.37) with the vector b_1. We get

$$\langle b_1, x\rangle + \langle b_1, a_1\rangle\langle b_1, x\rangle = \langle b_1, a\rangle, \tag{1.38}$$

and then

$$\langle b_1, x\rangle = \frac{\langle b_1, a\rangle}{1 + \langle b_1, a_1\rangle}$$

if the denominator $1 + \langle b_1, a_1\rangle$ is not zero. Using this value for $\langle b_1, x\rangle$ in (1.37), we find that

$$x = a - \frac{a_1\langle b_1, a\rangle}{1 + \langle b_1, a_1\rangle}. \tag{1.39}$$

We must now show that this expression for x actually does solve (1.37). It is conceivable that (1.37) does not have any solution. All that we have shown so far is that *if* (1.37) has a solution, it must be given by (1.38). However, by substituting (1.39) in (1.37), we see that the left-hand side of (1.37) is

$$a - \frac{a_1\langle b_1, a\rangle}{1 + \langle b_1, a_1\rangle} + a_1\langle b_1, a\rangle - \frac{a_1\langle b_1, a_1\rangle\langle b_1, a\rangle}{1 + \langle b_1, a_1\rangle},$$

and this reduces to a; consequently, if

$$1 + \langle b_1, a_1\rangle \neq 0,$$

then (1.39) is the unique solution of (1.37).

Suppose that

$$1 + \langle b_1, a_1\rangle = 0;$$

then (1.38) shows that we must have

$$\langle b_1, a\rangle = 0.$$

If $\langle b_1, a\rangle \neq 0$, (1.37) has no solution, since it leads to a contradiction. If $\langle b_1, a\rangle = 0$, then $x = a$ is one solution of (1.37). It is readily seen that every solution of (1.37) is given by the formula

$$x = a + \alpha a_1,$$

where α is any scalar.

Suppose now that D is the k-term dyad

$$a_1\rangle\langle b_1 + \cdots + a_k\rangle\langle b_k.$$

The equation $Lx = a$ becomes

$$Ix + Dx = x + a_1\langle b_1, x\rangle + \cdots + a_k\langle b_k, x\rangle = a. \tag{1.40}$$

Again, the method for solving this equation is to determine the values of the scalar products $\langle b_1, x\rangle, \cdots, \langle b_k, x\rangle$. Multiply (1.40) by the vectors

$b_1, \cdots, b_k$ in turn, and the following system of linear equations is obtained:

$$
\begin{array}{l}
(1 + \langle b_1, a_1\rangle)\langle b_1, x\rangle + \langle b_1, a_2\rangle\langle b_2, x\rangle + \cdots + \langle b_1, a_k\rangle\langle b_k, x\rangle = \langle b_1, a\rangle, \\
\langle b_2, a_1\rangle\langle b_1, x\rangle + (1 + \langle b_2, a_2\rangle)\langle b_2, x\rangle + \cdots + \langle b_2, a_k\rangle\langle b_k, x\rangle = \langle b_2, a\rangle \\
\quad \cdot \quad \cdot \quad \cdot \quad \cdot \quad \cdots \quad \cdot \quad \cdot \\
\langle b_k, a_1\rangle\langle b_1, x\rangle + \langle b_k, a_2\rangle\langle b_2, x\rangle + \cdots + (1 + \langle b_k, a_k\rangle)\langle b_k, x\rangle = \langle b_k, a\rangle.
\end{array}
\tag{1.41}
$$

After the values of the scalar products $\langle b_1, x\rangle, \cdots, \langle b_k, x\rangle$ are obtained by the usual methods for solving simultaneous equations, we find from (1.40) that

$$x = a - a_1\langle b_1, x\rangle - \cdots - a_k\langle b_k, x\rangle.$$

As an illustration of this method, consider the following integral equation:

$$u(t) + \lambda\int_0^1 su(s)\, ds = f(t), \tag{1.42}$$

where $f(t)$ is any function in $\mathcal{L}_2$ over (0, 1). This equation may be written $Lu = f$, where L is the identity plus the one-term dyad $\lambda\rangle\langle s$. Here λ corresponds to the vector a_1, and t (or s, since the variable of integration is immaterial) corresponds to the vector b_1.

To find u, multiply (1.42) by t and integrate from 0 to 1. We get

$$\int_0^1 tu(t)\, dt + \lambda\int_0^1 t\, dt \int_0^1 su(s)\, ds = \int_0^1 tf(t)\, dt$$

or

$$\int_0^1 su(s)\, ds \left(1 + \frac{\lambda}{2}\right) = \int_0^1 tf(t)\, dt.$$

Solving for the integral of su and substituting in (1.42), we find that

$$u(t) = f(t) - \lambda\left(1 + \frac{\lambda}{2}\right)^{-1} \int_0^1 tf(t)\, dt,$$

if $\lambda \neq -2$. If $\lambda = -2$, (1.42) does not have a solution unless

$$\int_0^1 tf(t)\, dt = 0,$$

in which case $u(t) = f(t) + \alpha$ is a solution for any value of α.

We may summarize the discussion of this section in the following

Rule. *To solve*

$$(I + D)x = a \tag{1.40}$$

where D is the k-term dyad

$$a_1\rangle\langle b_1 + a_2\rangle\langle b_2 + \cdots + a_k\rangle\langle b_k,$$

take the scalar product of (1.40) *with each of the vectors $b_1, \cdots, b_k$ and then* *solve the resulting set of simultaneous linear equations* (1.41) *for the scalar products* $\langle b_i, x\rangle$, $i = 1, 2, \cdots, k$. *Finally,*

$$x = a - Dx.$$

There is one special, but nevertheless important, case in which the simultaneous equations (1.41) have an immediate solution. In this case

$$\langle b_i, a_j\rangle = 0 \tag{1.43}$$

if $i \neq j$; for then the matrix of the coefficients in (1.41) reduces to its diagonal terms only, and we have

$$\langle b_i, x\rangle = \frac{\langle b_i, a\rangle}{1 + \langle b_i, a_i\rangle}, \quad i = 1, 2, \cdots, k.$$

A dyad D such that (1.43) holds is said to be in *diagonal* form. Similarly, if an operator L is written as an infinite sum of dyads,

$$L = a_1\rangle\langle b_1 + a_2\rangle\langle b_2 + \cdots$$

such that (1.43) holds for all $i \neq j$, then L is said to be represented in *diagonal* form. The advantages of the diagonal representation are many. For example, to solve the equation $Lx = a$ when L is in diagonal form, we assume that a solution exists and put

$$x = \Sigma\xi_j a_j.$$

From (1.43) we have

$$\xi_j = \frac{\langle b_j, x\rangle}{\langle b_j, a_j\rangle}.$$

Taking the scalar product of the equation $Lx = a$ with the vectors $b_1, b_2, \cdots$, in succession, we get

$$\langle b_j, a_j\rangle\langle b_j, x\rangle = \langle b_j, a\rangle;$$

consequently,

$$\xi_j = \frac{\langle b_j, a\rangle}{\langle b_j, a_j\rangle^2}$$

and

$$x = \sum\frac{\langle b_j, a\rangle}{\langle b_j, a_j\rangle^2}a_j.$$

One of our main goals will be to determine how to diagonalize operators. This question will be discussed in detail later. At present, we just note that if L is in diagonal form, then

$$La_j = \lambda_j a_j, \quad j = 1, 2, \cdots, \tag{1.44}$$

where

$$\lambda_j = \langle b_j, a_j\rangle.$$

A vector a_j such that (1.44) holds is called an *eigenvector* of L corresponding to the *eigenvalue* λ_j. We shall see that the problem of diagonalizing an operator is equivalent to the problem of finding all its eigenvectors.

PROBLEMS

1.30. Discuss the following integral equation for $u(s)$:

$$\int_0^1 \sin k(s - t)u(t)\, dt = f(s).$$

(*Hint.* Since $\sin k(s - t) = \sin ks \cos kt - \cos ks \sin kt$, the operator is a two-term dyad.)

1.31. Discuss the following integral equation:

$$\int_0^1 \frac{1 - s^n t^n}{1 - st} u(t)\, dt = f(s).$$

(*Hint.* The operator is an n-term dyad since

$$\frac{1 - s^n t^n}{1 - st} = 1 + st + s^2t^2 + \cdots + s^{n-1}t^{n-1}.)$$

1.32. If $f(s)$ is any function in $\mathcal{L}_2$ over (0, 1), solve the following integral equations:

$$(a)\quad u(s) + \int_0^1 \sin k(s - t)u(t)\, dt = f(s),$$

$$(b)\quad u(s) + \int_0^1 \frac{\sin (2n + 1)\pi(s - t)}{2 \sin \pi(s - t)} u(t)\, dt = f(s).$$

(*Hint.* The integral operator in (*b*) is a dyad because

$$\frac{\sin (2n + 1)\pi(s - t)}{2 \sin \pi (s - t)} = \frac{1}{2} + \sum_1^n \cos 2k\pi(s - t).)$$

1.33. Consider the equation $Dx = a$ where $D = a_1\rangle\langle b_1 + \cdots + a_k\rangle\langle b_k$ and where

$$\langle a_i, a_j\rangle = \langle b_i, b_j\rangle = 0$$

if $i \neq j$. Show that if a is in the manifold spanned by the vectors $a_1, \cdots, a_k$,

$$x = \sum_1^k \frac{\langle a, a_j\rangle b_j}{\langle a_j, a_j\rangle\langle b_j, b_j\rangle}.$$

1.34. Use Problem 1.33 to solve the infinite set of equations

$$a_m = \sum_0^\infty \varepsilon_n \xi_n \left(\frac{1}{\dfrac{2m + 1}{2} - n} + \frac{1}{\dfrac{2m + 1}{2} + n} \right), \quad m = 0, 1, 2, \cdots,$$

where $\varepsilon_0 = 2$, and $\varepsilon_n = 1$ for $n = 1, 2, \cdots$. (*Hint.* The operator is $\sum_0^\infty \cos n\phi\rangle\langle \cos \frac{2n + 1}{2}\phi$ on the space of $\mathcal{L}_2$ functions over $(0, \pi)$. The matrix representation is obtained by introducing $\cos n\phi$, $n = 0, 1, 2, \cdots$, as a basis for the space.)

Inverting the Identity plus a Small Operator

The fundamental problem of inverting the operator equation $Lx = a$ may be considered from the viewpoint of finding an operator inverse to L. This *inverse* operator, written L^{-1}, is a bounded linear operator such that

$$LL^{-1} = L^{-1}L = I,$$

the identity operator. In many cases, the inverse operator does not exist. In some cases, only one of the above equalities is satisfied. For example, consider the following operators in E_∞:

If

$$x = (\xi_1, \xi_2, \cdots);$$

then

$$Cx = (0, \xi_1, \xi_2, \cdots),$$
$$Dx = (\xi_2, \xi_3, \xi_4, \cdots).$$

We shall call C a "creation" operator because it creates an extra component, and analogously we shall call D a "destruction" operator. Now

$$DC = I,$$

but

$$CD \neq I.$$

We say D is a left-sided inverse of C but C has no inverse.

There is one important case in which the operator L can be proved to have an inverse. Write $L = I + M$; then the notation for the inverse operator suggests that

$$L^{-1} = (I + M)^{-1} = I - M + M^2 - M^3 + \cdots.$$

We shall show that this formal result will be correct if the operator M satisfies the expected condition for convergence of the right-hand side, namely, that M be, in some sense, less than one. More precisely, we shall prove

Theorem 1.2. *If for all vectors x, there exists a constant γ, independent of x, such that $0 < \gamma < 1$ and such that*

$$|Mx| < \gamma |x|;$$

then

$$(I + M)^{-1} = I - M + M^2 - M^3 + \cdots.$$

The proof of this theorem will consist of two parts. First, we prove that the partial sums of the infinite series on the right-hand side converge to a limit and, second, we show that this operator is the inverse of $I + M$.

For the first part, put

$$S_n = I - M + M^2 - \cdots (-)^{n-1} M^{n-1}.$$

For any vector x, we have (assume $n \geq m$):

$$|S_n x - S_m x| = |M^m x - M^{m+1} x + \cdots (-)^{n-m-1} M^{n-1} x|$$
$$\leq |M^m x| + |M^{m+1} x| + \cdots + |M^{n-1} x|$$

by the triangle inequality (Problem 1.8*b*). Since

$$|M^2 x| = |M(Mx)| < \gamma |Mx| < \gamma^2 |x|,$$

and, similarly,

$$|M^k x| < \gamma^k |x|,$$

we find that

$$|S_n x - S_m x| < (\gamma^m + \gamma^{m+1} + \cdots + \gamma^{n-1}) |x| < \frac{\gamma^m}{1 - \gamma} |x|.$$

By hypothesis, $\gamma < 1$; therefore, we can find an integer m_0 such that for $m > m_0$, we have

$$\gamma^m < \varepsilon(1 - \gamma).$$

This proves that the sequence $S_n x$ ($n = 1, 2, \cdots$) converges in the Cauchy sense. Since our space is complete, the sequence will converge to some limit vector y. We define an operator K by the equation

$$y = Kx = \lim S_n x.$$

It is readily seen that K is a linear operator and that K is bounded by

$$1 + \gamma + \gamma^2 + \cdots = (1 - \gamma)^{-1}.$$

For the second part of the proof, consider the product

$$S_n(I + M) = I + (-)^n M^n.$$

For any vector x, we have

$$S_n(I + M)x = x + (-)^n M^n x.$$

Since $|M^n x| < \gamma^n |x|$ and since $\gamma < 1$, we find that

$$\lim S_n(I + M)x = x.$$

All we need show now is that $\lim S_n(I + M) = K(I + M)$. This fact becomes evident from the following results:

$$|S_n(I + M)x - K(I + M)x| = |(M^n + M^{n+1} + \cdots)(I + M)x|$$
$$< (\gamma^n + \gamma^{n+1} + \cdots) |(I + M)x| < \frac{\gamma^n}{1 - \gamma}(1 + \gamma)|x|.$$

Similarly, we may show that

$$(I + M)Kx = \lim (I + M)S_n x = x;$$

consequently, we have proved Theorem 1.2, that

$$K = \lim (I - M + M^2 - \cdots (-)^{n-1} M^{n-1})$$

is the inverse of $I + M$.

The infinite series in Theorem 1.2 is called the *Neumann* series for the inverse operator $(I + M)^{-1}$. This series may be obtained also from the equation

$$(I + M)x = a$$

by an iteration or a perturbation method. With this approach, we consider Mx as a "small" term and write the equation as follows:

$$x = a - Mx.$$

To obtain an initial approximation, we neglect the term Mx and put $x_0 = a$. For the next approximation, we replace Mx by Mx_0 and get $x_1 = a - Mx_0$. Similarly, we put

$$x_{n+1} = a - Mx_n.$$

We must now determine whether $\lim x_n$ exists and whether it equals x. By successive iteration of this equation, we find that $x_n = S_{n+1}a$. If M satisfies the condition of Theorem 1.2, then the $\lim S_n$, and consequently the $\lim x_n$, exists and is equal to x.

There is another possible approach to the Neumann series. For the sake of convenience, consider the equation $x = a - \varepsilon Mx$ where ε is a scalar, instead of the equation $x = a - Mx$. Assume that

$$x = a_0 + \varepsilon a_1 + \varepsilon^2 a_2 + \cdots,$$

where $a_0, a_1, \cdots$ are unknown vectors independent of ε. Substituting this series in the equation for x, we get

$$a_0 + \varepsilon a_1 + \varepsilon^2 a_2 + \cdots = a - \varepsilon Ma_0 - \varepsilon^2 Ma_1 - \varepsilon^3 Ma_2 - \cdots,$$

and, after equating corresponding powers of ε, we get

$$a_0 = a$$

$$a_{n+1} = -Ma_n, \quad n = 0, 1, 2, 3, \cdots$$

or

$$a_n = (-)^n M^n a_0 = (-)^n M^n a.$$

Using this result, we obtain

$$x = a - \varepsilon Ma + \varepsilon^2 M^2 a - \cdots,$$

which reduces to the Neumann series when we put $\varepsilon = 1$.

To illustrate this discussion, consider again the integral equation solved in the preceding section, that is,

$$u(t) + \lambda \int_0^1 su(s)\, ds = f(t). \tag{1.42}$$

Assume that

$$u(t) = u_0(t) + \lambda u_1(t) + \lambda^2 u_2(t) + \cdots.$$

Substituting this in (1.42) and equating the coefficients of corresponding powers of λ, we get

$$u_n(t) + \int_0^1 su_{n-1}(s)\,ds = 0, \quad n = 1, 2, 3 \cdots,$$
$$u_0(t) \qquad\qquad = f(t).$$

Solving these equations, we find that

$$u_1(t) = -\int_0^1 sf(s)\,ds$$
$$u_2(t) = \tfrac{1}{2}\int_0^1 sf(s)\,ds, \text{ etc.}$$

Finally,

$$u(t) = f(t) - \lambda\left(1 - \frac{\lambda}{2} + \frac{\lambda^2}{4} - \cdots\right)\int_0^1 sf(s)\,ds$$
$$= f(t) - \frac{\lambda}{1 + \lambda/2}\int_0^1 sf(s)\,ds,$$

the result we obtained originally.

Notice that the iteration method can be applied only if the infinite series converges. Theorem 1.2 states that the series converges if the bound of the operator M is less than one. Now, from (1.42),

$$|Mx| = \left|\lambda\int_0^1 su(s)\,ds\right| \leq |\lambda|\left(\int_0^1 s^2\,ds\right)^{1/2}\left(\int_0^1 u^2\,ds\right)^{1/2}$$
$$= 3^{-1/2}|\lambda|\,|x|$$

by Schwarz's inequality; consequently, Theorem 1.2 will apply if $|\lambda| < 3^{1/2}$. However, inspection of the Neumann series for this case shows that the series converges when $|\lambda| < 2$ but diverges when $|\lambda| > 2$. This example is an illustration of the fact that the Neumann series may be used in cases where Theorem 1.2 does not apply, but it also shows that there are cases where the Neumann series cannot be used.

PROBLEMS

1.35. Show that the Neumann series converges if there exists an integer k, a constant C, and a constant γ, where $0 < \gamma < 1$, such that the bound of M^{k+n} is not greater than $C\gamma^n$ $(n = 1, 2, 3, \cdots)$.

1.36. Show that, if $Mu = \lambda\int_0^1 su(s)\,ds$, the bound of M^k is not greater than $|\lambda|^k 2^{-k+1/2}$, $(k = 1, 2, \cdots)$; then use Problem 1.35 to show that the Neumann series converges for $|\lambda| < 2$.

1.37. Show that, if

$$Mu(t) = \int_0^t q(s, t)u(s)\,ds,$$

where $q(s, t)$ is a continuous function of both s and t, the Neumann series for $(I + M)^{-1}$ converges. (*Hint.* Let $\mu_k(t)$ be the absolute value of $M^k u(t)$ for $0 \leq t \leq t_0$ and let $\int_0^t q(s, t)^2\, ds \leq k(t)^2$ for $0 \leq t \leq t_0$; then $\mu_k(t)^2 \leq k(t)^2 \int_0^t \mu_{k-1}(s)^2\, ds$ by Schwarz's inequality. By induction, show that

$$\mu_n(t)^2 \leq \frac{k(t)}{(n-1)!}\left(\int_0^t k(s)^2\, ds\right)^{n-1} \int_0^{t_0} u(s)^2\, ds.)$$

Completely Continuous Operators

Suppose that the operator L may be written as the sum of an operator V which has an inverse and an operator Y which is a k-term dyad; then the equation

$$Lx = (V + Y)x = a$$

may be solved as follows.

Since V has an inverse V^{-1}, we apply it to the above equation, and it becomes

$$(I + V^{-1}Y)x = V^{-1}a. \tag{1.45}$$

Suppose, now, that

$$Y = a_1\rangle\langle b_1 + a_2\rangle\langle b_2 + \cdots + a_k\rangle\langle b_k;$$

then

$$V^{-1}Y = V^{-1}a_1\rangle\langle b_1 + V^{-1}a_2\rangle\langle b_2 + \cdots + V^{-1}a_k\rangle\langle b_k$$

also is a k-term dyad. Consequently, the operator in (1.45) is the identity plus a k-term dyad, and therefore (1.45) may be solved by the methods discussed on pages 28–31.

As an illustration of this method, consider the following integral-differential equation:

$$u'(t) + \lambda \int_0^1 u(s)\, ds = f(t),$$

with the initial condition $u(0) = 0$. We invert the differential operator by integrating from 0 to t, and we get

$$u(t) + \lambda t \int_0^1 u(s)\, ds = \int_0^t f(s)\, ds. \tag{1.46}$$

The operator on the left-hand side is the identity plus the one-term dyad $\lambda t\rangle\langle 1$. To solve this equation by the method of pages 28–31, we must first find the value of the scalar product $\int_0^1 u(s)\, ds$. This can be done by integrating (1.46) from 0 to 1. We find that

$$\int_0^1 u(t)\, dt\left(1 + \frac{\lambda}{2}\right) = \int_0^1 dt \int_0^t f(s)\, ds = \int_0^1 (1 - s) f(s)\, ds.$$

Consequently,

$$\int_0^1 u(t)\, dt = \left(1 + \frac{\lambda}{2}\right)^{-1} \int_0^1 (1 - s) f(s)\, ds,$$

and, finally,

$$u(t) = \int_0^t f(s)\,ds - \lambda t\left(1 + \frac{\lambda}{2}\right)^{-1}\int_0^1 (1 - s)f(s)\,ds.$$

The methods of this section can be applied also to the case where $L = I + K$ if K is an operator which can be uniformly approximated by dyads. This means that there exists a sequence of dyads K_n such that

$$\lim \frac{|(K - K_n)x|}{|x|} = 0$$

uniformly for all vectors x or, in other words, given a positive number ε, there exists an integer n_0, independent of x, such that

$$\frac{|Kx - K_n x|}{|x|} < \varepsilon \tag{1.47}$$

for all values of $n > n_0$. Such an operator K is said to be *completely continuous.*†

Put $R_n = K - K_n$; then we may write

$$L = I + K = I + R_n + K_n.$$

If we take $\varepsilon < 1$ in (1.47), the bound of R_n is less than one, and therefore Theorem 1.2 may be used to show that $I + R_n$ has an inverse. Consequently, since L is the sum of the invertible operator $I + R_n$ and the dyad K_n, we may use the methods given in this section to solve the equation $(I + K)x = a$.

There is a simple condition which will ensure that an operator K can be uniformly approximated by dyads. The condition is stated in the following

Lemma. Let $e_1, e_2 \cdots$ *be an O.N. basis for the space. If* K *is bounded, and if*

$$\Sigma|Ke_n|^2 < \infty,$$

K *is completely continuous.*

To prove this lemma, note that every vector x may be represented as follows:

$$x = e_1\langle e_1, x\rangle + e_2\langle e_2, x\rangle + \cdots,$$

where

$$\langle x, x\rangle = \langle e_1, x\rangle^2 + \langle e_2, x\rangle^2 + \cdots. \tag{1.48}$$

For any operator K, we have formally

$$Kx = Ke_1\langle e_1, x\rangle + Ke_2\langle e_2, x\rangle + \cdots.$$

† The usual definition of a completely continuous operator K is as follows: If x_n is a bounded set of vectors, then the set Kx_n has a convergent subsequence. It can be shown that the definition of the text is equivalent to this one.

We show first that the series on the right-hand side converges if K is a bounded operator. Consider the vector

$$y_{mn} = \sum_{i=m}^{n} Ke_i\langle e_i, x\rangle = K\left(\sum_{m}^{n} e_i\langle e_i, x\rangle\right).$$

Since K is bounded,

$$|y_{mn}| < \gamma\left|\sum_{m}^{n} e_i\langle e_i, x\rangle\right| = \gamma\left(\sum_{m}^{n}\langle e_i, x\rangle^2\right)^{1/2}.$$

Since the series

$$\sum_{1}^{\infty}\langle e_i, x\rangle^2$$

converges, we can find an integer m_0 such that for $m > m_0$, $n > m_0$

$$\sum_{m}^{n}\langle e_i, x\rangle^2 < \varepsilon/\gamma$$

for any $\varepsilon > 0$; consequently, we have

$$|y_{mn}| < \varepsilon$$

for all $n > m_0$, $m > m_0$. This proves that the infinite series for Kx converges.

We must still prove that the convergence is uniform. The above proof does not do so because the value of m_0 obviously depends on the vector x. Of course, the reason for the non-uniformity is that so far we have used only the fact that K is a bounded operator. We must now use the hypothesis that

$$\sum_{1}^{\infty}|Ke_i|^2 < \infty. \tag{1.49}$$

Consider the vector

$$Kx - \sum_{1}^{n} Ke_i\langle e_i, x\rangle = \sum_{n+1}^{\infty} Ke_i\langle e_i, x\rangle.$$

Using the triangle inequality and the Cauchy-Schwarz inequalities,† we have

$$\left|\sum_{n+1}^{\infty} Ke_i\langle e_i, x\rangle\right| \le \sum_{n+1}^{\infty}|Ke_i|\cdot|\langle e_i, x\rangle| \le \left(\sum_{n+1}^{\infty}|Ke_i|^2\sum_{n+1}^{\infty}\langle e_i, x\rangle^2\right)^{1/2}$$

$$\le \left(\sum_{n+1}^{\infty}|Ke_i|^2\right)^{1/2}|x|$$

† The Cauchy-Schwarz inequality states that if a_i, b_i are real numbers, then $\left|\sum_{1}^{\infty} a_i b_i\right| \le (\Sigma a_i^2 \cdot \Sigma b_i^2)^{1/2}$. This is the content of Problem 1.2.

from (1.48). However, the convergence of the series in (1.49) implies that, given $\varepsilon > 0$, there exists an integer m_0, independent of x, such that for $n > m_0$

$$\left(\sum_{n+1}^{\infty} |Ke_i|^2\right)^{1/2} < \varepsilon;$$

therefore, we have

$$|Kx - \sum_{1}^{n} Ke_i\langle e_i, x\rangle| < \varepsilon|x|.$$

This proves that K is uniformly approximated by the partial sums of the series of dyads

$$\sum_{1}^{\infty} Ke_i\rangle\langle e_i;$$

consequently, K is completely continuous, and the lemma is proved.

We may immediately apply this lemma to discuss the integral equation

$$u(t) + \int_0^1 k(t, s)u(s)\, ds = f(t), \tag{1.50}$$

where the functions $u(t)$ and $f(t)$ are assumed to belong to $\bar{\mathcal{L}}_2$ over (0, 1). The operator K is here an integral operator with the kernel $k(s, t)$.

We assume a fact which will be proved later, namely, that the functions $\exp(2\pi int)$, $n = 0, \pm 1, \pm 2, \cdots$, form an O.N. basis for the space $\bar{\mathcal{L}}_2$. Then in this case the vectors Ke_n become

$$\int_0^1 k(t, s) \exp 2\pi ins\, ds,$$

and we have

$$\sum_{-\infty}^{\infty} |Ke_n|^2 = \sum_{-\infty}^{\infty} \int_0^1 \left| \int_0^1 k(t, s) \exp 2\pi ins\, ds \right|^2 dt \tag{1.51}$$

$$= \int_0^1 dt \sum_{-\infty}^{\infty} \int_0^1 \left| k(t, s) \exp 2\pi ins\, ds \right|^2.$$

Suppose we keep t fixed and consider $k(t, s)$ as a function of s in $\bar{\mathcal{L}}_2$. Equation (1.48) applied to this case gives

$$\int_0^1 |k(t, s)|^2\, ds = \sum_{-\infty}^{\infty} \left| \int_0^1 k(t, s) \exp 2\pi ins\, ds \right|^2;$$

consequently, (1.51) becomes

$$\Sigma |Ke_n|^2 = \int_0^1 dt \int_0^1 |k(t, s)|^2\, ds.$$

If, now,

$$\int_0^1 \int_0^1 |k(t, s)|^2 \, ds \, dt < \infty,$$

the lemma states that the integral operator in (1.50) is completely continuous; therefore, we may use the methods of this section to solve (1.50). Of course, often this is more a theoretical than a practical help since a sufficiently close approximation of the completely continuous operator may require a dyad of a large number of terms, and this computation would entail a prohibitive amount of algebra. Later we shall find easier methods for solving (1.50) in certain cases, but in most cases the methods of this section must be used.

PROBLEMS

1.38. Solve

$$u''(t) + \int_0^1 \sin k(s - t) u(s) \, ds = f(t)$$

with the conditions $u(0) = u'(0) = 0$.

1.39. Suppose that the operator K is diagonalized by the O.N. basis $e_1, e_2, \cdots$, that is, suppose $Ke_i = \lambda_i e_i$. Show that K is completely continuous if the numbers λ_i converge to zero. (*Hint.*

$$\left|\sum_{n+1}^{\infty} Ke_i \langle e_i, x\rangle\right|^2 = \sum_{n+1}^{\infty} \lambda_i^2 \langle e_i, x\rangle^2 < \varepsilon \sum_{1}^{\infty} \langle e_i, x\rangle^2$$

since λ_i converge to zero.)

1.40. Consider the integral operator $Ku = \int_{-1}^{1} k(t - s)u(s) \, ds$, where $k(t) = t^{-p}$ for $0 \leq t \leq 1$, and $= 0$ for all other values of t. Show that K is completely continuous in $\mathcal{L}_2$ over $(-1, 1)$ if $0 < p < 1$. (*Hint.* Introduce the basis exp πint and use Problem 1.39.)

1.41. Show that the sum of two completely continuous operators and the product of a completely continuous operator by a bounded operator are completely continuous. Finally, show that if $k(t, s)$ is the kernel of an integral operator K, and if $k(t, s) = (t - s)^{-p} k_1(s, t)$ where $0 < p < 1$ and where $\int_{-1}^{1} \int_{-1}^{1} |k_1(s, t)|^2 \, ds \, dt < \infty$, then K is completely continuous. (*Hint.* Use Problem 1.40. Write $(t - s)^{-p} = k(t - s) + k_2(t - s)$ and show $k_2(t - s)$ is completely continuous.)

The Adjoint of an Operator

All the methods we have used for solving operator equations, except that for the Neumann series, have been based on the assumption that the equation has a solution. Before discussing under what conditions the equation has a solution and under what conditions the solution is unique, we introduce a useful tool, the *adjoint* operator. A linear operator L^* is said to be the adjoint of L if, for all x and y in $\mathcal{S}$,

$$\langle y, Lx\rangle = \langle L^*y, x\rangle. \tag{1.52}$$

Because of the symmetry of the scalar product, it follows that L is also the adjoint of L^*, that is, $(L^*)^* = L$. Using (1.52), we see that

$$(L_1 + L_2)^* = L_1^* + L_2^*, \quad (L_1L_2)^* = L_2^* L_1^*.$$

If $L^* = L$, then L is said to be *self-adjoint*. In that case,

$$\langle y, Lx\rangle = \langle Ly, x\rangle = \langle x, Ly\rangle. \tag{1.53}$$

We shall prove:

Theorem 1.3. *Every bounded linear operator L has an adjoint.*

Consider the scalar product $\langle y, Lx\rangle$, where y is a fixed vector and x an arbitrary vector. Since L is bounded, the scalar product will be a bounded linear functional of x and therefore, by Theorem 1.1, there exists a vector w depending on y such that

$$\langle y, Lx\rangle = \langle w, x\rangle.$$

We now define an operator L^* to be the correspondence between w and y, and we write

$$L^*y = w. \tag{1.54}$$

It is easily seen that this correspondence is linear and that

$$\langle y, Lx\rangle = \langle L^*y, x\rangle,$$

and therefore that the operator, as defined by (1.54), is the adjoint of L.

This proof for the existence of an adjoint suggests the method by which the adjoint of a given operator can be found. Form the scalar product of y with Lx and then try to rewrite it as the scalar product of the vector x with a vector depending on y. Consider, for example, an operator A which is represented by the matrix in (1.25). Let y be a vector with components η_1, η_2, η_3; then the scalar product of y with Ax is

$$\begin{aligned}
&\eta_1(\alpha_{11}\xi_1 + \alpha_{12}\xi_2 + \alpha_{13}\xi_3) + \eta_2(\alpha_{21}\xi_1 + \alpha_{22}\xi_2 + \alpha_{23}\xi_3) \\
&\qquad\qquad + \eta_3(\alpha_{31}\xi_1 + \alpha_{32}\xi_2 + \alpha_{33}\xi_3) \\
&= \xi_1(\alpha_{11}\eta_1 + \alpha_{21}\eta_2 + \alpha_{31}\eta_3) + \xi_2(\alpha_{12}\eta_1 + \alpha_{22}\eta_2 + \alpha_{32}\eta_3) \\
&\qquad\qquad + \xi_3(\alpha_{13}\eta_1 + \alpha_{23}\eta_2 + \alpha_{33}\eta_3) \\
&= (x, A^*y),
\end{aligned}$$

where, clearly, A^* is the transpose matrix of A, that is, the matrix with the elements α_{ji} in the ith row and the jth column.

As another illustration of the adjoint, consider the operator defined in (1.27). Let $h(t)$ be an arbitrary function in $\mathcal{L}_2$; then

$$\langle h(t), Lf(t)\rangle = \langle h(t), g(t)\rangle = \int_\alpha^\beta dt\, h(t)\int_\alpha^\beta K(t, s)f(s)\, ds.$$

If the order of integration is interchanged, this becomes

$$\int_\alpha^\beta ds\, f(s)\int_\alpha^\beta K(t, s)h(t)\, dt = \langle f(t), L^*h(t)\rangle,$$

where

$$L^*h(t) = \int_\alpha^\beta K(s, t)h(s)\, ds.$$

Note that the kernel of the adjoint integral operator is the transpose of the kernel in (1.27).

As we have remarked before, the differential operator in (1.28) is unbounded so that the proof for the existence of an adjoint does not apply. Nevertheless, by means of an integration by parts we can construct an adjoint. For simplicity, we shall consider the differential operator, not in $\mathcal{S}$, but in the linear manifold $\mathcal{M}$ of all functions $f(t)$ in $\mathcal{S}$ which have second derivatives belonging to $\mathcal{S}$ and are such that

$f(\alpha) = f(\beta) = 0$. Now

$$\langle h(t), Lf(t)\rangle = \int_\alpha^\beta h(t)\left[\frac{d^2f}{dt^2} + t^2f(t)\right] dt$$

$$= \left[h(t)f'(t) - h'(t)f(t)\right]_\alpha^\beta + \int_\alpha^\beta f(t)\left[\frac{d^2h}{dt^2} + t^2h(t)\right] dt$$

$$= \int_\alpha^\beta f(t)\left[\frac{d^2}{dt^2} + t^2\right]h(t)\, dt = \langle f(t), L^*h(t)\rangle$$

if we assume that $h(\alpha) = h(\beta) = 0$. This shows that the differential operator of (1.28) is self-adjoint in the manifold $\mathcal{M}$.

PROBLEMS

1.42. Show that $\dfrac{d^{2n}}{dt^{2n}}$ is self-adjoint in the manifold of functions $f(t)$ in $\mathcal{L}_2$ which have their $2n$th derivatives in $\mathcal{L}_2$ and for which $f(\alpha) = f(\beta) = f'(\alpha) = f'(\beta) = \cdots = f^{(n)}(\alpha) = f^{(n)}(\beta) = 0$.

1.43. In $\mathcal{L}_2$ over $(-1, 1)$ find the adjoint of the reflection operator R which is such that $Rf(t) = f(-t)$.

1.44. Show that if L is bounded, then L^* has the same bound. (*Hint.* Use Problem 1.23 and the fact that $\langle Lx, y\rangle = \langle x, L^*y\rangle$.)

1.45. Prove that the operator defined by (1.54) is linear.

1.46. Prove that the adjoint of L defined by (1.31) is

$$L^* = y_1\rangle\langle x_1 + \cdots + y_k\rangle\langle x_k.$$

1.47. In a complex vector space the adjoint is still defined by (1.52). Show that the adjoint of an n-rowed square matrix in $\bar{E}_n$ is the complex conjugate of the transposed matrix.

1.48. In $\bar{\mathscr{L}}_2$, the space of all complex-valued functions such that the square of their absolute values is integrable, find the adjoint of the operators in (1.27) and (1.28).

The Existence and Uniqueness of the Solution of $Lx = a$

There are two distinct parts to the question of solving the equation $Lx = a$. First, does a solution exist and, second, is it unique? The *uniqueness* of the solution depends upon the existence of a non-zero solution of the related homogeneous equation

$$Lx = 0. \tag{1.55}$$

The precise relationship is given by

Theorem 1.4. *If the homogeneous equation* (1.55) *has a non-zero solution, the solution of the corresponding non-homogeneous equation is not unique. Conversely, if the solution of the non-homogeneous equation is not unique, there exists a non-trivial solution of the homogeneous equation.*

To prove this theorem, first suppose that $Lx_0 = 0$, where $x_0 \neq 0$; then $L(\alpha x_0) = 0$ for any α. And, if $Lx_1 = a$, then

$$L(x_1 + \alpha x_0) = Lx_1 + \alpha Lx_0 = a;$$

thus $x_1 + \alpha x_0$ also is a solution of $Lx = a$. Conversely, suppose that x_1 and x_2 are two distinct solutions of $Lx = a$; then $L(x_1 - x_2) = a - a = 0$, and thus $x_1 - x_2$ is a non-trivial solution of (1.55).

Notice that the proof that a solution, if it exists, is unique does not guarantee its existence. A simple example will clarify this point. In E_∞ consider the Creation operator C defined on page 34.

Now $Cx = 0$ implies that $x = 0$; therefore, the solution of

$$Cx = a,$$

if it exists, is unique. Suppose that

$$a = (1, 0, 0, \cdots);$$

it is clear, then, that there is no solution of the equation

$$Cx = (1, 0, 0, \cdots).$$

The existence of a solution of (1.55) will depend upon the existence of a non-trivial solution of the *adjoint* homogeneous equation. We shall prove

Theorem 1.5. *If the range of L is closed, the non-homogeneous equation*

$$Lx = a \tag{1.56}$$

has a solution for a given vector a if, and only if, a is orthogonal to every solution of the adjoint homogeneous equation

$$L^*z = 0. \tag{1.57}$$

From (1.56) and the definition of the adjoint, it follows that, for any x,

$$0 = \langle x, L^*z\rangle = \langle a, z\rangle,$$

which proves that a must be orthogonal to every solution of (1.57). To prove the converse, suppose that a were orthogonal to every solution of (1.57), and yet there did not exist any x satisfying (1.56). Consider the *range* of L, that is, the set of all vectors in $\mathcal{S}$ which can be written as Lx. The range is a linear manifold because if $Lx_1 = a_1$ and $Lx_2 = a_2$, then $L(\alpha x_1 + \beta x_2) = \alpha a_1 + \beta a_2$. By hypothesis, the range is closed and therefore a subspace. If a does not belong to the range, the Projection Theorem asserts that a has a projection a_r on the range, such that

$$z = a - a_r \tag{1.58}$$

is orthogonal to the range; that is,

$$\langle z, Lx\rangle = 0 \tag{1.59}$$

for all x. Since

$$\langle z, Lx\rangle = \langle L^*z, x\rangle = 0$$

for all x, it follows that

$$\langle L^*z, L^*z) = 0,$$

which implies that

$$L^*z = 0;$$

hence z as defined in (1.58) is a solution of (1.57). Now, from (1.58) and (1.59) it follows that

$$\langle z, a\rangle = \langle z, z + a_r\rangle = \langle z, z\rangle \neq 0,$$

which contradicts the hypothesis that a is orthogonal to every solution of (1.57). Therefore, a must be in the range, and a solution of (1.56) must exist.

As an illustration of Theorems 1.4 and 1.5, consider the case where $L = I + a_1\rangle\langle b_1$. On page 30, we discussed the equation

$$Lx = x + a_1\langle b_1, x\rangle = a, \tag{1.60}$$

and we found that this equation has a unique solution for every value of a if $1 + \langle b_1, a_1\rangle \neq 0$. However, if $1 + \langle b_1, a_1\rangle = 0$, this equation has a solution and, in fact, an infinite number of solutions only if a satisfies the condition that $\langle b_1, a\rangle = 0$. We shall show that these results are exactly those that would be obtained by an application of Theorems 1.4

and 1.5. Consider the homogeneous equation corresponding to (1.60), namely,

$$x + a_1\langle b_1, x\rangle = 0.$$

The only possible solution of this is when x is a multiple of a_1. Put $x = \alpha a_1$; then we get

$$\alpha a_1 + \alpha a_1\langle b_1, a_1\rangle = 0,$$

which implies that

$$1 + \langle b_1, a_1\rangle = 0.$$

We conclude that if $1 + \langle b_1, a_1\rangle$ does not equal zero, the homogeneous equation has no solution other than the zero vector, and then the solution, if any, of (1.60) is unique. However, if $1 + \langle b_1, a_1\rangle$ does equal zero, the solution, if any, of (1.60) is not unique.

To decide whether a solution of (1.60) exists, we must consider the adjoint homogeneous equation. The adjoint operator is $I + b_1\rangle\langle a_1$, and the equation to be studied is

$$z + b_1\langle z, a_1\rangle = 0.$$

The only possible solution of this occurs when z is a multiple of b_1. Put $z = \beta b_1$; then we get

$$\beta b_1 + \beta b_1\langle b_1, a_1\rangle = 0,$$

which implies that again

$$1 + \langle b_1, a_1\rangle = 0.$$

We conclude that, if $1 + \langle b_1, a_1\rangle$ does not equal zero, the adjoint homogeneous equation has no solution other than the zero vector, and then equation (1.60) always has a solution. However, if $1 + \langle b_1, a_1\rangle$ does equal zero, then by Theorem 1.5, equation (1.60) has a solution if and only if

$$\langle b_1, a\rangle = 0.$$

This last statement assumes the fact that L has a closed range. We shall prove this fact in the next section.

The operator L in (1.60) is one of the large class of operators which have the following property:

Either equation (1.60) *always has a unique solution or the corresponding homogeneous equation has a non-trivial solution.*

Such operators will be said to have the *Fredholm alternative* property. Not all operators have this property. Consider, for example, the Creation operator C that we defined on page 34. The equation $Cx = a$ does not always have a solution, and yet the equation $Cx = 0$ has only the trivial solution $x = 0$.

An operator which has the Fredholm alternative property will be called a *Fredholm operator*. For such an operator the uniqueness of a solution implies the existence. If the range of L is closed and if the fact that $L^*y = 0$ has a non-trivial solution implies the existence of a non-trivial solution of $Lx = 0$, then by Theorems 1.4 and 1.5 we see that L is a Fredholm operator.

Note that in Theorem 1.5 the condition that the range be closed is necessary for the validity of the theorem. For consider the following operator in E_∞:

$$Gx = (\xi_1, \tfrac{1}{2}\xi_2, \tfrac{1}{3}\xi_3, \cdots) \text{ if } x = (\xi_1, \xi_2, \cdots).$$

The operator G is bounded, linear, and self-adjoint, but its range is not closed. The only solution of

$$G^*x = Gx = 0$$

is $x = 0$ so we might expect the equation

$$Gx = y$$

always to have a solution. However, if we take

$$y = (1, \tfrac{1}{2}, \tfrac{1}{3}, \cdots),$$

we would get

$$x = (1, 1, 1, \cdots),$$

which is not a vector in E_∞ since its length is not finite.

We use the term *null space* of L^* for the subspace of all z such that $L^*z = 0$. Theorem 1.5 may be formulated in terms of this concept as follows:

The range of L is always orthogonal to the null space of L^*. If L has a closed range, the whole space $\mathfrak{S}$ is the direct sum of the range of L and the null space of L^*; in other words the range of L is the orthogonal complement of the null space of L^*.

PROBLEMS

1.49. Discuss these equations in E_∞:

$$(a)\ Cx = y;\ (b)\ Dx = y;\ (c)\ Gx = y.$$

For what values of y does x exist? When is the solution unique?

1.50. Let x and a be vectors in E_3, and let L be a matrix such that the element in the ith row and the jth column is $\delta_{i,j+1} + \delta_{i,j-1}$, where δ_{ij} is the Kronecker delta. Discuss the existence and solution of the equation $Lx = a$.

1.51. Suppose that L is an operator acting on E_n and that the dimension of the range of L is less than n; then show that there exists a non-trivial solution of $Lx = 0$. (*Hint.* If $e_1, e_2, \cdots, e_n$ is a basis for E_n, the vectors $Le_1, Le_2, \cdots, Le_n$ are linearly dependent.)

1.52. Show that if there exists a non-trivial solution of $L^*z = 0$ in E_n, there exists a non-trivial solution of $Lx = 0$; consequently, every operator in E_n is a Fredholm operator. (*Hint.* Use the fact that $\langle z, Ly\rangle = 0$ for all y in E_n and Problem 1.51.)

1.53. An operator L that commutes with its adjoint L^*, that is $LL^* = L^*L$, is called a *normal* operator. Show that every normal operator is a Fredholm operator. (*Hint.* For any x, $\langle Lx, Lx\rangle = \langle x, L^*Lx\rangle = \langle L^*x, L^*x\rangle$; therefore $Lx = 0$ implies $L^*x = 0$.)

Operators with Closed Range

In this section we shall show that most of the operators we have considered so far have a closed range; consequently, Theorems 1.4 and 1.5 apply, and the questions of existence or uniqueness of the solution to the operator equation can be decided. Our conclusions will be based on the following theorem, which we prove in the Appendix to this chapter:

Theorem 1A.II. *If the linear operator V and its adjoint V^* are operators with closed ranges, then the operator V plus a dyad has a closed range.*

We shall use this theorem to prove

Theorem 1.6. *If the linear operator $L = I + K$, where K is a completely continuous operator, then L has a closed range.*

First, note that the operator $I + M$ when M has a bound less than one is an operator with a closed range. This follows from Theorem 1.3, which shows that the inverse of $I + M$ exists for every vector in the space. Since the bound of M^* is the same as that of M, it also follows that $I + M^*$ has a closed range.

Now, because K is completely continuous, we may write $K = K_n + R_n$, where K_n is an n-term dyad and R_n is an operator with a bound less than one. Then the operator $V = I + R_n$ has a closed range and we have $L = V + K_n$. Since V^* also has a closed range, we conclude from an n-fold repetition of Theorem 1A.II that L has a closed range. This proves Theorem 1.6.

We may apply Theorems 1.4, 1.5, and 1.6 to the discussion of the existence and the uniqueness of the solutions of the integral equation

$$u(t) + \int_0^1 k(t, s)u(s)\, ds = f(t) \tag{1.60}$$

where

$$\int_0^1 \int_0^1 k(t, s)^2\, dt\, ds < \infty. \tag{1.61}$$

In the discussion of equation (1.50), it was shown that the integral operator is completely continuous; consequently, by Theorem 1.6, the range of the operator on the left-hand side of (1.50) is closed. From Theorems 1.4 and 1.5 we draw the following conclusions:

If (1.61) *is satisfied, then* (1.60) *has a solution if and only if*

$$\int_0^1 f(s)v(s)\, ds = 0$$

where $v(s)$ *is any solution of the equation*

$$v(s) + \int_0^1 k(t, s)v(t)\, dt = 0.$$

The solution of (1.60) *is unique if and only if there is no non-zero solution of the equation*

$$u(t) + \int_0^1 k(t, s)u(s)\, ds = 0. \tag{1.62}$$

From Problem 1.55 we conclude that *equation* (1.60) *always has a unique solution if, and only if,* (1.62) *has only the zero solution.*

PROBLEMS

1.54. If $L = I + D$ where D is a k-term dyad, prove that L is a Fredholm operator. (*Hint.* If a is in the range of D, then so is La. Suppose $L^*y = 0$. Let $a_1, a_2, \cdots, a_k$ be a basis for the range of D. If the vectors $La_1, \cdots, La_k$ are linearly independent, they also form a basis for the range of D. If y is orthogonal to the range of D, then $y = 0$; therefore, if $y \neq 0$, the vectors $La_1, \cdots, La_k$ are not linearly independent.)

1.55. Prove that, if $L = V + D$ where V and V^* have bounded inverses or if $L = I + K$ where K is completely continuous, then L is a Fredholm operator. (*Hint.* If $L = V + D$, then $V^{-1}L = I + D'$ where D' is a dyad. Use Problem 1.54 and the method of proof in Theorem 1.6.)

APPENDIX I

THE PROJECTION THEOREM

The Projection Theorem states essentially that through any point outside a subspace $\mathcal{M}$ there exists a perpendicular to $\mathcal{M}$. More precisely, the Projection Theorem states the following:

Theorem 1A.I. *If y is any vector not in a subspace $\mathcal{M}$, there exists in $\mathcal{M}$ a vector w, called the projection of y on $\mathcal{M}$, such that $y - w$ is orthogonal to $\mathcal{M}$; that is*

$$\langle y - w, x\rangle = 0$$

for every x in $\mathcal{M}$.

The vector $y - w$ is the perpendicular mentioned above. Since in Euclidean space the perpendicular is the shortest distance from a point to a subspace, we shall prove that a vector such as w exists by finding the minimum distance from y to the subspace $\mathcal{M}$. Therefore, we consider the values of

$$|y - x|^2 = \langle y - x, y - x\rangle$$

where x is any vector in $\mathcal{M}$. We assume a real-type scalar product over the field of real numbers; then, since y is not in $\mathcal{M}$, the value of $|y - x|$ is greater than zero.

Let δ be the greatest lower bound of the set of values of $|y - x|^2$ and let x_n be a sequence of vectors in $\mathcal{M}$ such that the limit of $|y - x_n|^2$ is δ. We shall show that the sequence x_n converges. For arbitrary $\varepsilon > 0$ we may find n_0 such that, for all $n > n_0$, we have

$$|y - x_n|^2 < \delta + \varepsilon.$$

Suppose that $m > n_0$; then by Problem 1.8*a*

$$\begin{aligned}|(y - x_n) + (y - x_m)|^2 &+ |(y - x_n) - (y - x_m)|^2 \\ &= 2|y - x_n|^2 + 2|y - x_m|^2 < 4(\delta + \varepsilon),\end{aligned}$$

or

$$|x_m - x_n|^2 < 4(\delta + \varepsilon) - 4\left|y - \frac{x_m + x_n}{2}\right|^2. \tag{1A.1}$$

Since $\mathcal{M}$ is a space, the vector $(x_m + x_n)/2$ is in it if the vectors x_m and x_n are; consequently, by the definition of the greatest lower bound, we have

$$\left|y - \frac{x_m + x_n}{2}\right|^2 \geq \delta.$$

Using this in (1A.1), we get

$$|x_m - x_n|^2 < 4\varepsilon.$$

Since ε was arbitrary, this proves that the sequence x_n of vectors in $\mathcal{M}$ converges to a limit. Denote this limit by w. By hypothesis, $\mathcal{M}$ is a subspace and consequently a closed manifold; therefore, w belongs to $\mathcal{M}$, and we have

$$\delta = |y - w|^2 \leq |y - x|^2 \tag{1A.2}$$

for all x in $\mathcal{M}$. If x and w belong to $\mathcal{M}$, so does the vector $w + \alpha x$ for any value of α. In (1A.2) replace x by $w + \alpha x$. We get

$$|y - w|^2 \leq |(y - w) + \alpha x|^2 = |y - w|^2 + 2\alpha\langle y - w, x\rangle + \alpha^2|x|^2$$

or

$$0 \leq \alpha^2|x|^2 + 2\alpha\langle y - w, x\rangle. \tag{1A.3}$$

Put

$$\alpha = -\frac{\langle y - w, x\rangle}{|x|^2}$$

and (1A.3) becomes the following:

$$0 \leq -\left(\frac{\langle y - w, x\rangle}{|x|}\right)^2.$$

This result implies

$$\langle y - w, x\rangle = 0,$$

which proves the theorem.

APPENDIX II

OPERATOR OF CLOSED RANGE PLUS A DYAD

We shall prove

Theorem 1A.II. *If the linear operator V and its adjoint V^* are operators with closed ranges, the operator V plus a dyad has a closed range.*

Put

$$L = V + a\rangle\langle b. \tag{1A.4}$$

The statement that L is an operator with a closed range $\mathcal{R}$ means that if there exists a sequence of elements y_n in $\mathcal{R}$ converging to a limit y, then y is in $\mathcal{R}$. Since y_n is in $\mathcal{R}$, there exists an element x_n such that $Lx_n = y_n$. What must be established is the existence of an element x such that $Lx = y$. Note that x need not be the limit of the sequence x_n. In fact, we shall give an example of an operator for which the sequence x_n does not have a limit.

To prove Theorem 1A.II, first assume that b is in the range of V^*; therefore, there exists an element b' such that

$$V^*b' = b.$$

We may now write

$$Lx_n = Vx_n + a\langle b, x_n\rangle = Vx_n + a\langle b', Vx_n\rangle = y_n. \tag{1A.5}$$

Put $Vx_n = t_n$, then from (1A.5) we have

$$t_n + a\langle b', t_n\rangle = y_n. \tag{1A.6}$$

Suppose that

$$1 + \langle b', a\rangle \neq 0. \tag{1A.7}$$

Multiply (1A.6) by b' and solve for $\langle b', t_n\rangle$. We get

$$\langle b', t_n\rangle = \frac{\langle b', y_n\rangle}{1 + \langle b', a\rangle}$$

and after substitution in (1A.6) we find that

$$t_n = y_n - \frac{a\langle b', y_n\rangle}{1 + \langle b', a\rangle}.$$

Since y_n converges to y, the right-hand side of this equation must converge to a limit t, and we have

$$t = y - \frac{a\langle b', y\rangle}{1 + \langle b', a\rangle}. \tag{1A.8}$$

By definition t_n is in the range of V. We have just proved that t_n converges to t. Since by assumption the range of V is closed, this means that there exists an element x such that

$$Vx = t = y - \frac{a\langle b', y\rangle}{1 + \langle b', a\rangle}. \tag{1A.9}$$

Multiply this equation by b' and solve for $\langle b', y\rangle$. We find

$$\langle b', y\rangle = \langle b', Vx\rangle[1 + \langle b', a\rangle].$$

Substituting this result in (1A.9) and simplifying, we get

$$y = Vx + a\langle b', Vx\rangle = Vx + a\langle b, x\rangle = Lx,$$

which was to be proved.

If (1A.7) is not satisfied, then

$$1 + \langle b', a\rangle = 0. \tag{1A.10}$$

Take the scalar product of (1A.6) with b', and using (1A.10), we find that

$$\langle b', y_n\rangle = 0$$

and going to the limit

$$\langle b', y\rangle = 0. \tag{1A.11}$$

Consider the element a. Either a is in the range of V or it is not. If a is in the range of V, there exists an element a' such that

$$Va' = a$$

and then we may write (1A.5) as

$$V[x_n + a'\langle b', Vx_n\rangle] = y_n.$$

This shows that y_n is in the range of V. Since by assumption the range of V is closed, there exists an element x such that

$$Vx = y.$$

From (1A.11) we conclude also that

$$\langle b', Vx\rangle = 0;$$

therefore,

$$Vx + a\langle b, x\rangle = Vx + a\langle b', Vx\rangle = y,$$

which was to be proved.

If a is not in the range of V, then from Theorem 1.5 it follows that there must exist a solution c of the equation

$$V^*c = 0$$

such that

$$\langle c, a\rangle \neq 0.$$

Put

$$b'' = b' + c.$$

We have

$$V^*b'' = b$$

and

$$1 + \langle b'', a\rangle = \langle c, a\rangle \neq 0.$$

If we replace b' by b'', we may repeat the proof given above for the case where (1A.7) was satisfied and thus again establish the existence of an element x such that

$$Lx = y.$$

Second, assume that b is not in the range of V^*. Since the range of V^* is closed, then by Theorem 1.5 there exists a solution d of the equation

$$Vd = 0 \tag{1A.12}$$

such that

$$\langle b, d\rangle \neq 0. \tag{1A.13}$$

Again, consider the element a. Either it is in the range of V or it is not. If a is not in the range of V, then, just as before, there exists a solution c of the equation

$$V^*c = 0 \tag{1A.14}$$

such that

$$\langle c, a\rangle \neq 0.$$

Note that because of (1A.14)

$$\langle c, Vx_n\rangle = 0. \tag{1A.15}$$

The scalar product of (1A.5) with c gives the result

$$\langle c, Vx_n\rangle + \langle c, a\rangle\langle b, x_n\rangle = \langle c, y_n\rangle.$$

With the help of (1A.15) we find that

$$\langle b, x_n\rangle = \frac{\langle c, y_n\rangle}{\langle c, a\rangle}.$$

Since y_n converges to a limit y, this equation shows that $\langle b, x_n\rangle$ converges to a limit β. We have

$$Vx_n = y_n - a\langle b, x_n\rangle \to y - \beta a.$$

By assumption the range of V is closed; therefore, there exists an element x' such that

$$Vx' = y - \beta a. \tag{1A.16}$$

Let α' be a scalar such that

$$\langle b, x' - \alpha' d\rangle = \beta. \tag{1A.17}$$

Put

$$x = x' - \alpha' d;$$

then because of (1A.12), (1A.16), and (1A.17) we have

$$Vx + a\langle b, x\rangle = y.$$

Finally, we consider the case in which a is in the range of V. As before, we assume

$$Va' = a$$

and we write (1A.5) as

$$V[x_n + a'\langle b, x_n\rangle] = y_n.$$

This shows that y_n is in the range of V. Since the range of V is closed, there exists an element x' such that

$$Vx' = y. \tag{1A.18}$$

Let α be the scalar such that

$$\langle b, x' - \alpha d\rangle = 0. \tag{1A.19}$$

Put

$$x = x' - \alpha d;$$

then from (1A.18) and (1A.19) we see that

$$Vx + a\langle b, x\rangle = y.$$

Thus, in all cases we have shown that y is in the range of L. This proves Theorem 1A.II.

We now construct the promised example of an operator for which the sequence x_n does not converge. Let $b_1, b_2, \cdots$ be an orthonormal basis in $\mathcal{S}$ and consider the operator

$$Dx = a\langle b_1, x\rangle.$$

Put

$$x_n = \frac{1}{n} b_1 + b_n, \quad n = 1, 2, \cdots,$$

then

$$Dx_n = y_n = \frac{1}{n} a.$$

The sequence y_n converges to zero, but the sequence x_n does not converge and no subsequence of the x_n converges.

2

SPECTRAL THEORY OF OPERATORS

Introduction

It was pointed out, in the previous chapter, that operator equations could be solved easily if the operator could be diagonalized, that is, if we could find a basis $e_1, e_2, \cdots$ and its reciprocal basis $f_1, f_2, \cdots$ such that the operator L could be represented as follows:

$$L = \lambda_1 e_1\rangle\langle f_1 + \lambda_2 e_2\rangle\langle f_2 + \cdots,$$

where $\lambda_1, \lambda_2, \cdots$ are scalars. As we have remarked before on pages 32 and 33, this representation implies that

$$Le_i = \lambda_i e_i, \quad i = 1, 2, \cdots. \tag{2.1}$$

We call the vectors e_i *eigenvectors* of L and the scalars λ_i *eigenvalues* of L. We shall say that the scalars λ_i are in the *spectrum* of L.

We see then that in order to diagonalize an operator, we must use its eigenvectors as a basis. In this chapter we shall discuss the eigenvectors and eigenvalues of an operator. We shall find that for some operators the eigenvectors do not span the space, and we shall discuss what can be done in such cases to obtain a simple representation for the operator. Finally, we shall apply this simple representation to solve various operator equations.

Because of the difficulties encountered when dealing with operators on infinite-dimensional spaces, most of our results will be for operators on finite-dimensional spaces, that is, essentially for matrices. However, when no restrictions are stated in the theorem, the results will be true for operators in general spaces.

Invariant Manifolds (Subspaces)

Equation (2.1) states that the one-dimensional space determined by the eigenvector e_i is invariant under L, that is, when L acts on any vector in the space, the result is still in the space. We generalize this concept of invariant vectors by considering *invariant manifolds* or *invariant subspaces.*

A manifold (subspace) $\mathcal{M}$ is called an invariant manifold (subspace) if, whenever m is a vector in $\mathcal{M}$, then Lm is also in $\mathcal{M}$. We prove

Theorem 2.1. *The whole space $\mathcal{S}$, the space containing only the zero vector, the null space of L, and the range of L are invariant manifolds of L. If $\mathcal{S}$ is finite-dimensional, the dimension of the null space plus the dimension of the range equals the dimension of $\mathcal{S}$.*

That the whole space, the space containing only the zero vector, and the null space are invariant manifolds is clear from the definition of invariant manifolds. Consider any vector r in the range. By the definition of the range, Lr is in the range, and consequently the range is also an invariant manifold.

Suppose that n is the dimension of $\mathcal{S}$, that ν is the dimension of the null space, and that ρ is the dimension of $L\mathcal{S}$, the range of L. Let $x_1, x_2, \cdots, x_\nu$ be a basis for the null space. Complete this basis to a basis for $\mathcal{S}$ by adjoining the vectors $x_{\nu+1}, x_{\nu+2}, \cdots, x_n$. We shall show that the vectors $Lx_{\nu+1}, \cdots, Lx_n$ form a basis for the range. First, consider any vector x in $\mathcal{S}$. We may write

$$x = \sum_1^n \xi_k x_k;$$

then

$$Lx = \sum_{\nu+1}^n \xi_k L x_k.$$

This shows that the vectors $Lx_{\nu+1}, \cdots, Lx_n$ span the range. Second, if

$$\alpha_{\nu+1} L x_{\nu+1} + \cdots + \alpha_n L x_n = 0,$$

then

$$L(\alpha_{\nu+1} x_{\nu+1} + \cdots + \alpha_n x_n) = 0,$$

and the vector $\alpha_{\nu+1}x_{\nu+1} + \cdots + \alpha_n x_n$ would be in the null space. However, this contradicts the assumptions that the vectors $x_1, \cdots, x_\nu$ form a basis for the null space and that $x_1, \cdots, x_n$ are linearly independent. The contradiction shows that the vectors $Lx_{\nu+1}, \cdots, Lx_n$ are linearly independent. Since they also span the range, they form a basis for the range. Since the number of elements in a basis is invariant, we conclude that

$$\rho = n - \nu,$$

which proves the theorem.

We shall call a set of vectors $x_{\nu+1}, \cdots, x_n$, such that the vectors $Lx_{\nu+1}, \cdots, Lx_n$ form a basis for the range of L, a set of *progenitors* for the range. The proof of Theorem 2.1 now implies the following

Corollary. *The finite-dimensional space $\mathcal{S}$ is the direct sum of the null space of L and the space spanned by any set of progenitors of the range of L.*

Incidentally, note that the space spanned by the set of progenitors is not necessarily invariant under L.

Theorem 2.2. *If $\mathcal{M}$ is a finite-dimensional invariant subspace of L, the effect of L on $\mathcal{M}$ may be represented by a matrix.*

In $\mathcal{M}$, choose a basis $m_1, m_2, \cdots m_k$. Consider Lm_j $(1 \leq j \leq k)$. Since $\mathcal{M}$ is invariant under L, Lm_j is in $\mathcal{M}$ and may therefore be expressed as a linear combination of the basis vectors. Put

$$Lm_j = \sum_{i=1}^{k} a_{ij} m_i. \tag{2.2}$$

Let x be any arbitrary vector in $\mathcal{M}$ with components $\xi_1, \xi_2, \cdots, \xi_k$ relative to the basis $m_1, \ldots, m_k$ so that

$$x = \xi_1 m_1 + \xi_2 m_2 + \cdots + \xi_k m_k.$$

Since Lx is also in $\mathcal{M}$, it can be expressed in terms of the basis $m_1, \cdots, m_k$. Suppose that

$$y = Lx = \eta_1 m_1 + \cdots + \eta_k m_k.$$

From (2.2) we find that

$$y = Lx = L\sum_{j=1}^{k} \xi_j m_j = \sum_{j=1}^{k} \xi_j L m_j = \sum_{j=1}^{k} \xi_j \sum_{i=1}^{k} \alpha_{ij} m_i = \sum_{i=1}^{k} m_i \sum_{j=1}^{k} \alpha_{ij} \xi_j \ ;$$

hence

$$\eta_i = \sum_{j=1}^{k} \alpha_{ij} \xi_j. \tag{2.3}$$

This shows that in $\mathcal{M}$, L can be represented by the k-rowed square matrix A whose elements are α_{ij}.

Suppose that the whole space $\mathcal{S}$ itself has finite dimension n and that $\mathcal{M}$ is a subspace of dimension k and is invariant under L; then we know that $\mathcal{S}$ can be expressed as the direct sum of $\mathcal{M}$ and a subspace $\mathcal{N}$ whose dimension is $n - k$. In general, $\mathcal{N}$ will not be invariant under L. The representation of L in $\mathcal{S}$ will be given by

Theorem 2.3. *If $\mathcal{M}$ is invariant under L and if $\mathcal{M}$ is a subspace of a finite-dimensional space $\mathcal{S}$, the operator L may be represented in $\mathcal{S}$ by the matrix*

$$\begin{pmatrix} A & B \\ 0 & D \end{pmatrix}$$

where A is the matrix of Theorem 2.2, B is a k by $n - k$ matrix, and D is an $(n - k)$-rowed square matrix.

Let $m_1, \cdots, m_k$ be a basis for $\mathcal{M}$, and $n_1, n_2, \cdots, n_{n-k}$ a basis for $\mathcal{N}$. Since $\mathcal{S}$ is the direct sum of $\mathcal{M}$ and $\mathcal{N}$, every vector x in $\mathcal{S}$ can be expressed uniquely in terms of m_i and n_j. We write

$$x = \begin{pmatrix} \mu \\ \nu \end{pmatrix}$$

where μ denotes the k components in the subspace $\mathcal{M}$ and ν denotes the $n - k$ components in the subspace $\mathcal{N}$.

To find the representation of L we assume

$$L = \begin{pmatrix} A & B \\ C & D \end{pmatrix} \begin{matrix} k \text{ rows} \\ n-k \text{ rows} \end{matrix}$$
$$k \text{ columns} \qquad n-k \text{ columns}$$

where A, B, C, D are submatrices of the matrix for L; then

$$Lx = \begin{pmatrix} A & B \\ C & D \end{pmatrix} \begin{pmatrix} \mu \\ \nu \end{pmatrix} = \begin{pmatrix} A\mu + B\nu \\ C\mu + D\nu \end{pmatrix}.$$

Suppose that x is in $\mathcal{M}$ and hence $\nu = 0$; then Lx should also be in $\mathcal{M}$ since $\mathcal{M}$ is invariant under L. This implies that $C\mu = 0$ for arbitrary μ, and therefore $C = 0$. This implies that Lx on $\mathcal{M}$ is just $A\mu$. Hence, A is the matrix described in Theorem 2.2. This proves Theorem 2.3.

Corollary. *If $\mathcal{M}$ and $\mathcal{N}$ are both invariant subspaces of L and if $\mathcal{S}$ is the direct sum of $\mathcal{M}$ and $\mathcal{N}$, the effect of L on $\mathcal{S}$ is represented by the diagonal block matrix*

$$L = \begin{pmatrix} A & 0 \\ 0 & D \end{pmatrix}.$$

This follows from the fact that if x is in $\mathcal{N}$, $\mu = 0$; and since Lx is also in $\mathcal{N}$, then $B\nu = 0$ for all ν.

In the case described in the corollary, L may be considered as the sum of two operators, L_1 acting only on $\mathcal{M}$ and L_2 acting only on $\mathcal{N}$. We put

$$L_1 = \begin{pmatrix} A & 0 \\ 0 & 0 \end{pmatrix}, \quad L_2 = \begin{pmatrix} 0 & 0 \\ 0 & D \end{pmatrix},$$

so that

$$L = L_1 + L_2,$$
$$L_1 L_2 = L_2 L_1 = 0.$$

The most important invariant subspaces are those which are one-dimensional; that is, $\mathcal{M}$ contains only vectors of the form αm, where m is a fixed vector and α an arbitrary scalar. By the definition of invariant subspace, it follows that

$$Lm = \lambda m \tag{2.4}$$

where λ is a scalar. The vector m is said to be a *characteristic vector* or *eigenvector* of L, and λ is a *characteristic value* or *eigenvalue* of L. Geometrically, (2.4) means that, under the transformation L, vectors in the m direction keep their direction fixed but have their lengths multiplied by λ. Note that $\mathcal{M}$ is contained in the null space of $L - \lambda I$, where I is the identity operator. Every vector in the null space of $L - \lambda I$ is an eigenvector of L corresponding to the eigenvalue λ. The dimension of the null space will be called the *multiplicity* of the eigenvalue λ.

We prove

Theorem 2.4. *If $\mathcal{M}$ is a finite-dimensional subspace invariant under L, there exists an eigenvector of L in $\mathcal{M}$.*

By Theorem 2.2., L acting in $\mathcal{M}$ can be represented by a matrix A relative to a fixed basis; hence if x is any vector in $\mathcal{M}$,

$$Lx = Ax.$$

Now for x to be an eigenvector, we must have

$$(A - \lambda I)x = \sum_{j=1}^{k} (\alpha_{ij} - \lambda \delta_{ij})\xi_j = 0. \tag{2.5}$$

A necessary and sufficient condition for an eigenvector to exist is that there should be a solution of (2.5) for which not all ξ_j are zero. Such a solution will exist if and only if the determinant of the coefficients, $\det |\alpha_{ij} - \lambda\delta_{ij}|$ is zero. Since the determinant is a polynomial of the kth degree in λ, it will certainly have a zero, real or complex. For any zero, there will exist a solution of (2.5) for which not all ξ_j are zero. The vector with these scalars ξ_j as components will be an eigenvector.

As an illustration, suppose L is an operator that has the following matrix representation in E_2:

$$L = \begin{pmatrix} 3 & 4 \\ 1 & 3 \end{pmatrix}.$$

If the vector with components (ξ_1, ξ_2) is an eigenvector of L, there exists a real or complex number λ such that

$$L\begin{pmatrix} \xi_1 \\ \xi_2 \end{pmatrix} = \lambda \begin{pmatrix} \xi_1 \\ \xi_2 \end{pmatrix}$$

or

$$(L - \lambda I)\begin{pmatrix} \xi_1 \\ \xi_2 \end{pmatrix} = 0. \tag{2.6}$$

Equation (2.6) implies the following set of linear equations:

$$\begin{aligned} (3 - \lambda)\xi_1 + 4\xi_2 &= 0, \\ \xi_1 + (3 - \lambda)\xi_2 &= 0. \end{aligned} \tag{2.7}$$

These equations have a non-trivial solution if and only if

$$\begin{vmatrix} 3-\lambda & 4 \\ 1 & 3-\lambda \end{vmatrix} = 0. \tag{2.8}$$

The roots of (2.8) are $\lambda = 1$ and $\lambda = 5$. Put $\lambda = 1$ in (2.7) and we find that the ratio of ξ_1 to ξ_2 is $-2/1$. Hence, any vector x_1 with its components proportional to $-2, 1$ is then an eigenvector of L corresponding to the eigenvalue $\lambda = 1$. Similarly, when we put $\lambda = 5$ in (2.7), the ratio of ξ_1 to ξ_2 is $2/1$; thus, a vector x_2 with components proportional to 2, 1 is an eigenvector of L corresponding to the eigenvalue $\lambda = 5$. Note that the two vectors x_1 and x_2 are *not* orthogonal.

Since x_1 and x_2 are linearly independent, any vector x in E_2 can be written as follows:

$$x = \eta_1 x_1 + \eta_2 x_2,$$

where η_1 and η_2 are scalars. From the linearity of L and the properties of x_1 and x_2 it follows that

$$Lx = \eta_1 L x_1 + \eta_2 L x_2 = \eta_1 x_1 + 5\eta_2 x_2.$$

This result may be used to solve the equation

$$Lx = a. \tag{2.9}$$

Write $a = \alpha_1 x_1 + \alpha_2 x_2$ where α_1 and α_2 are known scalars; then (2.9) becomes

$$\eta_1 x_1 + 5\eta_2 x_2 = \alpha_1 x_1 + \alpha_2 x_2,$$

and consequently

$$\eta_1 = \alpha_1, \quad \eta_2 = \frac{\alpha_2}{5}.$$

This shows that the solution of (2.9) is

$$x = \alpha_1 x_1 + \frac{\alpha_2 x_2}{5}.$$

We see then that there are two essentially different methods for solving (2.9). One way is to find the inverse operator L^{-1}. The other way is to find the eigenvalues and eigenvectors of L and then to solve the equation, using these. Both techniques will be important in later work.

PROBLEMS

2.1. Consider the linear vector space whose elements $p(x, y, z)$ are homogeneous polynomials in x, y, z, and consider the permutation operator P such that

$$Pp(x, y, z) = p(y, x, z).$$

Show that the subspace of polynomials of a fixed degree is invariant under P. What is the matrix representation of P in the subspace of polynomials of degree one? What are the eigenvalues and eigenvectors of P in that subspace? (*Hint.* The three polynomials x, y, and z form a basis for the subspace.)

2.2. Consider the operator P of the preceding problem in the subspace of polynomials of degree two. Find the matrix that represents P, and determine its eigenvalues and eigenvectors.

2.3. Let (x, y, z) represent the coordinates of a point in E_3. Consider the operator defined by the following equations:

$$x' = 2x + y + z,$$
$$y' = -3x - y + 2z,$$
$$z' = x - 3z.$$

Show that the subspace defined by $x + y + z = 0$ is invariant under this operator. Find the representation of the operator in this subspace and find the eigenvalues of the representation.

Commuting Operators

We digress for a moment from the problem of finding the eigenvectors of an operator. Instead, we shall prove some theorems which will be interesting in themselves and will enable us also to clarify the distinction between an operator and its representation.

Two operators L and K are said to *commute* if $LK = KL$. We prove

Theorem 2.5.† *If L and K commute, then the range and null space of L are invariant manifolds for K. Similarly, the range and null space of K are invariant manifolds of L.*

To prove this, let x be in the null space of L; that is, let $Lx = 0$; then

$$0 = KLx = LKx.$$

Consequently, Kx is in the null space of L, and therefore the null space of L is invariant under K.

Suppose now that y is in the range of L; that is, there exists a vector z such that $Lz = y$. Then

$$Ky = KLz = L(Kz);$$

consequently, Ky is in the range of L, and therefore the range of L is invariant under K. Similarly, we can prove that the range and null space of K are invariant under L.

An important consequence of this theorem is given by the following

Corollary. *If L and K commute and if one of the operators has an eigenvalue of finite multiplicity, both operators have a common eigenvector; that is, there exists a vector x such that*

$$Lx = \lambda x, \quad Kx = \kappa x,$$

where λ and κ are scalars.

† This theorem contains the essential part of Schur's lemma. See Murnaghan, *Theory of Group Representations*, Johns Hopkins Press, Baltimore, 1938.

The proof depends on noticing that, since $Ly = \lambda y$ implies $(L - \lambda I)y = 0$, the eigenvectors of L corresponding to a given eigenvalue λ are elements of the null space for the operator $L - \lambda I$. Now, if L commutes with K, so will $L - \lambda I$; consequently, by Theorem 2.5, the null space of $L - \lambda I$ is invariant under K. Since by hypothesis this null space is finite-dimensional, Theorem 2.2 shows that K can be represented as a matrix on this space. It follows from Theorem 2.4 that K has an eigenvector x in this null space; consequently, we have

$$Kx = \kappa x \quad \text{and} \quad (L - \lambda I)x = 0,$$

which proves the corollary.

We shall discuss some applications of this corollary to the theory of differential equations. Consider the differential operator

$$L = -\frac{d^2}{dt^2} + q(t)$$

acting on the linear manifold of twice-differentiable functions $u(t)$ such that $u(1) = u(-1) = 0$. Suppose that $q(t)$ is an *even* function of t; then if R is the operator such that

$$Ru(t) = u(-t),$$

we can show that R and L commute. First, the manifold of functions $u(t)$, such that $u(1) = u(-1) = 0$, is invariant under R; second,

$$R(Lu) = R[-u''(t) + q(t)u(t)] = -u''(-t) + q(-t)u(-t)$$

and

$$L(Ru) = L[u(-t)] = -u''(-t) + q(t)u(-t).$$

Since by assumption $q(t) = q(-t)$, we have $RL = LR$.

The corollary tells us that R and L must have common eigenvectors, that is, there exist functions $u(t)$ such that

$$Lu = \lambda u \quad \text{and} \quad Ru = \rho u.$$

The eigenvalues of R may be readily found. From

$$Ru(t) = u(-t) = \rho u(t)$$

we get

$$R^2u(t) = Ru(-t) = u(t) = \rho Ru(t) = \rho^2 u(t).$$

This equation implies that $\rho^2 = 1$, or, $\rho = \pm 1$. The eigenfunctions corresponding to $\rho = 1$ are functions $u(t)$ such that

$$Ru(t) = u(-t) = u(t),$$

that is, even functions of t, whereas the eigenfunctions corresponding to $\rho = -1$ are functions $v(t)$ such that

$$Rv(t) = v(-t) = -v(t),$$

that is, odd functions of t. We conclude then that the eigenfunctions of L may be separated into either even or odd functions of t.

As a final application, consider the differential operator

$$L = -\frac{d^2}{dt^2} + q(t)$$

acting on the linear manifolds of functions $u(t)$ which are twice-differentiable from $-\infty$ to ∞. Suppose that $q(t)$ is periodic with period h, that is, $q(t + h) = q(t)$ for all t. We shall define a translation operator T as the operator such that

$$Tu(t) = u(t + h),$$

and we shall prove that T commutes with L.

We have

$$TLu(t) = T[-u''(t) + q(t)u(t)] = -u''(t + h) + q(t + h)u(t + h)$$

and $$LTu(t) = L[u(t + h)] = -u''(t + h) + q(t)u(t + h).$$

Since by assumption $q(t)$ has period h, we see that

$$TL = LT.$$

From Theorem 2.5 the null space of L is invariant under T, that is, the space of all functions $u(t)$ which are solutions of the equation

$$-u''(t) + q(t)u(t) = 0 \tag{2.10}$$

is invariant under T. This means that if $u(t)$ is a solution of (2.10), then so is $u(t + h)$. From the theory of linear differential equations we know that there exist two fundamental solutions $u_1(t)$ and $u_2(t)$ of equation (2.10) such that any solution $u(t)$ of (2.10) is expressible as a linear combination, that is,

$$u(t) = \alpha u_1(t) + \beta u_2(t),$$

where α and β are scalars; consequently, the null space of L is two-dimensional.

By Theorem 2.2 the action of T on this null space can be represented by a 2×2 matrix. To obtain this matrix we may proceed as follows.

Since $u_1(t + h)$ is a solution of (2.10), we may write

$$u_1(t + h) = \alpha_{11}u_1(t) + \alpha_{21}u_2(t).$$

Similarly, we have

$$u_2(t + h) = \alpha_{12}u_1(t) + \alpha_{22}u_2(t).$$

If we put

$$u(t + h) = \gamma u_1(t) + \delta u_2(t),$$

we get

$$\begin{aligned}\gamma u_1(t) + \delta u_2(t) &= \alpha u_1(t + h) + \beta u_2(t + h)\\ &= (\alpha\alpha_{11} + \beta\alpha_{12})u_1(t) + (\alpha\alpha_{21} + \beta\alpha_{22})u_2(t).\end{aligned}$$

We see then that

$$\begin{pmatrix}\gamma\\ \delta\end{pmatrix} = \begin{pmatrix}\alpha_{11} & \alpha_{12}\\ \alpha_{21} & \alpha_{22}\end{pmatrix}\begin{pmatrix}\alpha\\ \beta\end{pmatrix}.$$

The matrix

$$A = \begin{pmatrix}\alpha_{11} & \alpha_{12}\\ \alpha_{21} & \alpha_{22}\end{pmatrix}$$

is the representation of the operator T acting on the two-dimensional space of solutions of (2.10). Note that the operator T is still a translation operator and is not the matrix A but that its effect may be represented by the matrix A. The matrix representation is not unique, for if we use two other fundamental solutions of (2.10) instead of $u_1(t)$ and $u_2(t)$ we shall represent T by a matrix different from A.

Later, we shall investigate the relationship between the different matrices that may represent the same operator. At present, we note that by Theorem 2.4 every matrix has at least one eigenvector. In general, a two-dimensional matrix such as A will have two eigenvectors. Suppose the components of these two eigenvectors are (α_1, β_1) and (α_2, β_2), corresponding to the eigenvalues λ_1 and λ_2, respectively. Each eigenvector will define an eigenfunction of T as follows:

$$v_1(t) = \alpha_1 u_1(t) + \beta_1 u_2(t),$$

$$v_2(t) = \alpha_2 u_1(t) + \beta_2 u_2(t).$$

We shall have

$$v_1(t + h) = Tv_1(t) = \lambda_1 v_1(t), \quad v_2(t + h) = Tv_2(t) = \lambda_2 v_2(t).$$

Put

$$\rho_1 = h^{-1} \log \lambda_1,$$

then

$$v_1(t + h) = e^{h\rho_1} v_1(t)$$

or

$$e^{-\rho_1(t+h)} v_1(t + h) = e^{-\rho_1 t}\, v_1(t).$$

This shows that the function $w_1(t) = e^{-\rho_1 t}\, v_1(t)$ is a periodic function of t with a period h. A similar result holds for the function $w_2(t) = e^{-\rho_2 t}\, v_2(t)$, where $\rho_2 = h^{-1} \log \lambda_2$. Therefore, we conclude that equation (2.10) has two linearly independent solutions $v_1(t)$, $v_2(t)$ such that

$$v_1(t) = e^{\rho_1 t} w_1(t), \quad v_2(t) = e^{\rho_2 t} w_2(t),$$

where $w_1(t)$ and $w_2(t)$ are periodic functions of t with period h.†

† This result is known as Floquet's theorem. See Ince, *Ordinary Differential Equations*, pp. 381–384, Dover, New York, 1944.

If we now introduce $v_1(t)$ and $v_2(t)$ as a basis for the space of solutions $u(t)$ of (2.10), we see that we get

$$u(t) = \alpha' v_1(t) + \beta' v_2(t),$$
$$u(t+h) = Tu(t) = \alpha'\lambda_1 v_1(t) + \beta'\lambda_2 v_2(t).$$

From these considerations it is easy to discuss the possibility of finding solutions of (2.10) which have the period h. Clearly, $v_1(t)$ will be such a periodic function if $\lambda_1 = 1$. Similarly, the possibility of finding solutions of (2.10) which have a period that is any multiple of h may be discussed.

PROBLEMS

2.4. If T_h is the translation operator through a distance h, show that, for all values of h, T_h commutes with any differential operator with constant coefficients.

2.5 If R_θ is the operator which rotates $E_2(\xi, \eta)$ through an angle θ, show that it commutes with the Laplacian $\dfrac{\partial^2}{\partial \xi^2} + \dfrac{\partial^2}{\partial \eta^2}$.

2.6. Suppose that L_1 and L_2 are operators which have no common invariant manifolds except the zero vector and the whole space. If K is an operator which commutes with both L_1 and L_2, prove that either K maps every element onto zero or K has a unique inverse. (*Hint.* By Theorem 2.5 the range and null space of K are invariant manifolds of both L_1 and L_2.)

2.7 Suppose that $K = L_1L_2$, where $L_1L_2 - L_2L_1 = I$. Show that if x is an eigenvector of K corresponding to the eigenvalue λ, then L_1x is an eigenvector of K corresponding to the eigenvalue $\lambda - 1$, and L_2x is an eigenvector of K corresponding to the eigenvalue $\lambda + 1$. (*Hint.* We have $L_1(L_1L_2x) - L_1L_2(L_1x) = L_1x$ and also $L_1L_2(L_2x) - L_2(L_1L_2x) = L_2x$.)

• **2.8.** Use Problem 2.7 to find the eigenvalues and eigenfunctions of the operator $K = -\dfrac{d^2}{dt^2} + \dfrac{t^2}{4}$ in the space of square integrable functions over $-\infty$ to ∞. (*Hint.* Put $L_1 = \dfrac{t}{2} - \dfrac{d}{dt}$, $L_2 = \dfrac{t}{2} + \dfrac{d}{dt}$ and note that $\exp(-t^2/4)$ is an eigenfunction corresponding to the eigenvalue $1/2$.)

Generalized Eigenvectors

The use of eigenvectors to solve the equation $Lx = a$ is justified only if the eigenvectors of L span the whole space. It is easy to give cases where the eigenvectors of L do not span the whole space. For example, let L be represented by the following matrix in E_3:

$$L = \begin{pmatrix} 1 & 1 & 2 \\ 0 & 1 & 3 \\ 0 & 0 & 2 \end{pmatrix}.$$

We have det $|L - \lambda I| = (\lambda - 1)^2(\lambda - 2)$; hence $\lambda = 1$ and $\lambda = 2$. For $\lambda = 1$, the equation $(L - \lambda)x = 0$ becomes

$$\begin{aligned} \xi_2 + 2\xi_3 &= 0, \\ 3\xi_3 &= 0, \\ 2\xi_3 &= 0, \end{aligned}$$

and these equations imply $\xi_2 = \xi_3 = 0$; therefore, the eigenvector x_1 corresponding to $\lambda = 1$ has components proportional to 1, 0, 0. For $\lambda = 2$, the equation $(L - \lambda)x = 0$ becomes

$$\begin{aligned} -\xi_1 + \xi_2 + 2\xi_3 &= 0, \\ -\xi_2 + 3\xi_3 &= 0, \\ 0 &= 0, \end{aligned}$$

and then $\xi_2 = 3\xi_3$, $\xi_1 = 5\xi_3$; therefore, the eigenvector x_3 corresponding to $\lambda = 2$ has components proportional to 5, 3, 1. These two eigenvectors x_1 and x_3 obviously do not span the three-dimensional space. To be able to span the space we use a vector x_2 for which

$$(L - \lambda I)x_2 \neq 0$$

but for which

$$(L - \lambda I)^2 x_2 = 0.$$

It is easy to see that when $\lambda = 1$ one such vector x_2 has components proportional to 0, 1, 0. However, when $\lambda = 2$, there is no such eigenvector.

A vector x_k, for which

$$(L - \lambda_0 I)^{k-1} x_k \neq 0, \tag{2.11}$$

but for which

$$(L - \lambda_0 I)^k x_k = 0, \tag{2.12}$$

will be called a *generalized eigenvector* or an *eigenvector of rank k* corresponding to the eigenvalue λ_0. Put

$$\begin{aligned} (L - \lambda_0 I)x_k &= x_{k-1}, \\ (L - \lambda_0 I)^2 x_k &= x_{k-2}, \\ &\cdot \quad \cdot \quad \cdot \quad \cdot \quad \cdot \\ (L - \lambda_0 I)^{k-1} x_k &= x_1. \end{aligned} \tag{2.13}$$

Since from (2.12)

$$(L - \lambda_0 I)^j x_j = (L - \lambda_0 I)^j (L - \lambda_0 I)^{k-j} x_k = (L - \lambda_0 I)^k x_k = 0, \tag{2.14}$$

and from (2.11)

$$(L - \lambda_0 I)^{j-1} x_j = (L - \lambda_0 I)^{j-1} (L - \lambda_0 I)^{k-j} x_k = (L - \lambda_0 I)^{k-1} x_k \neq 0 \tag{2.15}$$

we see that x_j is a generalized eigenvector of rank j corresponding to the eigenvalue λ_0.

Eigenvectors of different rank will be linearly independent. Suppose that

$$\alpha_1 x_1 + \alpha_2 x_2 + \cdots + \alpha_j x_j = 0;$$

then by (2.14)

$$(L - \lambda_0 I)^{j-1}(\alpha_1 x_1 + \alpha_2 x_2 + \cdots + \alpha_j x_j) = \alpha_j (L - \lambda_0 I)^{j-1} x_j = 0.$$

But from (2.15) this result implies $\alpha_j = 0$. Similarly, we can show that $\alpha_{j-1}, \alpha_{j-2}, \cdots, \alpha_1$ are all zero, and hence the vectors $x_1, \cdots, x_k$ are linearly independent.

Consider the set $\mathcal{N}$ of all vectors x in $\mathcal{S}$ for which there exists an integer k such that

$$(L - \lambda_0 I)^k x = 0. \tag{2.16}$$

We shall call this set $\mathcal{N}$ the *generalized null space* of the operator $L - \lambda_0$. $\mathcal{N}$ is a linear subspace of $\mathcal{S}$, for if x_1 and x_2 belong to $\mathcal{N}$ with exponents k_1 and k_2, respectively, then

$$(L - \lambda_0)^k(\alpha_1 x_1 + \alpha_2 x_2) = 0,$$

where k is equal to the larger of k_1 and k_2.

If $\mathcal{S}$ has the finite dimension n, the smallest integer k for which (2.16) holds must be not greater than n. For, if this were not so, there would exist a vector x_k such that

$$(L - \lambda_0)^k x_k = 0,$$
$$(L - \lambda_0)^{k-1} x_k \neq 0,$$

with $k > n$. Then consider the vectors $x_1, x_2, \cdots, x_k$, defined in (2.13). We have shown that these k vectors are linearly independent. But a space of dimension n has at most n linearly independent vectors; consequently, k is not greater than n.

We see then that

$$(L - \lambda_0)^n x = (L - \lambda_0)^{n-k}(L - \lambda_0)^k x = 0$$

for every x in $\mathcal{N}$. We write this result as

$$(L - \lambda_0)^n \mathcal{N} = 0,$$

and we say that the operator $(L - \lambda_0)^n$ *annihilates* $\mathcal{N}$. There may exist an integer m smaller than n such that $(L - \lambda_0)^m$ annihilates $\mathcal{N}$. The smallest such integer will be called the *index* of the eigenvalue λ_0. More precisely, the integer m is the index of the eigenvalue λ_0 if

$$(L - \lambda_0)^m \mathcal{N} = 0; \tag{2.17}$$

but

$$(L - \lambda_0)^{m-1} \mathcal{N} \neq 0. \tag{2.18}$$

This last formula means that there exists a vector x in $\mathcal{N}$ such that

$$(L - \lambda_0)^{m-1}x \neq 0.$$

The generalized null space $\mathcal{N}$ is an invariant subspace of L, for if x is in $\mathcal{N}$, then by (2.17) $(L - \lambda_0)^m x = 0$. Consequently, Lx is also in $\mathcal{N}$ because

$$(L - \lambda_0)^m Lx = L(L - \lambda_0)^m x = 0.$$

By Theorem 2.2 the action of L on $\mathcal{N}$ can be represented by a matrix. In the next section we shall obtain a simple representation for L by introducing an appropriate basis in $\mathcal{N}$.

PROBLEMS

2.9. In E_∞ consider the "Destruction" operator D of Chapter 1 defined as follows: $Dx = (\xi_2, \xi_3, \cdots)$ if $x = (\xi_1, \xi_2, \cdots)$. Show that the only eigenvector corresponding to the eigenvalue zero is $x_1 = (1, 0, 0, \cdots)$. Show also that zero is an eigenvalue of infinite index. (*Hint.* $D^j x_j = 0$ when x_j is a vector all of whose components are zero except the jth component.)

2.10. Consider the following operators in E_4:

$$L_1 = \begin{pmatrix} 0 & 1 & 0 & 0 \\ 0 & 0 & 1 & 0 \\ 0 & 0 & 0 & 1 \\ 0 & 0 & 0 & 0 \end{pmatrix}, \quad L_2 = \begin{pmatrix} 0 & 1 & 0 & 0 \\ 0 & 0 & 0 & 0 \\ 0 & 0 & 0 & 1 \\ 0 & 0 & 0 & 0 \end{pmatrix}, \quad L_3 = \begin{pmatrix} 0 & 1 & 0 & 0 \\ 0 & 0 & 1 & 0 \\ 0 & 0 & 0 & 0 \\ 0 & 0 & 0 & 0 \end{pmatrix}, \quad L_4 = \begin{pmatrix} 0 & 1 & 0 & 0 \\ 0 & 0 & 0 & 0 \\ 0 & 0 & 0 & 0 \\ 0 & 0 & 0 & 0 \end{pmatrix}.$$

Find the eigenvectors and generalized eigenvectors of these operators. (*Hint.* Show that L_1 has a generalized eigenvector of rank 4, L_2 has two generalized eigenvectors of rank 2, L_3 has one generalized eigenvector of rank 3, and L_4 has one generalized eigenvector of rank 2.)

2.11. Show that for a completely continuous operator K the generalized null space corresponding to a non-zero eigenvalue λ is finite-dimensional. (*Hint.* Suppose that the generalized null space $\mathcal{N}$ has infinite dimension. Put $K = K_n + R$ where K_n is an n-term dyad and where the bound of R is $\varepsilon < \lambda/2$. Let $e_1, e_2, \cdots$ be linearly independent vectors in $\mathcal{N}$ such that $(K - \lambda)e_{m+1}$ is a linear combination of $e_1, \cdots, e_m$ for all values of m. By using projections we construct from these vectors the vectors $g_1, g_2, \cdots$ such that for all m the length of g_m is one, the distance between g_{m+1} and any linear combination of $g_1, \cdots, g_m$ is not less than one, and $(K - \lambda)g_{m+1}$ is a linear combination of $g_1, \cdots, g_m$. Then for $m > m'$, we have $K_n(g_m - g_{m'}) = \lambda(g_m - h_m) - R(g_m - g_{m'})$, where h_m is a linear combination of $g_1, \cdots, g_{m-1}$. From the definition of the g-vectors, this implies $|K_n(g_m - g_{m'})| > \lambda/2$. Since the vectors $K_n g_m$ belong to an n-dimensional subspace, there exists a subsequence of the g-vectors such that $K_n g_m$ converges to a limit.)

Representation of an Operator in a Generalized Null Space

We wish to find a basis for $\mathcal{N}$, the generalized null space of $L - \lambda_0$, such that the representation in terms of this basis will be as simple as possible. In the preceding section we saw that all the elements of $\mathcal{N}$ are eigenvectors or generalized eigenvectors of L. We saw also that repeated applications of $L - \lambda_0$ to a generalized eigenvector of rank r generates a *chain* of r generalized eigenvectors. The basis we want will be made up of maximal chains of vectors, that is, chains which contain the largest possible numbers of vectors.

A specific example will clarify the method. Consider the following operator in E_6:

$$L = \begin{pmatrix} 0 & 0 & 0 & 0 & 1 & 0 \\ 0 & 0 & 1 & 1 & 0 & -1 \\ 0 & 0 & 0 & 0 & 0 & 0 \\ 0 & 0 & 0 & 0 & 0 & 0 \\ 0 & 0 & 1 & 0 & 0 & -1 \\ 0 & 0 & 0 & 0 & 0 & 0 \end{pmatrix}.$$

If we denote the column vector which has one in the jth row and zero in the other rows by a_j ($j = 1, 2, \cdots, 6$), an arbitrary vector x in E_6 will be represented by

$$x = \sum_1^6 \xi_j a_j$$

and we find that

$$\begin{aligned} Lx &= \xi_5 a_1 + (\xi_3 + \xi_4 - \xi_6)a_2 + (\xi_3 - \xi_6)a_5, \\ L^2x &= (\xi_3 - \xi_6)a_1, \\ L^3x &= 0. \end{aligned} \tag{2.19}$$

The last equation shows that E_6 is a generalized null space of index three for the eigenvalue zero. We wish to simplify the representation of L by finding a new basis in E_6. Consider the effect of L^2 on E_6. From the second equation in (2.19) we see that the null space $\mathcal{N}_2$ of L^2 is the space of all vectors in E_6 such that $\xi_3 = \xi_6$, and that the range of L^2 is the space spanned by a_1. By the corollary to Theorem 2.1, the space E_6 will be the direct sum of $\mathcal{N}_2$ and the space $\mathcal{P}_2$ of the progenitors of the range of L^2. As a progenitor of the range space we may take any vector x such that $L^2x = a_1$; for example, put $x = a_3$. The vector a_3 will be a basis for $\mathcal{P}_2$ and also one vector in the new basis.

Now we shall find a basis for the null space $\mathcal{N}_2$ of L^2, that is, the space of all vectors x in E_6 such that $\xi_3 = \xi_6$. Consider the effect of L on $\mathcal{N}_2$. From the first equation in (2.19) we see that the range of L on $\mathcal{N}_2$ is two-dimensional (note $\xi_3 = \xi_6$) and that it is spanned by the vectors a_1 and a_2. Note that the null space of L acting on E_6 is the same as the null space of L acting on $\mathcal{N}_2$. Again by the corollary to Theorem 2.1, this time applied to $\mathcal{N}_2$ and the operator L, we see that $\mathcal{N}_2$ will be the direct sum of the null space $\mathcal{N}_1$ of L acting on $\mathcal{N}_2$ and the space $\mathcal{P}_1$ of progenitors of the range of L acting on $\mathcal{N}_2$. We try to find a basis for $\mathcal{P}_1$. Since the vector a_3 is already in $\mathcal{P}_2$, we consider the vector $b_2 = La_3$. We have

$$b_2 = La_3 = a_2 + a_5.$$

Since $L^2 b_2 = 0$, the vector b_2 is in $\mathcal{N}_2$ and since $Lb_2 = a_1$, we see that b_2 is in $\mathcal{P}_1$. To complete a basis for $\mathcal{P}_1$ we may take any vector x in $\mathcal{N}_2$ such that $Lx = a_2$; for example, take $x = a_4$. The vectors a_3, b_2, and a_4 will be part of the new basis for E_6.

The rest of the basis will be found by obtaining a basis for $\mathcal{N}_1$. The space $\mathcal{N}_1$ is the set of vectors x such that

$$\xi_5 = \xi_3 + \xi_4 - \xi_6 = \xi_3 - \xi_6 = 0.$$

Consider the effect of L^0, that is, the identity operator, on $\mathcal{N}_1$. The range of L^0 acting on $\mathcal{N}_1$ is exactly the three-dimensional space $\mathcal{N}_1$, and the null space of L^0 acting on $\mathcal{N}_1$ is the zero vector. We shall have a basis for E_6 if we adjoin to the vectors we already have, namely, a_3, b_2, and a_4, a set of progenitors of the range of L^0 acting on $\mathcal{N}_1$. To find these progenitors, consider L acting on the basis for $\mathcal{P}_1$, that is, on b_2 and a_4. We have

$$Lb_2 = L(La_3) = a_1$$
$$La_4 = a_2.$$

Because b_2 and a_4 are in $\mathcal{N}_2$, the vectors Lb_2 and La_4 are in $\mathcal{N}_1$ and can be used as progenitors of the range of L^0 acting on $\mathcal{N}_1$. To get a complete set of progenitors for this range, we take any vector x in $\mathcal{N}_1$ which is independent of b_2 and La_4; for example, take the vector $x = a_3 - a_6$.

To sum up, we have obtained a new basis for E_6. Denote the vectors of this basis by $b_1, b_2, \cdots, b_6$, where

$$b_1 = L^2 a_3 = a_1,$$
$$b_2 = La_3 = a_2 + a_5,$$
$$b_3 = a_3,$$
$$b_4 = La_4 = a_2,$$
$$b_5 = a_4,$$
$$b_6 = a_3 - a_6.$$

Note that this basis contains a chain of length three, a chain of length two, and a chain of length one. In terms of this basis, we find the following simple representation for L:

$$L = \begin{pmatrix} 0 & 1 & 0 & 0 & 0 & 0 \\ 0 & 0 & 1 & 0 & 0 & 0 \\ 0 & 0 & 0 & 0 & 0 & 0 \\ 0 & 0 & 0 & 0 & 1 & 0 \\ 0 & 0 & 0 & 0 & 0 & 0 \\ 0 & 0 & 0 & 0 & 0 & 0 \end{pmatrix}.$$

We shall prove

Theorem 2.6. *If $\mathcal{N}$ is a generalized null space of index ν for L, there exists a basis for $\mathcal{N}$ such that in terms of this basis L has the following representation:*

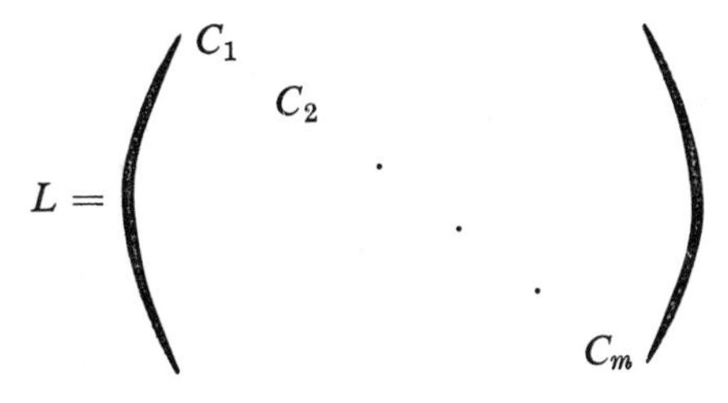

$$L = \begin{pmatrix} C_1 & & & & \\ & C_2 & & & \\ & & \cdot & & \\ & & & \cdot & \\ & & & & \cdot \\ & & & & & C_m \end{pmatrix}$$

where C_j $(1 \leq j \leq m)$ is a square matrix of order δ_j all of whose elements are zero except that the elements in the diagonal above the main diagonal are all one. The numbers δ_j are independent of the basis chosen, and we have

$$1 \leq \delta_j \leq \nu, \quad 1 \leq j \leq m.$$

The proof of this theorem will be based on the method used in the previous section. Let $\mathcal{N}_j$ $(0 \leq j \leq \nu)$ be the null space of L^j acting on $\mathcal{N}$ and let ν_j be its dimension. We see that $\mathcal{N}_0$ contains only the zero vector and that $\mathcal{N}_\nu$ is the whole space. We prove the following lemma:

Lemma. *The null space of L^k acting on $\mathcal{N}_{k+1}$ is the null space $\mathcal{N}_k$.*

The null space of L^k acting on $\mathcal{N}_{k+1}$ is the set of vectors x in $\mathcal{N}_{k+1}$ such that $L^k x = 0$. From this it follows that x is in $\mathcal{N}_k$. Conversely, if x is in $\mathcal{N}_k$, then

$$L^{k+1}x = LL^k x = 0,$$

which shows that x is also in $\mathcal{N}_{k+1}$ and therefore in the null space of L^k acting on $\mathcal{N}_{k+1}$. This proves the lemma and also the fact that $\mathcal{N}_k$ is contained in $\mathcal{N}_{k+1}$.

Let $\mathcal{P}_k$ be the space of progenitors of the range of L^k acting on $\mathcal{N}_{k+1}$, that is, $\mathcal{P}_k$ is the space spanned by the vectors p_j $(1 \leq j \leq \rho_k)$, where p_j are vectors in $\mathcal{N}_{k+1}$ such that the vectors $L^k p_j$ are linearly independent. By the corollary to Theorem 2.1, we know that the space $\mathcal{N}_{k+1}$ is the direct sum of the space $\mathcal{N}_k$ and the space $\mathcal{P}_k$; consequently, the dimension of $\mathcal{P}_k$ is

$$\rho_k = \nu_{k+1} - \nu_k.$$

We may write

$$\mathcal{N}_{k+1} = \mathcal{N}_k \oplus \mathcal{P}_k, \quad (0 \leq k \leq \nu - 1).$$

Since $\mathcal{N} = \mathcal{N}_\nu$ and $\mathcal{N}_0 = 0$, by induction we can show that

$$\mathcal{N} = \mathcal{P}_{\nu-1} \oplus \mathcal{P}_{\nu-2} \oplus \cdots \oplus \mathcal{P}_1 \oplus \mathcal{P}_0.$$

The proof will be completed by finding a suitable basis for the spaces $\mathcal{P}_k$ $(0 \leq k \leq \nu - 1)$. We start with a basis for $\mathcal{P}_{\nu-1}$. Then we show that L acting on this basis will give a set of linearly independent vectors in $\mathcal{P}_{\nu-2}$. Complete this set to a basis in $\mathcal{P}_{\nu-2}$. We find that L acting on this basis for $\mathcal{P}_{\nu-2}$ will give a set of linearly independent vectors in $\mathcal{P}_{\nu-3}$. Complete this last set to a basis in $\mathcal{P}_{\nu-3}$, and continue in the same way until we have a basis for all the $\mathcal{P}_k$. The following lemma will justify this procedure.

Lemma. *If the vectors p_j $(1 \leq j \leq \rho_k)$ form a basis for $\mathcal{P}_k$, the vectors Lp_j $(1 \leq j \leq \rho_k)$ are linearly independent vectors in $\mathcal{P}_{k-1}$.*

Note that this lemma implies that $\rho_{k-1} \geq \rho_k$.

From the definition of $\mathcal{P}_k$ we know that p_j is in $\mathcal{N}_{k+1}$ and therefore Lp_j is in $\mathcal{N}_k$. Again, from the definition of $\mathcal{P}_k$, if

$$\alpha_1 L^k p_1 + \cdots + \alpha_\rho L^k p_\rho = 0, \ (\rho = \rho_k)$$

then

$$\alpha_1 = \alpha_2 = \cdots = \alpha_\rho = 0, \ (\rho = \rho_k);$$

consequently, since a relation of the form

$$\alpha_1 L^j p_1 + \cdots + \alpha_\rho L^j p_\rho = 0, \quad (\rho = \rho_k)$$

by multiplication with L^{k-j} implies

$$\alpha_1 L^k p_1 + \cdots + \alpha_\rho L^k p_\rho = 0, \quad (\rho = \rho_k),$$

it follows that the vectors $Lp_1, \cdots, Lp_\rho$ are linearly independent and also that the vectors $L^k p_1, \cdots, L^k p_\rho$ are linearly independent. This proves first that the vectors $Lp_1, \cdots, Lp_\rho$, $(\rho = \rho_k)$ are linearly independent vectors and second that they are part of $\mathcal{P}_{k-1}$.

The basis is now found by induction. Let x_j $(1 \leq j \leq \rho_{\nu-1})$ be a basis for $\mathcal{P}_{\nu-1}$. By the above lemma, the vectors Lx_j are linearly independent vectors in $\mathcal{P}_{\nu-2}$. To these vectors, we can adjoin vectors x_j $(\rho_{\nu-1} < j \leq \rho_{\nu-2})$ in $\mathcal{P}_{\nu-2}$ so that the set of vectors Lx_j $(1 \leq j \leq \rho_{\nu-1})$ and x_j $(\rho_{\nu-1} < j \leq \rho_{\nu-2})$ forms a basis for $\mathcal{P}_{\nu-2}$. Similarly, if y_j $(1 \leq j \leq \rho_k)$ forms a basis for $\mathcal{P}_k$, then, using the lemma above, we can find vectors y_j $(\rho_k < j \leq \rho_{k-1})$ such that the set of vectors Ly_j $(1 \leq j \leq \rho_k)$ and y_j $(\rho_k < j \leq \rho_{k-1})$ forms a basis for $\mathcal{P}_{k-1}$.

This process is continued until we have found a basis for $\mathcal{P}_0$. Combining all the basis vectors, we obtain the following basis for $\mathcal{N}$:

$$
\begin{array}{ll}
x_j, Lx_j, \cdots, L^{\nu-1}x_j, & (1 \leq j \leq \rho_{\nu-1}) \\
x_j, Lx_j, \cdots, L^{\nu-2}x_j, & (\rho_{\nu-1} < j \leq \rho_{\nu-2}) \\
\cdots & \cdots \\
x_j, Lx_j, \cdots, L^{k}x_j, & (\rho_{k+1} < j \leq \rho_k) \\
\cdots & \cdots \\
x_j & (\rho_1 < j \leq \rho_0).
\end{array}
\tag{2.20}
$$

To obtain the representation given in Theorem 2.6, we re-arrange the vectors in (2.20) as follows:

Since x_1 is a generalized eigenvector of rank ν, we may put

$$y_1 = L^{\nu-1}x_1, \quad y_2 = L^{\nu-2}x_1, \cdots, \quad y_\nu = x_1;$$

these vectors span a space invariant under L. The representation of L on this space is exactly the matrix C_1. Similarly, if $\rho_{\nu-1} \geq 2$, the vector x_2 is a generalized eigenvector of rank ν and we put

$$y_{\nu+1} = L^{\nu-1}x_2, \quad y_{\nu+2} = L^{\nu-2}x_2, \cdots, y_{2\nu} = x_2;$$

however, if $\rho_{\nu-1} = 1$, the vector x_2 is a generalized eigenvector of rank $\nu - 1$, and we put

$$y_{\nu+1} = L^{\nu-2}x_2, \quad y_{\nu+2} = L^{\nu-3}x_2, \cdots, y_{2\nu-1} = x_2.$$

In either case, these vectors span a space invariant under L and the representation of L on this space is the matrix C_2. Continuing in this way, we obtain the representation given in Theorem 2.6. Note that

$$
\begin{aligned}
\delta_j &= \nu, & &(1 \leq j \leq \rho_{\nu-1}), \\
&= \nu - 1, & &(\rho_{\nu-1} < j \leq \rho_{\nu-2}), \\
&\ \ \vdots \\
&= 1, & &(\rho_1 < j \leq \rho_0),
\end{aligned}
$$

and that $m = \rho_0$.

PROBLEM

2.12. Simplify the representation of the following operators by using the appropriate basis:

$$L_1 = \begin{pmatrix} 0&0&0&1&0 \\ 0&0&0&1&0 \\ 0&0&0&1&0 \\ 0&0&0&0&0 \\ 0&0&0&0&0 \end{pmatrix}, \quad L_2 = \begin{pmatrix} 0&1&1&1&1 \\ 0&0&1&1&1 \\ 0&0&0&1&1 \\ 0&0&0&0&1 \\ 0&0&0&0&0 \end{pmatrix}, \quad L_3 = \begin{pmatrix} 0&1&1&1&1 \\ 0&0&0&0&0 \\ 0&0&0&1&1 \\ 0&0&0&0&0 \\ 0&0&0&0&0 \end{pmatrix}.$$

Canonical Form of an Operator in a Finite-Dimensional Space

Suppose that λ is an eigenvalue of index m for the operator L. The set of vectors in $\mathcal{S}$ which can be written as $(L - \lambda I)^m x$ will be called the *generalized range* of L for the eigenvalue λ. It is clear that the generalized range is a subspace. We prove

Theorem 2.7. *The space $\mathcal{S}$ is the direct sum of the generalized range and the generalized null space for any eigenvalue λ_0 of L.*

Let x be any vector in $\mathcal{S}$. Consider the vectors $(L - \lambda_0 I)^m x$, $(L - \lambda_0 I)^{2m} x$, $(L - \lambda_0 I)^{3m} x, \cdots$. Since $\mathcal{S}$ has dimension n, the first $n + 1$ of these vectors must be linearly dependent; that is, there exist scalars $\alpha_0, \alpha_1, \cdots, \alpha_n$ such that

$$\alpha_0 x + \alpha_1 (L - \lambda_0)^m x + \cdots + \alpha_n (L - \lambda_0)^{nm} x = 0.$$

If $\alpha_0 \neq 0$, then

$$x = -(L - \lambda_0)^m \left[\frac{\alpha_1}{\alpha_0} x + \cdots + \frac{\alpha_n}{\alpha_0} (L - \lambda_0)^{(n-1)m} x \right];$$

hence x would be in the generalized range. If $\alpha_0 = \alpha_1 = \cdots = \alpha_{k-1} = 0$ but $\alpha_k \neq 0$, then

$$\alpha_k (L - \lambda_0)^{km} (x - y) = 0, \tag{2.21}$$

where

$$-\alpha_k y = \alpha_{k+1} (L - \lambda_0)^m x + \cdots + \alpha_n (L - \lambda_0)^{(n-k)m} x.$$

The vector y is clearly in the generalized range. Since m is the index of λ_0, equation (2.21) implies that

$$(L - \lambda_0 I)^m (x - y) = 0.$$

If we write

$$x = y + (x - y), \tag{2.22}$$

x has been expressed as the sum of a vector in the generalized range and a vector in the generalized null space for λ_0.

The representation (2.22) is unique, for suppose that

$$x = y_1 + z_1 = y_2 + z_2$$

when y_1 and y_2 are in the generalized range and z_1 and z_2 are in the generalized null space. By subtraction we have

$$w = y_1 - y_2 = z_2 - z_1,$$

which shows that the vector w is in both the generalized range and in the generalized null space. This is impossible unless w is the zero vector because, if w is in the generalized range, then

$$w = (L - \lambda_0)^m x_1$$

for some x_1; and, if w is in the generalized null space, then

$$(L - \lambda_0)^m w = (L - \lambda_0)^{2m} x_1 = 0.$$

Since m is the index of λ_0, the last relation implies that

$$(L - \lambda_0)^m x_1 = w = 0;$$

consequently, both $y_1 = y_2$ and $z_1 = z_2$. This proves the uniqueness of the representation in (2.22) and completes the proof of Theorem 2.7.

Theorem 2.1 can easily be extended to show that the generalized range and the generalized null space are invariant subspaces of L. The corollary of Theorem 2.3 shows that L may be represented as follows:

$$L = \begin{pmatrix} A & 0 \\ 0 & D \end{pmatrix}.$$

Here A represents the effect of L on the generalized null space, and D represents the effect of L on the generalized range. In the preceding section, we have shown that A can be reduced to a simple form with non-zero elements only on the main diagonal and the diagonal above it. We now proceed to simplify D.

The matrix D represents the effect of L on an invariant subspace, namely, the generalized range. By Theorem 2.4, L has an eigenvalue λ_1 in this subspace. Determine the generalized null space and generalized range of L for the eigenvalue λ_1. The direct sum of these invariant subspaces will be the generalized range for λ_0. Consequently, D can be reduced to the following form:

$$D = \begin{pmatrix} A_1 & 0 \\ 0 & D_1 \end{pmatrix},$$

where A_1 represents the effect of D on the generalized null space for λ_1 and D_1 represents the effect of D on the generalized range for λ_1.

It is clear that this process may be continued with D_1. Eventually, we find that

$$L = \begin{pmatrix} A & 0 & \cdot & \cdot & \cdot & 0 \\ 0 & A_1 & \cdot & \cdot & \cdot & \cdot \\ \cdot & \cdot & \cdot & \cdot & \cdot & \cdot \\ \cdot & \cdot & \cdot & \cdot & \cdot & \cdot \\ \cdot & \cdot & \cdot & \cdot & \cdot & \cdot \\ 0 & \cdot & \cdot & \cdot & \cdot & A_k \end{pmatrix}, \tag{2.23}$$

where A_j represents the effect of L on the generalized null space for the eigenvalue λ_j. The previous section shows that A_j can be represented as follows:

$$A_j = \begin{pmatrix} \lambda_j & 1 & \cdot & \cdot & \cdot & \cdot \\ \cdot & \lambda_j & 1 & \cdot & \cdot & \cdot \\ \cdot & \cdot & \cdot & \cdot & \cdot & \cdot \\ \cdot & \cdot & \cdot & \cdot & \cdot & \cdot \\ \cdot & \cdot & \cdot & \cdot & \lambda_j & 1 \\ \cdot & \cdot & \cdot & \cdot & \cdot & \lambda_j \end{pmatrix}. \tag{2.24}$$

The representation (2.23) with blocks for each eigenvalue appearing as in (2.24) is called the *Jordan canonical form* of a matrix.

The preceding discussion has proved

Theorem 2.8. *If a suitable basis is used, every operator on a finite-dimensional space may be represented by a matrix in the Jordan canonical form.*

We shall illustrate the above proof by finding the Jordan canonical form of the following matrix:

$$L = \begin{pmatrix} 5 & -1 & 1 & 1 & 0 & 0 \\ 1 & 3 & -1 & -1 & 0 & 0 \\ 0 & 0 & 4 & 0 & 1 & 1 \\ 0 & 0 & 0 & 4 & -1 & -1 \\ 0 & 0 & 0 & 0 & 3 & 1 \\ 0 & 0 & 0 & 0 & 1 & 3 \end{pmatrix}.$$

We find that $\det |L - \lambda I| = (\lambda - 4)^5(\lambda - 2)$; therefore, 4 is an eigenvalue of multiplicity five and 2 is an eigenvalue of multiplicity one. Let us now determine the generalized null space of $L - 4I$. The successive powers of $L - 4I$ are as follows:

$$L - 4I = \begin{pmatrix} 1 & -1 & 1 & 1 & 0 & 0 \\ 1 & -1 & -1 & -1 & 0 & 0 \\ 0 & 0 & 0 & 0 & 1 & 1 \\ 0 & 0 & 0 & 0 & -1 & -1 \\ 0 & 0 & 0 & 0 & -1 & 1 \\ 0 & 0 & 0 & 0 & 1 & -1 \end{pmatrix},$$

$$(L - 4I)^2 = \begin{pmatrix} 0 & 0 & 2 & 2 & 0 & 0 \\ 0 & 0 & 2 & 2 & 0 & 0 \\ 0 & 0 & 0 & 0 & 0 & 0 \\ 0 & 0 & 0 & 0 & 0 & 0 \\ 0 & 0 & 0 & 0 & 2 & -2 \\ 0 & 0 & 0 & 0 & -2 & 2 \end{pmatrix},$$

$$(L - 4I)^3 = \begin{pmatrix} 0 & 0 & 0 & 0 & 0 & 0 \\ 0 & 0 & 0 & 0 & 0 & 0 \\ 0 & 0 & 0 & 0 & 0 & 0 \\ 0 & 0 & 0 & 0 & 0 & 0 \\ 0 & 0 & 0 & 0 & -4 & 4 \\ 0 & 0 & 0 & 0 & 4 & -4 \end{pmatrix}.$$

It is easy to see that $(L - 4I)^4$, $(L - 4I)^5$, and still higher powers of $L - 4I$ will look like $(L - 4I)^3$ except for different numbers in the two-rowed square matrix in the lower-right-hand corner; therefore, any vector that is annihilated by some power of $L - 4I$ will be annihilated by $(L - 4I)^3$. This shows that 4 is an eigenvalue of index three.

To determine the generalized null space, consider $(L - 4I)^3x$, where x is an arbitrary vector with components $\xi_1, \xi_2, \cdots, \xi_6$. For convenience, we shall write the column vectors as row vectors. We have

$$(L - 4I)^3x = (0, 0, 0, 0, -4\xi_5 + 4\xi_6, 4\xi_5 - 4\xi_6);$$

therefore, the null space of $(L - 4I)^3$ is defined by the relation $\xi_5 = \xi_6$. Since the whole space is six-dimensional and since the null space of $(L - 4I)^3$ is defined by only one relation, it follows that this null space is five-dimensional. We reduce L to canonical form by finding a suitable basis in this null space.

As in the proof of Theorem 2.6, we consider the generalized null space of $(L - 4I)^0$, $(L - 4I)^1$, and $(L - 4I)^2$. We have

$$(L - 4I)^0 x = (\xi_1, \xi_2, \xi_3, \xi_4, \xi_5, \xi_6)$$

$$(L - 4I)^1 x = (\xi_1 - \xi_2 + \xi_3 + \xi_4, \xi_1 - \xi_2 - \xi_3 - \xi_4, \xi_5 + \xi_6, - \xi_5 - \xi_6, \\ - \xi_5 + \xi_6, \xi_5 - \xi_6)$$

$$(L - 4I)^2 x = (2\xi_3 + 2\xi_4, 2\xi_3 + 2\xi_4, 0, 0, 2\xi_5 - 2\xi_6, - 2\xi_5 + 2\xi_6).$$

We shall denote the generalized null space of $(L - 4I)^k$ $(1 \leq k \leq 3)$ by $\mathcal{N}_k$.

The space $\mathcal{N}_3$ is defined by $\xi_5 = \xi_6$. Using the equation above for $(L - 4I)^2 x$, we see that the effect of $(L - 4I)^2$ on any vector x in $\mathcal{N}_3$ is as follows:

$$(L - 4I)^2 x = (2\xi_3 + 2\xi_4, 2\xi_3 + 2\xi_4, 0, 0, 0, 0) \\ = 2(\xi_3 + \xi_4)(1, 1, 0, 0, 0, 0).$$

The range of this operator is clearly one-dimensional. As a progenitor we may take the vector

$$x_1 = (0, 0, 1, 0, 0, 0).$$

The space $\mathcal{N}_2$ is defined by the relations

$$\xi_5 = \xi_6 \text{ and } \xi_3 = - \xi_4.$$

Using the equation for $(L - 4I)x$, we see that the effect of $(L - 4I)$ on any vector x in $\mathcal{N}_2$ is as follows:

$$(L - 4I)x = (\xi_1 - \xi_2, \xi_1 - \xi_2, 2\xi_5, - 2\xi_5, 0, 0) \\ = (\xi_1 - \xi_2)(1, 1, 0, 0, 0, 0) + 2\xi_5(0, 0, 1, - 1, 0, 0).$$

This range is two-dimensional. As one progenitor we take the vector

$$x_2 = (L - 4I)x_1 = (1, - 1, 0, 0, 0, 0).$$

As another progenitor, we take the vector

$$x_3 = (0, 0, 0, 0, 1, 1).$$

The space $\mathcal{N}_1$ is defined by the relations

$$\begin{aligned} \xi_1 - \xi_2 + \xi_3 + \xi_4 &= 0, \\ \xi_1 - \xi_2 - \xi_3 - \xi_4 &= 0, \\ \xi_5 + \xi_6 &= 0, \\ \xi_5 - \xi_6 &= 0. \end{aligned}$$

These relations imply

$$\begin{aligned} \xi_5 &= \xi_6 = 0, \\ \xi_2 &= \xi_1, \\ \xi_4 &= - \xi_3; \end{aligned}$$

consequently, the effect of $(L - 4I)^0 = I$ on any vector x in $\mathcal{N}_1$ is as follows:

$$\begin{aligned} Ix &= (\xi_1, \xi_1, \xi_3, -\xi_3, 0, 0) \\ &= \xi_1(1, 1, 0, 0, 0, 0) + \xi_3(0, 0, 1, -1, 0, 0). \end{aligned}$$

Again, the range is two-dimensional. Since the vectors

$$x_4 = (L - 4I)x_2 = 2(1, 1, 0, 0, 0, 0)$$

and

$$x_5 = (L - 4I)x_3 = 2(0, 0, 1, -1, 0, 0)$$

are progenitors of the range, we are finished. The vectors x_1, x_2, x_3, x_4, x_5 form a basis of $\mathcal{N}_3$. We arrange them as follows:

$$y_1 = x_4,\ y_2 = x_2,\ y_3 = x_1,\ y_4 = x_5,\ y_5 = x_3.$$

If we adjoin to these vectors the vector y_6 which is the eigenvector for the eigenvalue $\lambda = 2$, then in terms of this basis L takes the following form:

$$L = \begin{pmatrix} 4 & 1 & 0 & 0 & 0 & 0 \\ 0 & 4 & 1 & 0 & 0 & 0 \\ 0 & 0 & 4 & 0 & 0 & 0 \\ 0 & 0 & 0 & 4 & 1 & 0 \\ 0 & 0 & 0 & 0 & 4 & 0 \\ 0 & 0 & 0 & 0 & 0 & 2 \end{pmatrix}.$$

PROBLEM

2.13. Express the following matrices in the Jordan canonical form:

(*a*)
$$\begin{pmatrix} 7 & -1 & -3 & 1 \\ -1 & 7 & 1 & -3 \\ -3 & 1 & 7 & -1 \\ 1 & -3 & -1 & 7 \end{pmatrix},$$

(*b*)
$$\begin{pmatrix} 8 & -2 & -2 & 0 \\ 0 & 6 & 2 & -4 \\ -2 & 0 & 8 & -2 \\ 2 & -4 & 0 & 6 \end{pmatrix}.$$

Similarity Transformations

Consider the operator L acting on the finite-dimensional space $\mathcal{S}$. If we introduce a set of basis vectors e_i $(1 \leq i \leq n)$, the operator will be represented by a matrix (λ) with elements λ_{ij} $(1 \leq i, j \leq n)$. If we use a different set of basis vectors e_i', the operator L will be represented by a different matrix (λ') with elements λ'_{ij}. The matrix (λ') is said to be *similar* to the matrix (λ). We shall derive a relationship connecting the two matrices.

Consider first the unprimed basis e_i. Since any vector x in $\mathcal{S}$ may be written as

$$x = \sum_1^n \xi_j e_j, \tag{2.25}$$

where the ξ_i are scalars, we have

$$Lx = \sum_1^n \xi_j L e_j. \tag{2.26}$$

This shows that to find the matrix representation for L we need to express the vectors Le_j in terms of the basis e_i. Put

$$Le_j = \sum_{i=1}^n \lambda_{ij} e_i, \quad (j = 1, 2, \cdots, n), \tag{2.27}$$

where the λ_{ij} are scalars; then, substituting in (2.26), we get

$$Lx = \sum_{j=1}^n \xi_j \sum_{i=1}^n \lambda_{ij} e_i = \sum_{i=1}^n e_i \sum_{j=1}^n \lambda_{ij} \xi_j. \tag{2.28}$$

This equation shows that the components of Lx relative to the e_i basis are obtained from the components of x in the same basis by multiplying them with the matrix (λ) whose elements are λ_{ij}.

By using the basis reciprocal to e_i, we shall obtain an explicit expression for λ_{ij}. Denote the vectors of this reciprocal basis by f_i $(1 \leq i \leq n)$. From the properties of the reciprocal basis (Chapter 1), we have

$$I = \sum_1^n e_i \rangle\langle f_i, \tag{2.29}$$

where I is the identity operator. Since

$$x = Ix = \sum_1^n e_i \langle f_i, x\rangle,$$

we see by comparison with (2.25) that the components of x in the e_i basis are given by the formula

$$\xi_i = \langle f_i, x\rangle, \quad (i = 1, 2, \cdots, n). \tag{2.30}$$

From (2.27) it follows that the scalars λ_{ij} are the components of the vectors Le_j; therefore, by (2.30) we get

$$\lambda_{ij} = \langle f_i, Le_j\rangle.$$

We state this result as a

Lemma. *Relative to the e_i basis, the operator L is represented by a matrix (λ) whose elements are*

$$\lambda_{ij} = \langle f_i, Le_j\rangle.$$

Another way to obtain the matrix for L is as follows:

The components of Lx relative to the e_i basis are given by $\langle f_i, Lx\rangle$. By using (2.29), we have

$$L = LI = \sum_{j=1}^{n} Le_j\rangle\langle f_i;$$

therefore,

$$\langle f_i, Lx\rangle = \sum_{j=1}^{n}\langle f_i, Le_j\rangle\langle f_i, x\rangle,$$

which is equivalent to (2.28).

Suppose that we now consider the primed basis e_i' and its corresponding reciprocal basis f_i'. The operator L will be represented in terms of this basis by the matrix (λ') with elements

$$\lambda_{ij}' = \langle f_i', Le_j'\rangle.$$

We shall find the relationship between λ_{ij}' and λ_{ij} by expressing the vectors of the primed bases in terms of the vectors of the unprimed bases. Using (2.29), we get

$$e_j'\rangle = Ie_j'\rangle = \sum_{k=1}^{n} e_k\langle f_k, e_j'\rangle,$$

and

$$\langle f_i' = \langle f_i'I = \sum_{m=1}^{n}\langle f_i', e_m\rangle\langle f_m.$$

Substituting these results in the expression for λ_{ij}', we find that

$$\lambda_{ij}' = \sum_{m=1}^{n}\sum_{k=1}^{n}\langle f_i', e_m\rangle\langle f_m, Le_k\rangle\langle f_k, e_j'\rangle;$$

or, if we put (σ) for the matrix with elements

$$\sigma_{im} = \langle f_i', e_m\rangle$$

and (τ) for the matrix with elements

$$\tau_{kj} = \langle f_k, e_j'\rangle,$$

we find the following relation between matrices:

$$(\lambda') = (\sigma)(\lambda)(\tau).$$

This expression can be simplified because we shall show that the matrices (σ) and (τ) are the inverse of each other. Consider the matrix product $(\sigma)(\tau)$ whose elements are given by

$$\sum_{k=1}^{n}\langle f'_i, e_k\rangle\langle f_k, e'_j\rangle = \langle f'_i, e'_j\rangle$$

by the use of (2.29). Since the f'_i are the vectors of the basis reciprocal to the e'_i basis, we have

$$\langle f'_i, e'_j\rangle = \delta_{ij};$$

therefore,

$$(\sigma)(\tau) = I,$$

the identity matrix, and

$$(\sigma) = (\tau)^{-1}.$$

Using this result, we obtain the following relationship between similar matrices:

$$(\lambda') = (\tau)^{-1}(\lambda)(\tau).$$

Since

$$e'_j\rangle = \sum_{i=1}^{n} e_i\rangle\langle f'_i, e'_j\rangle = \sum_{i=1}^{n}\tau_{ij}e_i\rangle,$$

we see that (τ) is the matrix which expresses the primed basis e'_i in terms of the unprimed basis e_i.

The above discussion has proved the following

Lemma. *If (λ') and (λ) are similar matrices, that is, if they are matrix representations of the same operator, there exists a matrix (τ) such that*

$$(\lambda') = (\tau)^{-1}(\lambda)(\tau). \tag{2.31}$$

We shall call the transformation by which (λ) becomes (λ') a *similarity transformation.*

Suppose that, instead of an operator, we are given an $n \times n$ matrix (μ), with elements μ_{ij}. This matrix may be used to define an operator M in E_n. Introduce a basis e_j $(1 \le j \le n)$, where e_j is the column vector which has one as its jth component and zero for all the other components. We put

$$Me_j = \sum_{i=1}^{n}\mu_{ij}e_i; \tag{2.32}$$

and if

$$x = \sum_{1}^{n}\xi_j e_j,$$

we put

$$Mx = \sum_{1}^{n} \xi_j M e_j. \tag{2.33}$$

The operator M so defined is clearly a linear operator on the finite-dimensional space E_n. Theorem 2.8 states that, if a suitable basis is used, then M may be represented by a matrix (μ') in the Jordan canonical form; consequently, from (2.31) there exists a matrix (τ) such that

$$(\mu') = (\tau)^{-1}(\mu)(\tau).$$

This shows that an arbitrary matrix is always similar to a matrix in the Jordan canonical form. We state this result as

Theorem 2.9. *Every matrix may be transformed into the Jordan canonical form by means of a similarity transformation.*

It is readily seen that the Jordan canonical form for an operator is unique except for the order of arrangement of the eigenvalues; it follows that two matrices are similar if and only if their Jordan canonical forms are the same except for the order of arrangement of the eigenvalues. This implies that the two matrices must have the same eigenvalues with the same multiplicity and with the same number of generalized eigenvectors.

Right and Left Eigenvectors of a Matrix

The discussion of the preceding section has shown that an arbitrary matrix (μ) can be transformed into the Jordan canonical form by a similarity transformation with matrix (τ). This matrix (τ) was the matrix which expressed the eigenvectors and generalized eigenvectors of M in terms of the original basis e_i $(1 \leq i \leq n)$. Let x_1 be an eigenvector of M corresponding to the eigenvalue λ_1; then

$$Mx_1 = \lambda_1 x_1.$$

Put

$$x_1 = \Sigma \xi_i e_i;$$

from (2.32) and (2.33) we get

$$\sum e_i \sum_j \mu_{ij} \xi_j = \lambda_1 \sum_i \xi_i e_i$$

or

$$\Sigma \mu_{ij} \xi_j = \lambda_1 \xi_i.$$

This result shows that the column vector whose components are ξ_j will be an eigenvector of the matrix (μ) corresponding to the eigenvalue λ_1. Similar results will hold for all the eigenvectors and generalized eigenvectors of M. When expressed in terms of the e_i basis, they will be eigenvectors and generalized eigenvectors of the matrix (μ). We see then that (τ) will be the matrix whose columns are the vectors and generalized eigenvectors of the matrix (μ).

An example will clarify the situation. Consider the six-dimensional matrix whose generalized eigenvectors were found in the illustrative example (page 78). We have

$$(\mu) = \begin{pmatrix} 5 & -1 & 1 & 1 & 0 & 0 \\ 1 & 3 & -1 & -1 & 0 & 0 \\ 0 & 0 & 4 & 0 & 1 & 1 \\ 0 & 0 & 0 & 4 & -1 & -1 \\ 0 & 0 & 0 & 0 & 3 & 1 \\ 0 & 0 & 0 & 0 & 1 & 3 \end{pmatrix}.$$

The generalized eigenvectors corresponding to the eigenvalue 4 are as follows:

$$y_1 = \begin{pmatrix} 2 \\ 2 \\ 0 \\ 0 \\ 0 \\ 0 \end{pmatrix} \quad y_2 = \begin{pmatrix} 1 \\ -1 \\ 0 \\ 0 \\ 0 \\ 0 \end{pmatrix} \quad y_3 = \begin{pmatrix} 0 \\ 0 \\ 1 \\ 0 \\ 0 \\ 0 \end{pmatrix} \quad y_4 = \begin{pmatrix} 0 \\ 0 \\ 2 \\ -2 \\ 0 \\ 0 \end{pmatrix} \quad y_5 = \begin{pmatrix} 0 \\ 0 \\ 0 \\ 0 \\ 1 \\ 1 \end{pmatrix}.$$

There is also one eigenvector corresponding to the eigenvalue 2:

$$y_6 = \begin{pmatrix} 0 \\ 0 \\ 0 \\ 0 \\ 1 \\ -1 \end{pmatrix}.$$

If we form the matrix with these vectors as columns, we have

$$(\tau) = \begin{pmatrix} 2 & 1 & 0 & 0 & 0 & 0 \\ 2 & -1 & 0 & 0 & 0 & 0 \\ 0 & 0 & 1 & 2 & 0 & 0 \\ 0 & 0 & 0 & -2 & 0 & 0 \\ 0 & 0 & 0 & 0 & 1 & 1 \\ 0 & 0 & 0 & 0 & 1 & -1 \end{pmatrix}.$$

Inverting this matrix, we get

$$(\tau)^{-1} = \tfrac{1}{2}\begin{pmatrix} 1/2 & 1/2 & 0 & 0 & 0 & 0 \\ 1 & -1 & 0 & 0 & 0 & 0 \\ 0 & 0 & 2 & 2 & 0 & 0 \\ 0 & 0 & 0 & -1 & 0 & 0 \\ 0 & 0 & 0 & 0 & 1 & 1 \\ 0 & 0 & 0 & 0 & 1 & -1 \end{pmatrix}.$$

We let e_i, $(1 \leq i \leq 6)$, represent the column vector all of whose components are zero except the ith component which is one, and similarly we let f_i, $(1 \leq i \leq 6)$, represent the row vector all of whose components are zero except the ith which is one. Then we may write

$$(\tau) = y_1\rangle\langle f_1 + \cdots + y_6\rangle\langle f_6$$

and

$$(\tau)^{-1} = e_1\rangle\langle u_1 + \cdots + e_6\rangle\langle u_6,$$

where $u_1, \cdots, u_6$ are the row vectors of the matrix $(\tau)^{-1}$. Note that

$$\langle f_i, e_j\rangle = \langle u_i, y_j\rangle = \delta_{ij}.$$

It is easy to check that

$$(2.34) \qquad (\tau)^{-1}(\mu)(\tau) = \begin{pmatrix} 4 & 1 & 0 & 0 & 0 & 0 \\ 0 & 4 & 1 & 0 & 0 & 0 \\ 0 & 0 & 4 & 0 & 0 & 0 \\ 0 & 0 & 0 & 4 & 1 & 0 \\ 0 & 0 & 0 & 0 & 4 & 0 \\ 0 & 0 & 0 & 0 & 0 & 2 \end{pmatrix} = (\mu'),$$

the Jordan canonical form. We have

$$\begin{aligned}(\mu') = e_1\rangle\langle(4f_1 + f_2) + e_2\rangle\langle(4f_2 + f_3) + 4e_3\rangle\langle f_3 \\ + e_4\rangle\langle(4f_4 + f_5) + 4e_5\rangle\langle f_5 + 2e_6\rangle\langle f_6.\end{aligned}$$

If we rewrite (2.34) as $(\mu)(\tau) = (\tau)(\mu')$ we get

$$(\mu)y_1\rangle\langle f_1 + \cdots + (\mu)y_6\rangle\langle f_6 = y_1\rangle\langle(4f_1 + f_2) + \cdots + 2y_6\rangle\langle f_6.$$

From this result we have the following equations which show again that $y_1, y_2, y_3, y_4, y_5, y_6$ are generalized eigenvectors:

$$(2.35) \qquad \begin{aligned} (\mu)y_1 &= 4y_1, \\ (\mu)y_2 &= y_1 + 4y_2, \\ (\mu)y_3 &= y_2 + 4y_3, \\ (\mu)y_4 &= 4y_4, \\ (\mu)y_5 &= y_4 + 4y_5, \\ (\mu)y_6 &= 2y_6. \end{aligned}$$

Suppose now that we rewrite (2.34) as follows:

$$(\tau)^{-1}(\mu) = (\mu')(\tau)^{-1}.$$

Then we find that

$$e_1\rangle\langle u_1(\mu) + \cdots + e_6\rangle\langle u_6(\mu) = 4e_1\rangle\langle u_1 + (e_1 + 4e_2)\rangle\langle u_2 + \cdots + 2e_6\rangle\langle u_6.$$

This implies that

$$\begin{aligned} u_1(\mu) &= 4u_1 + u_2, \\ u_2(\mu) &= 4u_2 + u_3, \\ u_3(\mu) &= 4u_3, \\ u_4(\mu) &= 4u_4 + u_5, \\ u_5(\mu) &= 4u_5, \\ u_6(\mu) &= 2u_6. \end{aligned}$$

These equations are very similar to the equations in (2.35) for the eigenvectors and generalized eigenvectors of (μ). We shall say that the vectors $u_1, \cdots, u_6$ are *left eigenvectors* or *left generalized eigenvectors* of the matrix (μ). The vectors $y_1, \cdots, y_6$ we considered previously are called *right eigenvectors* and *right generalized eigenvectors of* (μ).

Notice that the vectors $u_1, \cdots, u_6$ are the right eigenvectors or right generalized eigenvectors of the matrix (μ^T), with elements μ_{ij}^T, where

$$\mu_{ij}^T = \mu_{ji}.$$

The matrix (μ^T) is called the *transpose* matrix of the matrix (μ).†

We shall show that the results of this example are valid for an arbitrary matrix (μ). Let $e_1', e_2', \cdots, e_n'$ be the eigenvectors and generalized eigenvectors of (μ) corresponding to the eigenvalues $\lambda_1, \lambda_2, \cdots, \lambda_n$, respectively. We suppose that the vectors $e_1', \cdots, e_n'$ are so chosen that the matrix (μ) becomes the Jordan canonical form. Then we have

$$(\mu)e_j' = \lambda_j e_j' + \alpha_j e_{j-1}', \quad (1 \leq j \leq n) \tag{2.36}$$

where $\alpha_j = 0$ if e_j' is an eigenvector of rank one, but $\alpha_j = 1$ if e_j' is a generalized eigenvector of rank greater than one. Let $f_1', f_2', \cdots, f_n'$ be the basis reciprocal to $e_1', e_2', \cdots, e_n'$; therefore

$$\langle f_k', e_j'\rangle = \delta_{kj}, \quad (1 \leq k, j \leq n). \tag{2.37}$$

Multiplying (2.36) by $\langle f_k'$, we get

$$\langle f_k', (\mu)e_j'\rangle = \lambda_j\langle f_k', e_j'\rangle + \alpha_j\langle f_k', e_{j-1}'\rangle. \tag{2.38}$$

† The transpose of a matrix is the matrix with its rows and columns interchanged. We note that the transpose of the matrix product $(\mu)(\lambda)$ is $(\lambda^T)(\mu^T)$.

Since $(\lambda_k - \lambda_j)\delta_{jk} = 0$ for all values of j and k, we see from (2.37) that

(2.39a) $$\lambda_j\langle f_k', e_j'\rangle = \lambda_k\langle f_k', e_j'\rangle$$

and

(2.39b) $$\alpha_j\langle f_k', e_{j-1}'\rangle = \alpha_j\langle f_{k+1}', e_j'\rangle = \alpha_{k+1}\langle f_{k+1}', e_j'\rangle.$$

Note that if $k = n$, we have

$$\langle f_k', e_{j-1}'\rangle = 0$$

for $1 \le j \le n$; therefore, we may put $\alpha_{n+1} = 0$ and $f_{n+1}' = f_n'$. Using (2.39a) and (2.39b) in (2.38), we get

$$\langle f_k', (\mu)e_j'\rangle = \lambda_k\langle f_k', e_j'\rangle + \alpha_{k+1}\langle f_{k+1}', e_j'\rangle.$$

Since this equation holds for every value of j, we may multiply it by $\langle f_j'$ and sum the resulting equations from $j = 1$ to $j = n$. In this way we find

(2.40) $$\langle f_k'(\mu) = \lambda_k\langle f_k' + \alpha_{k+1}\langle f_{k+1}'$$

since, by definition,

$$\Sigma e_j'\rangle\langle f_j' = I.$$

Equation (2.40) shows that f_k' is either a left eigenvector or a left generalized eigenvector of (μ), according as $\alpha_{k+1} = 0$ or $\alpha_{k+1} = 1$. In the preceding section, we have shown that (τ), the matrix which transforms (μ) into its canonical form, had the vectors $e_1', \cdots, e_n'$ as its columns; consequently,

$$(\tau) = e_1'\rangle\langle f_1 + \cdots + e_n'\rangle\langle f_n.$$

We also showed in the previous section that

$$(\tau)^{-1} = e_1\rangle\langle f_1' + \cdots + e_n\rangle\langle f_n';$$

consequently, the rows of $(\tau)^{-1}$ are the vectors $f_1', \cdots, f_n'$. We have thus proved

Theorem 2.10. *Let (μ) be an arbitrary matrix with $e_1', \cdots, e_n'$ as its right eigenvectors and right generalized eigenvectors. Let (τ) be the matrix whose columns are the vectors $e_1', \cdots, e_n'$; then $(\tau)^{-1}(\mu)(\tau)$ is a matrix in the Jordan canonical form and the rows of $(\tau)^{-1}$ are left eigenvectors or left generalized eigenvectors of (μ).*

PROBLEMS

2.14. Find the left and right eigenvectors and generalized eigenvectors of the matrices in Problem 2.13.

2.15. Suppose that (λ) and (μ) are matrices which commute with each other, that is, suppose that $(\lambda)(\mu) = (\mu)(\lambda)$. Show that there exists a matrix (τ) such that both $(\tau)^{-1}(\lambda)(\tau)$ and $(\tau)^{-1}(\mu)(\tau)$ are in the Jordan canonical form.

2.16. If B is a Jordan canonical matrix, show that B commutes with B^* if and only if B is a diagonal matrix.

Eigenvalues of the Adjoint Operator

The statements in the preceding section about the left eigenvectors of a matrix (μ) can be interpreted to give results about the eigenvectors of the transposed matrix (μ^T). From (2.40) we get

$$(\mu^T)f'_j\rangle = \lambda_j f'_j\rangle + \alpha_{j+1}f'_{j+1}\rangle.$$

This shows that, according as $\alpha_{j+1} = 0$ or $\alpha_{j+1} = 1$, the vector f'_j is an eigenvector or generalized eigenvector of (μ) corresponding to the eigenvalue λ_j.

If L is an arbitrary linear operator represented by a finite-dimensional matrix (μ), the adjoint operator L^* will be represented by the transposed matrix (μ^T). From Theorem 2.10 we conclude that *every eigenvalue of L is an eigenvalue of L^* and to every chain of length k for L there corresponds a chain of the same length k for L^*.*

Similar statements are not true for arbitrary operators in general spaces. We shall give two examples to show that an operator may have λ as an eigenvalue but its adjoint does not have λ as an eigenvalue. The first example is one we have already considered in another connection in Chapter 1. If $x = (\xi_1, \xi_2, \cdots)$ is any vector in E_∞, we have defined $Dx = (\xi_2, \xi_3, \cdots)$. The operator D has the adjoint C where $Cx = (0, \xi_1, \xi_2, \cdots)$. Zero is an eigenvalue of D but is not an eigenvalue of C.

In the second example, define the operator F as follows:

$$Fx = (\xi_2, \xi_2 + \xi_3, \xi_3 + \xi_4, \xi_4 + \xi_5, \cdots);$$

then it is readily seen that

$$F^*x = (0, \xi_1 + \xi_2, \xi_2 + \xi_3, \xi_3 + \xi_4, \cdots).$$

The vector $e = (1, 0, 0, \cdots)$ is an eigenvector of F corresponding to the eigenvalue zero. The operator F^*, on the other hand, does not have zero as an eigenvalue; for the equation $F^*x = 0$ implies that

$$\xi_1 + \xi_2 = 0, \quad \xi_2 + \xi_3 = 0, \quad \xi_3 + \xi_4 = 0, \cdots.$$

By solving these equations recursively, we have

$$\xi_n = (-)^{n+1}\xi_1;$$

but clearly the vector

$$(\xi_1, -\xi_1, \xi_1, -\xi_1, \cdots)$$

has infinite length; consequently, there is no vector x such that $F^*x = 0$, and zero is not an eigenvalue of F^*.

However, if we restrict the class of operators suitably, we shall get results similar to those we have already obtained for matrices. In the appendix to this chapter, we shall prove

Theorem 2.11. *Let λ be an eigenvalue of finite index for the operator*

L and let both $L - \lambda$ and $L^* - \lambda$ have closed ranges; then λ is an eigenvalue of L^* also. Moreover, to any chain of length k for L, namely, vectors $e_1, e_2, \cdots, e_k$ such that

$$e_j = (L - \lambda)^{j-1} e_1, \quad (j = 1, 2, \cdots, k),$$

there corresponds a chain of length k for L^*, namely, vectors $f_1, f_2, \cdots, f_k$ such that

$$f_j = (L^* - \lambda)^{k-j} f_k, \quad (j = 1, 2, \cdots, k),$$

and such that

$$\langle f_j, e_i \rangle = \delta_{ij}, \quad (i, j = 1, 2, \cdots, k).$$

The examples we discussed prior to stating Theorem 2.11 show that the conditions of the theorem are necessary for its validity. In the first example, zero was an eigenvalue of infinite index whereas, in the second example, the range of F was not closed. To see that the range of F is not closed, put $Fx_n = a_n$, where x_n is the vector whose first $2n$ components are alternatively $+1$ and -1, the next n components are given by the formula

$$\xi_j = (-1)^j \left(1 - \frac{j - 2n}{n}\right), \quad 2n + 1 \leq j \leq 3n,$$

and the remaining components are zero. We find that a_n is the vector whose first component is one and the rest zero except for those between the $2n$th and $3n$th places, which are alternately $1/n$ and $-1/n$. It is clear that the sequence a_n converges to a limit vector a, namely, the vector whose first component is one and all the rest zero. However, the equation $Fx = a$ has no solution. This proves that the range of F is not closed.

PROBLEMS

2.17. If a complex-type scalar product is used and the conditions of Theorem 2.11 are satisfied, show that, when λ is an eigenvalue of L, then $\bar{\lambda}$, the complex conjugate of λ, is an eigenvalue of L^*.

2.18. If $\int_0^1 \int_0^1 k(s, t)^2 \, ds \, dt < \infty$ and $\int_0^1 f(s)^2 \, ds < \infty$, show that the integral equation

$$u(s) + \lambda \int_0^1 k(s, t) u(t) \, dt = f(s)$$

has a unique solution belonging to $\mathcal{L}_2$ if, and only if, the homogeneous equation

$$u(s) + \lambda \int_0^1 k(s, t) u(t) \, dt = 0$$

has only the trivial solution $u(s) = 0$. (*Hint.* The integral operator is completely continuous. Use Theorem 2.11 and Problem 2.11.)

Characteristic Equation of a Matrix

In previous sections we have seen that the eigenvalues of a matrix A must be zeroes of the polynomial $c(\lambda) = \det |A - \lambda I|$. Conversely, every zero of the polynomial $c(\lambda)$ will be an eigenvalue of A. The equation $c(\lambda) = 0$ is called the *characteristic equation* of A. We shall prove the following

Theorem 2.12. *If B is a matrix similar to A, the characteristic equation of B is the same as the characteristic equation of A, namely, $c(\lambda) = 0$.*

If B is similar to A, there exists a non-singular matrix P such that $B = P^{-1}AP$. Obviously,

$$B - \lambda I = P^{-1}(A - \lambda I)P.$$

Using the well-known theorem that the determinant of a product of matrices is the product of the determinants of the matrices, we find that

$$\det |B - \lambda I| = \det P^{-1} \cdot \det |A - \lambda I| \cdot \det P.$$

Since

$$\det P^{-1} \cdot \det P = 1,$$

we conclude that

$$\det |B - \lambda I| = \det |A - \lambda I| = c(\lambda),$$

which proves the theorem.

By Theorem 2.9 the similarity transformation can be so chosen that B is in the Jordan canonical form. The characteristic equation for a matrix in the Jordan canonical form is clearly

$$c(\lambda) = (\lambda_1 - \lambda)^{n_1} \cdots (\lambda_k - \lambda)^{n_k},$$

where $\lambda_1, \cdots, \lambda_k$ are the eigenvalues of the matrix, and $n_1, \cdots, n_k$ are the dimensions of the generalized null spaces corresponding to the eigenvalues $\lambda_1, \cdots, \lambda_k$.

If λ_j is an eigenvalue of index m_j, by (2.13) the generalized null space for the eigenvalue λ_j contains a chain of m_j linearly independent vectors. This implies that n_j, the dimension of the generalized null space, is not less than m_j. We state this result as follows:

The multiplicity of λ_j as a root of the characteristic equation is not less than the index of λ_j as an eigenvalue of A.

We shall use this result to prove

Theorem 2.13.† *Every matrix A satisfies its characteristic equation, that is, the matrix $c(A)$ is identically zero.*

Consider any generalized eigenvector y corresponding to the eigenvalue λ_j. Since y belongs to the generalized null space of $A - \lambda_j$, we have

$$(A - \lambda_j)^{m_j} y = 0.$$

† This result is known as the Cayley-Hamilton theorem.

Since $n_j \geq m_j$, we conclude that

$$(A - \lambda_j)^{n_j} y = 0$$

and that

$$c(A)y = (\lambda_1 I - A)^{n_1} \cdots (\lambda_k I - A)^{n_k} y = 0. \tag{2.41}$$

In the proof of the existence of the Jordan canonical form, we showed that the eigenvectors and generalized eigenvectors of A formed a basis for the space. From (2.41), we see that the matrix $c(A)$ applied to any basis vector gives zero; consequently, $c(A)$ applied to any vector in the space gives zero. This implies that $c(A)$ must be identically zero.

As an illustration of Theorem 2.13, consider the matrix

$$L = \begin{pmatrix} 3 & 4 \\ 1 & 3 \end{pmatrix}.$$

The characteristic equation is

$$c(\lambda) = \begin{vmatrix} 3 - \lambda & 4 \\ 1 & 3 - \lambda \end{vmatrix} = \lambda^2 - 6\lambda + 5 = 0.$$

Now

$$L^2 = \begin{pmatrix} 13 & 24 \\ 6 & 13 \end{pmatrix}$$

and we see that

$$L^2 - 6L + 5I = \begin{pmatrix} 13 & 24 \\ 6 & 13 \end{pmatrix} - \begin{pmatrix} 18 & 24 \\ 6 & 18 \end{pmatrix} + \begin{pmatrix} 5 & 0 \\ 0 & 5 \end{pmatrix} = \begin{pmatrix} 0 & 0 \\ 0 & 0 \end{pmatrix}.$$

PROBLEMS

2.19. Show that $(\lambda - 4)^2(\lambda - 2) = 0$ is the characteristic equation for both the matrices

$$A = \begin{pmatrix} 6 & 2 & -2 \\ -2 & 2 & 2 \\ 2 & 2 & 2 \end{pmatrix} \quad \text{and} \quad B = \begin{pmatrix} 6 & 2 & 2 \\ -2 & 2 & 0 \\ 0 & 0 & 2 \end{pmatrix}.$$

Show that $A^2 - 6A + 8I = (A - 4I)(A - 2I) = 0$ but that $B^2 - 6B + 8I \neq 0$. Compute that $(B - 4I)^2(B - 2I) = 0$.

2.20. Show that the matrix A in Problem 2.19 has two linearly independent eigenvectors corresponding to $\lambda = 4$, but that the matrix B does not have two linearly independent eigenvectors corresponding to $\lambda = 4$; instead B has a generalized eigenvector of rank two. This is the reason that A satisfies a polynomial equation of lower degree than its characteristic equation.

2.21. The *trace* of a matrix is the sum of the elements on the main diagonal. Show that the trace of the matrix representing an operator L in terms of the basis $e_1, \cdots, e_n$ is $\sum_1^n \langle f_j, Le_j \rangle$, where $f_1, \cdots, f_n$ is the basis reciprocal to $e_1, \cdots, e_n$. Show that if A and B are arbitrary matrices, then $tr(AB) = tr(BA)$.

2.22. Show that if two matrices are similar, their traces are equal and their determinants are equal; therefore, from the Jordan canonical form it follows that the trace of a matrix is the sum of its eigenvalues and the determinant is the product of its eigenvalues.

2.23. By actual computation, show that for the matrix in Problem 2.19

$$A^3 = 28A - 48I.$$

Compare this result with the remainder when λ^3 is divided by $\lambda^2 - 6\lambda + 8$.

Self-adjoint Operators

In the section Scalar Product in Abstract Spaces in Chapter 1, we stated that we shall use a real-type scalar product but that our vector space will be taken over the field of complex numbers. It was also stated that this may cause difficulty with the concept of the length of a vector because the scalar product of a vector with itself might be negative or even complex. However, it is clear that vectors in E_n all of whose components are real have real non-negative lengths. Now, we shall generalize to arbitrary spaces the concept of a vector with all real components.

Suppose that our space has n linearly independent vectors, each such that the scalar product of the vector with itself is not zero. By using a slight modification of the Schmidt orthogonalization process, we can prove the existence of an O.N. basis B composed of vectors $x_1, \cdots, x_n$. If

$$x = \xi_1 x_1 + \cdots + \xi_n x_n,$$

we shall call the scalars $\xi_1, \cdots, \xi_n$ the *components* of x relative to B. We say the vector $\bar{x}$ is the *complex conjugate vector to* x relative to B if

$$\bar{x} = \bar{\xi}_1 x_1 + \cdots + \bar{\xi}_n x_n,$$

where $\bar{\xi}_1, \cdots, \bar{\xi}_n$ are the complex conjugate scalars to $\xi_1, \cdots, \xi_n$. We have

$$\overline{\alpha x} = \bar{\alpha}\,\bar{x},$$
$$\overline{x + y} = \bar{x} + \bar{y},$$
$$\langle \bar{x}, x\rangle = |\xi_1|^2 + \cdots + |\xi_n|^2 > 0$$

unless $x = 0$, in which case $\langle \bar{x}, x\rangle = 0$.

A vector x such that $x = \bar{x}$ is said to be a *real* vector relative to B. Note that the vector $x + \bar{x}$ is a real vector and that the scalar product of a real vector with itself is always non-negative.

The concepts of real and complex-conjugate vectors depend upon the O.N. basis. If we start with a different O.N. basis B' containing the vectors $y_1, \cdots, y_n$ and if we have

$$x = \eta_1 y_1 + \cdots + \eta_n y_n,$$

then the vector complex conjugate to x relative to B' will be the vector

$$\bar{\eta}_1 y_1 + \cdots + \bar{\eta}_n y_n.$$

In general this vector will not be equal to the vector $\bar{x}$ that we have defined before. Therefore, the results that we shall obtain will be relative to the fixed O.N. basis with which we start. Usually, we shall take for our fixed basis in E_n the vectors e_j, $(1 \leq j \leq n)$, where the components of e_j are zero except for the jth component which is one.

An operator L is said to be a *real* operator if

$$\overline{Lu} = L\bar{u}$$

where the bar indicates the *complex conjugate vector.* If L is self-adjoint and real, the theory of its canonical form becomes much simpler. We first prove

Theorem 2.14. *If L is a real self-adjoint operator, its eigenvalues are real and it has no eigenvectors of rank larger than one.*

Suppose that λ is an eigenvalue of L and u the corresponding eigenvector such that

$$Lu = \lambda u.$$

If we take the complex conjugate of this equation and use the fact that L is real, we have

$$\overline{Lu} = L\bar{u} = \bar{\lambda}\bar{u};$$

consequently, $\bar{\lambda}$ is an eigenvalue of L corresponding to the eigenvector $\bar{u}$. Now, since L is self-adjoint,

$$\lambda\langle\bar{u}, u\rangle = \langle\bar{u}, \lambda u\rangle = \langle\bar{u}, Lu\rangle = \langle L\bar{u}, u\rangle = \langle\bar{\lambda}\bar{u}, u\rangle = \bar{\lambda}\langle\bar{u}, u\rangle.$$

This result implies that either $\langle\bar{u}, u\rangle = 0$ or $\lambda = \bar{\lambda}$. The first alternative is impossible because $u \neq 0$; therefore, $\lambda = \bar{\lambda}$, which means that λ is real. Suppose that u is not real; then, as we have shown above,

$$L\bar{u} = \bar{\lambda}\bar{u} = \lambda\bar{u}.$$

Thus the real vector $u + \bar{u}$ will be an eigenvector since

$$L(u + \bar{u}) = \lambda(u + \bar{u}).$$

Consequently, if λ is real, the eigenvectors corresponding to it can be chosen to be real.

Suppose now that v is a generalized eigenvector of rank two corresponding to the eigenvalue λ. We may assume v to be a real vector; if it is not real, then, just as before, we use $v + \bar{v}$. Because L is self-adjoint, we have

$$0 = \langle v, (L - \lambda I)^2 v\rangle = \langle(L - \lambda I)v, (L - \lambda I)v\rangle = |(L - \lambda I)v|^2.$$

Since $(L - \lambda I)v$ is a real vector, this equation implies that

$$(L - \lambda I)v = 0;$$

therefore, v is an ordinary eigenvector of rank one.

If L had an eigenvector w of rank higher than two, then by applying $L - \lambda I$ to w a sufficient number of times we would obtain an eigenvector of rank two. The previous argument has shown this to be impossible; therefore, there are no generalized eigenvectors for L. This completes the proof of the theorem.

If we go back to the proof of Theorem 2.6, we see that, if there are no generalized eigenvectors, then $\nu = 1$ and all $\delta_j = 1$. This implies that the Jordan canonical form for such a matrix is a diagonal matrix, that is, a matrix whose only non-zero elements are on the main diagonal.

An immediate consequence of Theorem 2.14 is that the *Jordan canonical form for a real self-adjoint matrix is a diagonal matrix.* The elements on the diagonal are just the eigenvalues of L.

Another important fact about self-adjoint operators is given by

Theorem 2.15. *If L is self-adjoint, the eigenvectors of L which correspond to different eigenvalues are orthogonal.*

Let λ and μ be distinct eigenvalues of L, and let u and v be the corresponding eigenvectors; then, since L is self-adjoint,

$$\lambda\langle u, v\rangle = \langle \lambda u, v\rangle = \langle Lu, v\rangle = \langle u, Lv\rangle = \langle u, \mu v\rangle = \mu\langle u, v\rangle.$$

This result implies that

$$(\lambda - \mu)\langle u, v\rangle = 0.$$

Since $\lambda \neq \mu$, we must have $\langle u, v\rangle = 0$; this proves the theorem.

A slightly different arrangement of the proof of this theorem is worth noting. We have

$$Lu = \lambda u, \quad Lv = \mu v.$$

Form the scalar product of the first equation with v, the scalar product of the second with u, then subtract the second result from the first. We have

$$\langle v, Lu\rangle - \langle u, Lv\rangle = (\lambda - \mu)\langle u, v\rangle.$$

Since L is self-adjoint, the left side is zero and the conclusion follows as before.

PROBLEMS

2.24. Prove that if L is self-adjoint with a complex-type scalar product, the eigenvalues of L are real numbers. (*Hint.* Consider $\lambda\langle u, v\rangle$ and proceed as in Theorem 2.14.)

2.25. Prove that if L is self-adjoint with a complex-type scalar product, it has no eigenvectors of rank greater than one and its eigenvectors corresponding to different eigenvalues are orthogonal.

2.26. Prove that a symmetric matrix with real elements is self-adjoint. Show that the symmetric matrix

$$\begin{pmatrix} 3 & i \\ i & 1 \end{pmatrix}$$

has an eigenvector of rank two. Compare this result with Theorem 2.14. Note that the eigenvector is orthogonal to itself.

2.27. Find the eigenvalues and the eigenvectors of the following matrix:

$$\begin{pmatrix} 7 & -3 & -1 & 1 \\ -3 & 7 & 1 & -1 \\ -1 & 1 & 7 & -3 \\ 1 & -1 & -3 & 7 \end{pmatrix}.$$

Verify the fact that the eigenvectors corresponding to different eigenvalues are orthogonal.

Orthogonal Matrices

From the results of the preceding section and from Theorem 2.9, it follows that a self-adjoint matrix can be reduced to a diagonal matrix by means of a similarity transformation; that is, if L is a self-adjoint matrix and P is a matrix whose columns are the eigenvectors of L, then

$$P^{-1}LP = D,$$

where D is a diagonal matrix. We shall now show that as a consequence of Theorem 2.15, we may take $P^{-1} = P^*$ (P^* being the matrix transpose or adjoint to P).

Let $u_1, u_2, \cdots, u_n$ be the eigenvectors of L. If u_i and u_j correspond to different eigenvalues, by Theorem 2.15 we have

$$\langle u_i, u_j \rangle = 0.$$

Suppose that two or more eigenvectors correspond to the same eigenvalue; for example, suppose that $u_1, u_2, \cdots, u_k$ correspond to the eigenvalue λ; then any vector in the space spanned by $u_1, \cdots, u_k$ will be an eigenvector corresponding to the eigenvalue λ because

$$L(\alpha_1 u_1 + \cdots + \alpha_k u_k) = \alpha_1 L u_1 + \cdots + \alpha_k L u_k = \lambda(\alpha_1 u_1 + \cdots + \alpha_k u_k).$$

By the Schmidt orthogonalization process we may set up an orthonormal basis in this space. Denote these basis vectors again by $u_1, \cdots, u_k$. Do the same in all cases where there are multiple eigenvectors for one eigenvalue. Consequently, we have

$$\langle u_i, u_j \rangle = 0, \quad i \neq j$$

even if u_i and u_j belong to the same eigenvalue. Notice that we can always normalize the eigenvectors; that is, we can find a constant c_i such that the length of $c_i u_i$ is one. If we assume this done and use y_i to represent the normalized eigenvector, we have finally

$$\langle y_i, y_j \rangle = \delta_{ij}. \tag{2.42}$$

We state this result as

Theorem 2.16. *The eigenvectors of a real self-adjoint operator form an orthonormal basis for the space.*

As an illustration of this theorem, consider the matrix in Problem 2.27. Its eigenvalues are 12, 8, 4, and 4. The corresponding eigenvectors are $u_1 = (1, -1, -1, 1)$, $u_2 = (1, -1, 1, -1)$, $u_3 = (1, 1, 1, 1)$, and $u_4 = (1, 1, 0, 0)$. Any linear combination of the last two vectors is also an eigenvector corresponding to the eigenvalue 4. We find that $u_3 - u_4$ is orthogonal to u_4. We take u_1, u_2, u_4, and $u_3 - u_4$ as a set of orthogonal eigenvectors. Normalize these vectors and we have an O.N. basis:

$$y_1 = \tfrac{1}{2}u_1,\ y_2 = \tfrac{1}{2}u_2,\ y_3 = (\tfrac{1}{2})^{1/2}u_4,\ y_4 = (\tfrac{1}{2})^{1/2}(u_3 - u_4).$$

Consider now the matrix P whose columns are the vectors y_1, y_2, y_3, y_4; that is,

$$P = \begin{pmatrix} \frac{1}{2} & \frac{1}{2} & (\frac{1}{2})^{\frac{1}{2}} & 0 \\ -\frac{1}{2} & -\frac{1}{2} & (\frac{1}{2})^{\frac{1}{2}} & 0 \\ -\frac{1}{2} & \frac{1}{2} & 0 & (\frac{1}{2})^{\frac{1}{2}} \\ \frac{1}{2} & -\frac{1}{2} & 0 & (\frac{1}{2})^{\frac{1}{2}} \end{pmatrix}.$$

Note that P^*, which is the transpose of P, is the matrix whose rows are the eigenvectors y_1, y_2, y_3, y_4; that is,

$$P^* = \begin{pmatrix} \frac{1}{2} & -\frac{1}{2} & -\frac{1}{2} & \frac{1}{2} \\ \frac{1}{2} & -\frac{1}{2} & \frac{1}{2} & -\frac{1}{2} \\ (\frac{1}{2})^{\frac{1}{2}} & (\frac{1}{2})^{\frac{1}{2}} & 0 & 0 \\ 0 & 0 & (\frac{1}{2})^{\frac{1}{2}} & (\frac{1}{2})^{\frac{1}{2}} \end{pmatrix}.$$

Clearly, $P^*P = I$, the identity matrix. We shall show in general that this is just a consequence of the fact that the vectors y_1, y_2, y_3, y_4 form an O.N. basis.

Suppose that $y_1, y_2, \cdots, y_n$ are the O.N. eigenvectors for the real self-adjoint operator L. Consider the matrix P whose *columns* are the eigenvectors of L; we may write

$$P = y_1\rangle\langle e_1 + \cdots + y_n\rangle\langle e_n.$$

The adjoint P^* will then have the eigenvectors of L as its *rows*; we may write

$$P^* = f_1\rangle\langle y_1 + \cdots + f_n\rangle\langle y_n.$$

Now the product P^*P will be

$$P^*P = f_1\rangle\langle e_1 + \cdots + f_n\rangle\langle e_n = I, \tag{2.43}$$

the identity matrix because $\langle y_i, y_j\rangle = \delta_{ij}$.

A matrix P which satisfies equation (2.43) is called an *orthogonal* matrix. Since P is the matrix which transformed L into a diagonal matrix, we have proved

Theorem 2.17. *A real self-adjoint matrix can be transformed into a diagonal matrix by an orthogonal transformation.*

The importance of the *orthogonal transformations* lies in the fact that they *leave scalar products invariant.* Consider two vectors u, v and an orthogonal transformation P which transforms them into u' and v', respectively, so that

$$u' = Pu, \quad v' = Pv;$$

then by (2.43)

$$\langle u', v'\rangle = \langle Pu, Pv\rangle = \langle u, P^*Pv\rangle = \langle u, v\rangle.$$

Geometrically, *the orthogonal transformations correspond to a rotation of the coordinate axes around the origin.*

PROBLEMS

2.28. Prove that the rows or columns of an orthogonal matrix form an O.N. basis.

2.29. Find the most general form for a two-dimensional orthogonal matrix. Compare it with the matrix for a rotation of the coordinate axes.

2.30. Prove that the determinant of an orthogonal transformation is ± 1. Show that the orthogonal transformation in Theorem 2.17 can be chosen so that its determinant is plus one.

Unitary and Hermitian Matrices

In the discussion of the preceding sections, we have always used a real-type scalar product. For many applications, especially in quantum mechanics, it is necessary to consider a complex-type scalar product. In this section we shall point out how the previous theorems must be modified to cover such cases.

The main difference in the statements of the theorems arises from the fact that the scalar product is no longer symmetric, but instead we have

$$\langle u, v\rangle = \overline{\langle v, u\rangle},$$

where the bar indicates the complex conjugate. Consequently, A^*, the adjoint of the matrix A, is not the transpose of A but is the transpose of A with all elements replaced by their complex conjugates. A self-adjoint matrix H is a matrix which is equal to the complex conjugate of its transpose; for example,

$$H = \begin{pmatrix} 2 & i \\ -i & 3 \end{pmatrix}.$$

Such a matrix is said to be a *Hermitian* matrix.

As shown in Problems 2.24 and 2.25, a Hermitian matrix has real eigenvalues only and eigenvectors of rank one only. Consequently, when a Hermitian matrix is transformed by a similarity transformation to the Jordan canonical form, the result will be a diagonal matrix with real elements. We shall show that for a Hermitian matrix the similarity transformation may be replaced by a *unitary* transformation, that is, a transformation with matrix U such that the transpose of the complex conjugate of U is the inverse of U. Such a matrix is called a *unitary* matrix.

First, it is easy to see that for a Hermitian matrix H, eigenvectors corresponding to different eigenvalues are orthogonal. Just as in the real case, this follows from the self-adjointness of the Hermitian matrix. Second, if there are different eigenvectors corresponding to the same eigenvalue, they can be orthogonalized in the same manner as before. Finally, we see that the eigenvectors of H will form an O.N. basis for the space.

Let u_j, $j = 1, 2, \cdots, n$, be the O.N. eigenvectors of H, and let U be the matrix whose columns are the eigenvectors u_j. The adjoint of U will have the complex conjugate of the vectors u_j as its rows. Since the vectors u_j form an O.N. basis,

$$\langle u_i, u_j \rangle = \delta_{ij}.$$

Because of the complex scalar product, this formula shows that the complex conjugate of the transpose of U is the inverse of U. Therefore, U is unitary, and we have proved

Theorem 2.18. *A Hermitian matrix can be transformed by a unitary transformation into a diagonal matrix with real elements.*

Note also that a unitary matrix leaves the scalar product invariant. The proof is the same as that for an orthogonal matrix.

The results of this and the preceding sections are listed in the following summary:

(*a*) The types of scalar products are

Real scalar product	Complex scalar product

(*b*) A self-adjoint matrix is called

Symmetric	Hermitian

(*c*) It can be transformed into a diagonal form with real elements on the diagonal by means of a transformation which leaves the scalar product unchanged. Such a transformation is called

Orthogonal	Unitary

(*d*) The transformation has the property that its inverse is equal to its adjoint; that is, the inverse of the matrix of the transformation equals its

Transpose.	Complex conjugate transpose.

Quadratic Forms

If we use a real-type scalar product, a quadratic form may be considered abstractly as $\langle x, Ax\rangle$, where A is any self-adjoint matrix. For example, if the quadratic form is

$$4\xi_1^2 + 4\xi_2^2 + 2\xi_3^2 - 4\xi_1\xi_2 + 4\xi_1\xi_3 + 4\xi_2\xi_3,$$

the matrix A is

$$\begin{pmatrix} 4 & -2 & 2 \\ -2 & 4 & 2 \\ 2 & 2 & 2 \end{pmatrix}.$$

Note that since the coefficients of the cross terms such as $\xi_1\xi_3$ and $\xi_2\xi_3$ appear twice in the matrix, they are taken at half their value in the quadratic form.

By Theorem 2.17 we know that A can be transformed into diagonal form by means of an orthogonal transformation. If we use the technique of the preceding section, we may show that the orthogonal transformation is represented by the matrix

$$P = \begin{pmatrix} 2^{-1/2} & \frac{1}{2} & \frac{1}{2} \\ -2^{-1/2} & \frac{1}{2} & \frac{1}{2} \\ 0 & -2^{-1/2} & 2^{-1/2} \end{pmatrix}.$$

Let us now introduce the change of variables defined by this matrix P. We put

$$\begin{aligned} \xi_1 &= 2^{-1/2}\eta_1 + \tfrac{1}{2}\eta_2 + \tfrac{1}{2}\eta_3, \\ \xi_2 &= -2^{-1/2}\eta_1 + \tfrac{1}{2}\eta_2 + \tfrac{1}{2}\eta_3, \\ \xi_3 &= -2^{-1/2}\eta_2 + 2^{-1/2}\eta_3. \end{aligned}$$

Since P is an orthogonal matrix, the change from the (ξ_1, ξ_2, ξ_3) coordinates to the (η_1, η_2, η_3) coordinates will be the result of a rotation of the coordinate axes. After substituting these expressions for ξ_1, ξ_2, ξ_3, we find that the quadratic form becomes

$$6\eta_1^2 + 2(1 - 2^{1/2})\eta_2^2 + 2(1 + 2^{1/2})\eta_3^2.$$

It is interesting to note the geometrical interpretation of this result. Consider the following equation obtained by setting the original quadratic form equal to a constant:

$$4\xi_1^2 + 4\xi_2^2 + 2\xi_3^2 - 4\xi_1\xi_2 + 4\xi_1\xi_3 + 4\xi_2\xi_3 = c.$$

This equation represents a second-degree surface in three-dimensional Euclidean space which is so oriented that its principal axes are oblique to the coordinate axes. After a suitable rotation of the coordinate axes, the surface is represented by the equation

$$6\eta_1^2 + 2(1 - 2^{1/2})\eta_2^2 + 2(1 + 2^{1/2})\eta_3^2 = c;$$

now the surface is so oriented that its principal axes are the coordinate axes.

The fact that we can thus rotate the coordinate axes so that the principal axes of a second-degree surface in n dimensions become the coordinate axes turns out to be very important and has many applications. Note also that, as a result of this rotation, a quadratic form in n variables is written as the sum or difference of squares. We shall show that the method illustrated above in the case of three dimensions is applicable to any number of dimensions. This will prove the following

Theorem 2.19. *A real quadratic form can be transformed by means of an orthogonal transformation into a form having only square terms.*

To prove this, let

$$Q = \sum_{i=1}^{n} \sum_{j=1}^{n} a_{ij}\zeta_i\zeta_j, \quad a_{ij} = a_{ji},$$

be the quadratic form whose matrix A has the real elements a_{ij}. Note that A is symmetrical and is therefore self-adjoint. Suppose that P is the orthogonal matrix which diagonalizes A so that

$$P^*AP = D,$$

where D is a diagonal matrix whose diagonal elements $\lambda_1, \lambda_2, \cdots, \lambda_n$ are the eigenvalues of A. Let z be the vector whose components are $\zeta_1, \zeta_2, \cdots, \zeta_n$; then we may write

$$Q = \sum_{i=1}^{n} \zeta_i \sum_{j=1}^{n} a_{ij}\zeta_i = \langle z, Az \rangle.$$

Now introduce a new set of variables $\eta_1, \eta_2, \cdots, \eta_n$ by means of the transformation P, that is, put

$$z = Py,$$

where y is a vector with components $\eta_1, \eta_2, \cdots, \eta_n$. Under this transformation, the quadratic form becomes

$$Q = \langle Py, APy \rangle = \langle y, P^*APy \rangle = \langle y, Dy \rangle = \sum_{k=1}^{n} \lambda_k \eta_k^2,$$

which has only square terms. Thus the theorem is proved.

There are several points about the proof which are worth noticing. First, since P is an orthogonal matrix,

$$\begin{aligned} \zeta_1^2 + \cdots + \zeta_n^2 &= \langle z, z \rangle = \langle Py, Py \rangle = \langle y, P^*Py \rangle = \langle y, y \rangle, \\ &= \eta_1^2 + \cdots + \eta_n^2. \end{aligned}$$

This shows that although we have reduced the quadratic form Q into a linear combination of squares, the quadratic form that is a sum of squares has been left invariant. Second, when Q is so reduced the coefficients are exactly the eigenvalues of the matrix of the quadratic form. Third, since D is purely a diagonal matrix, the value of the determinant of D is $\lambda_1 \lambda_2 \cdots \lambda_n$. From the definition of D we have

$$\det D = \det P^* \cdot \det A \cdot \det P,$$

but since $P^*P = I$, $\det P^* \det P = 1$; consequently,

$$\det D = \det A.$$

This proves that

$$\det A = \lambda_1 \lambda_2 \cdots \lambda_n, \tag{2.44}$$

or, expressed verbally, *the determinant of a matrix is equal to the product of its eigenvalues.*

If we use a complex-type scalar product, the expression $\langle x, Ax \rangle$, where A is self-adjoint, is called a *Hermitian form* or a *Hermitian quadratic form.* For example, suppose that

$$A = \begin{pmatrix} 5 & 2i \\ -2i & 2 \end{pmatrix};$$

then the Hermitian form is $5\bar{\xi}_1\xi_1 + 2i\bar{\xi}_1\xi_2 - 2i\bar{\xi}_2\xi_1 + 2\bar{\xi}_2\xi_2$. By Theorem 2.18 we know that A can be transformed into a diagonal form by means of a unitary transformation. Since the eigenvalues of A are 6 and 1, and the normalized eigenvectors are $(-5^{-1/2} i, 2 \cdot 5^{-1/2})$ and $(2 \cdot 5^{-1/2} i, 5^{-1/2})$, the unitary transformation is given by the matrix

$$U = \begin{pmatrix} \dfrac{-i}{\sqrt{5}} & \dfrac{2i}{\sqrt{5}} \\ \dfrac{2}{\sqrt{5}} & \dfrac{1}{\sqrt{5}} \end{pmatrix}.$$

Now let us introduce a transformation with matrix U:

$$\xi_1 = -\,i\eta_1 5^{-1/2} + 2i\eta_2 5^{-1/2},$$
$$\xi_2 = 2\eta_1 5^{-1/2} + \eta_2 5^{-1/2}.$$

We find that the Hermitian form reduces to $\bar{\eta}_1\eta_1 + 6\,\bar{\eta}_2\eta_2$.

A similar procedure can be applied when A is a Hermitian matrix in n dimensions, with elements a_{ij}, where $a_{ij} = \bar{a}_{ji}$. The corresponding Hermitian form is

$$H = \sum_{i=1}^{n} \sum_{j=1}^{n} a_{ij}\bar{\xi}_i\xi_j.$$

By a unitary transformation, A becomes a diagonal matrix where the diagonal elements are the necessarily real eigenvalues of A; we will then have

$$H = \sum_{k=1}^{n} \lambda_k |\eta|_k^2.$$

The proof is similar to that of Theorem 2.19. We state the result in the following

Theorem 2.20. *A Hermitian form may be transformed by means of a unitary transformation into a form containing only squares of absolute values.*

PROBLEMS

2.31. Find the orthogonal matrix that will reduce the quadratic form $4(\xi_1^2 + \xi_2^2 + \xi_3^2 + \xi_4^2) - 2(\xi_1 + \xi_2)(\xi_3 - \xi_4)$ to a linear combination of square terms only.

2.32. Find the unitary matrix that will reduce the Hermitian form $41|\xi_1|^2 + 12i\bar{\xi}_2\xi_1 - 12i\bar{\xi}_1\xi_2 + 34|\xi_2|^2$ to a form without cross terms.

2.33. A real quadratic form $\langle x, Ax\rangle$ is called *positive-definite* if and only if the value of $\langle x, Ax\rangle$ is greater than zero, except when $x = 0$. Prove that $\langle x, Ax\rangle$ is positive-definite if and only if the eigenvalues of A are all greater than zero. (*Hint.* If x_0 is an eigenvector of A, then $\langle x_0, Ax_0\rangle = \lambda\langle x_0, x_0\rangle$.)

2.34. If A is positive-definite, put $\langle x, Ay\rangle = \langle x, y\rangle_1$. Show that $\langle x, y\rangle_1$ satisfies all the axioms for a scalar product. Use the Schwarz inequality for $\langle x, y\rangle_1$ to prove

$$\langle x, Ay\rangle^2 \le \langle x, Ax\rangle\langle y, Ay\rangle.$$

2.35. Prove that the minimum value of the ratio $\langle x, Ax\rangle/\langle x, x\rangle$ is λ_1, the eigenvalue with the smallest possible numerical value. Prove also that the maximum value of the ratio is λ_m, the eigenvalue with the largest possible numerical value. Suppose that x_1 is an eigenvector of A corresponding to the smallest eigenvalue λ_1 and that $\lambda_2 \ge \lambda_1$ is the next eigenvalue in order of increasing magnitude. Show that the minimum value of the ratio $\langle x, Ax\rangle/\langle x, x\rangle$ for all vectors x restricted by the condition $\langle x, x_1\rangle = 0$ will be λ_2. (*Hint.* Reduce the quadratic form to its diagonal form.)

2.36. Let y be a fixed vector and let $m(y)$ be the minimum value of the ratio $\langle x, Ax\rangle/\langle x, x\rangle$ for all vectors x such that $\langle x, y\rangle = 0$. Then the maximum value of $m(y)$ as y varies over all possible vectors will be λ_2. Prove this result. Consider its geometrical significance in three dimensions and then extend it to higher eigenvalues.† Note that, in contrast to Problem 2.35, this method does not require a knowledge of the vector x_1.

Evaluation of an Integral

In order to illustrate the application of the theorems that have been proved about quadratic forms, we shall evaluate the following n-dimensional integrals:

$$I_1 = \int_{-\infty}^{\infty} \cdots \int_{-\infty}^{\infty} \exp\left[i\sum_{k=1}^{n} t_k x_k - \sum_{i=1}^{n}\sum_{j=1}^{n} a_{ij}x_i x_j\right] dx_1 \cdots dx_n,$$

$$I_2 = \int_{-\infty}^{\infty} \cdots \int_{-\infty}^{\infty} F\left[\sum_{k=1}^{n} t_k x_k\right] \exp\left[-\sum_{i=1}^{n}\sum_{j=1}^{n} a_{ij}x_i x_j\right] dx_1 \cdots dx_n.$$

Here $F(u)$ is an arbitrary integrable function of the variable u over the interval $(-\infty, \infty)$. We assume that $a_{ij} = a_{ji}$ and, in order to insure that the integrals converge, we also assume that the quadratic form in the exponent is positive-definite. Let A be the matrix whose elements are a_{ij}, let x be the vector whose components are $x_1, x_2, \cdots, x_n$, and let t be the vector whose components are $t_1, t_2, \cdots, t_n$. Then the integrals I_1 and I_2 may be written as follows:

$$I_1 = \int \exp\,[i\langle t, x\rangle - \langle x, Ax\rangle]\, dx$$

$$I_2 = \int F[\langle t, x\rangle] \exp\,[-\langle x, Ax\rangle]\, dx,$$

where the integration is extended over the whole n-dimensional space.

Before evaluating I_1 and I_2, we shall derive the following well-known result:

$$J = \int_{-\infty}^{\infty} e^{-at^2} dt = (\pi/a)^{1/2}. \tag{2.45}$$

The proof of this follows from the fact that

$$J^2 = \int_{-\infty}^{\infty} e^{-at^2}\, dt \int_{-\infty}^{\infty} e^{-as^2}\, ds = \iint e^{-a(s^2+t^2)}\, ds\, dt,$$

and by a transformation to polar coordinates

$$J^2 = \int_0^{2\pi}\int_0^{\infty} e^{-ar^2} r\, dr\, d\theta = \pi/a.$$

† This is an illustration of the minimax principle. See Courant-Hilbert, *Methods of Mathematical Physics*, volume 1, chapter I, Interscience Publishers, New York, 1953.

Now, let P be the orthogonal transformation with determinant one (Problem 2.30) that diagonalizes A so that

$$P^*AP = D,$$

D being a diagonal matrix with elements $\lambda_1, \lambda_2, \cdots, \lambda_n$ all greater than zero. Put $x = Py$; then

$$I_1 = \int \exp [i\langle P^*t, y\rangle - \langle y, Dy\rangle]\, dy,$$
$$I_2 = \int F[\langle P^*t, y\rangle] \exp [-\langle y, Dy\rangle]\, dy,$$

where the integration is again extended over the whole n-dimensional space. Note that the Jacobian of the transformation from the x-coordinates to the y-coordinates is one, since the determinant of P is one. Put $y = D^{-1/2} z$†; then

$$I_1 = (\det D)^{-1/2} \int \exp [i\langle D^{-1/2}P^*t, z\rangle - \langle z, z\rangle]\, dz,$$
$$I_2 = (\det D)^{-1/2} \int F[\langle D^{-1/2}P^*t, z\rangle] \exp [-\langle z, z\rangle]\, dz.$$

The integration is still over the entire n-dimensional space.

In evaluating I_1, we shall try to complete the square in the exponent by putting $z = w + b$ where w is a variable vector and b is a fixed vector. The exponent in I_1 becomes

$$-\langle w, w\rangle - 2\langle w, b\rangle - \langle b, b\rangle + i\langle s, w\rangle + i\langle s, b\rangle.$$

Here we have written s for the vector $D^{-1/2}P^*t$. If we put $b = is/2$, the linear term in w will vanish and the exponent reduces to

$$-\langle w, w\rangle - \frac{\langle s, s\rangle}{4}.$$

Using (2.45), we obtain

$$I_1 = \det D^{-1/2} \exp [-\langle s, s\rangle/4] \int \exp [-\langle w, w\rangle]\, dw$$
$$= (\pi)^{n/2} \det D^{-1/2} \cdot \exp [-\langle s, s\rangle/4].$$

Now, by (2.44),

$$\det D^{-1/2} = (\lambda_1 \cdots \lambda_n)^{-1/2} = (\det A)^{-1/2},$$

and then

$$\langle s, s\rangle = \langle D^{-1/2}P^*t, D^{-1/2}P^*t\rangle = \langle t, PD^{-1}P^*t\rangle = \langle t, A^{-1}t\rangle \tag{2.46}$$

since, from the definition of D,

$$D^{-1} = P^{-1}A^{-1}(P^*)^{-1}.$$

We have, finally, the following result:

$$I_1 = (\pi)^{n/2}(\det A)^{-1/2} \exp [-\langle t, A^{-1}t\rangle/4]. \tag{2.47}$$

† The matrix $D^{-1/2}$ is the diagonal matrix whose elements are the reciprocal of the positive square root of the elements of D.

To complete the evaluation of I_2, we rotate the coordinate axes so that the first axis lies along the direction of the vector $s = D^{-1/2}P^*t$. This rotation is the result of a transformation by an orthogonal matrix of determinant one. Let Q be this matrix and $z = Qw$ the transformation; then

$$I_2 = \det D^{-1/2} \int F[\langle Q^*s, w\rangle] \exp[-\langle w, w\rangle]\, dw.$$

Suppose that $w_1, w_2, \cdots, w_n$ are the components of w; then by the definition of the transformation, Q^*s has only its first component (call it α) different from zero; and hence

$$\langle Q^*s, w\rangle = \alpha w_1.$$

Finally, by integration over $w_2, w_3, \cdots, w_n$ we find that

$$I_2 = (\pi)^{(n-1)/2} (\det A)^{-1/2} \int_{-\infty}^{\infty} F(\alpha u) e^{-u^2}\, du, \tag{2.48}$$

where

$$\alpha^2 = \langle Q^*s, Q^*s\rangle = \langle s, s\rangle = \langle t, A^{-1}t\rangle.$$

We combine our results into

Theorem 2.21. *If A is a positive-definite self-adjoint matrix, then*

$$\int \exp[i\langle t, x\rangle - \langle x, Ax\rangle]\, dx = (\pi)^{n/2} (\det A)^{-1/2} \exp[-\langle t, A^{-1}t\rangle/4]$$

and

$$\int F[\langle t, x\rangle] \exp[-\langle x, Ax\rangle]\, dx = \pi^{(n-1)/2} (\det A)^{-1/2} \int_{-\infty}^{\infty} F(\alpha u) e^{-u^2}\, du.$$

Here the integrals on the left are extended over the entire n-dimensional space, and

$$\alpha^2 = \langle t, A^{-1}t\rangle.$$

PROBLEM

2.37. Show that

$$\int_{-\infty}^{\infty} \cdots \int_{-\infty}^{\infty} x_k x_m \exp\left[-\sum_i \sum_j a_{ij} x_i x_j\right] dx_1 \cdots dx_n = \tfrac{1}{2}\pi^{n/2} (\det A)^{-3/2} D_{km}$$

when D_{km} is the minor of a_{km} in the matrix A. (*Hint.* Differentiate (2.47) with respect to t_k and t_m and then put $t_1 = t_2 = \cdots = t_n = 0$.)

Simultaneous Reduction of Two Quadratic Forms

Suppose that a mechanical system is determined by n coordinates $\zeta_1, \zeta_2, \cdots, \zeta_n$ and that the kinetic energy of the system is given by the real positive-definite quadratic form

$$T = \tfrac{1}{2} \sum_{i=1}^{n} \sum_{j=1}^{n} a_{ij} \dot{\zeta}_i \dot{\zeta}_j = \tfrac{1}{2}\langle \dot{z}, A\dot{z}\rangle,$$

where the dots mean derivatives with respect to time, whereas the potential energy of the system is given by the real quadratic form

$$V = \tfrac{1}{2}\sum_{i=1}^{n}\sum_{j=1}^{n} b_{ij}\zeta_i\zeta_j = \tfrac{1}{2}\langle z, Bz\rangle.$$

We show that a new set of coordinates $\eta_1, \eta_2, \cdots, \eta_n$ can be introduced such that

$$2T = \dot{\eta}_1^2 + \cdots + \dot{\eta}_n^2$$

and

$$2V = \lambda_1\eta_1^2 + \cdots + \lambda_n\eta_n^2.$$

The η-coordinates are called the *normal coordinates* of the system. In terms of these coordinates the motion of the system is determined by the following simple set of equations:

$$\frac{d^2\eta_k}{dt^2} = -\lambda_k\eta_k, \quad k = 1, 2, \cdots, n.$$

In order to determine the normal coordinates, we must first solve the following eigenvalue problem. Find the eigenvalues λ for which the equation

$$(B - \lambda A)x = 0 \tag{2.49}$$

has a non-trivial solution. When (2.49) is considered as a set of linear equations for the components of x, it is clear that λ is an eigenvalue if and only if

$$\det |B - \lambda A| = 0.$$

Since this determinantal equation is in general of the nth degree, it will give n values for λ. To every non-repeated value of λ there will exist an eigenvector x satisfying (2.49). We assume that n eigenvectors x_1, $x_2, \cdots, x_n$ exist corresponding to the eigenvalues $\lambda_1, \lambda_2, \cdots, \lambda_n$, respectively.

Now, since A and B are matrices of quadratic forms, they are self-adjoint; therefore

$$\begin{aligned} 0 = \langle x_k, Bx_j\rangle - \langle x_j, Bx_k\rangle &= \lambda_j\langle x_k, Ax_j\rangle - \lambda_k\langle x_j, Ax_k\rangle \\ &= (\lambda_j - \lambda_k)\langle x_k, Ax_j\rangle. \end{aligned}$$

If $\lambda_j \neq \lambda_k$, this result implies that

$$\langle x_k, Ax_j\rangle = 0,$$

or, expressed verbally, that eigenvectors corresponding to different eigenvalues are orthogonal with respect to the matrix A. Just as for real self-adjoint matrices, the following argument will show that the orthogonality of the eigenvectors implies that the eigenvalues and the eigenvectors are real. Suppose that λ_1 is complex; then $\bar{\lambda}_1$ is also an eigenvalue and $\bar{x}_1$ the corresponding eigenvector. By the orthogonality property, $\langle x_1, Ax_1\rangle = 0$. Since A is positive-definite, this implies that $x_1 = 0$, and hence λ_1 is not an eigenvalue. This contradiction shows that λ_1 is real and then x_1 must be real.

Because A is positive-definite, we have $\langle x_k, Ax_k\rangle > 0$. We may therefore multiply x_k by a suitable constant to ensure that $\langle x_k, Ax_k\rangle = 1$. Consequently, the eigenvectors will form an orthonormal set with respect to A; that is,

$$\langle x_k, Ax_j\rangle = \delta_{jk}. \tag{2.50}$$

Note that

$$\langle x_k, Bx_j\rangle = \langle x_k, \lambda_j Ax_j\rangle = \lambda_j \delta_{jk}. \tag{2.51}$$

Now let Q be the matrix whose columns are the eigenvectors $x_1, x_2, \cdots, x_n$; we may write

$$Q = x_1\rangle\langle f_1 + \cdots + x_n\rangle\langle f_n$$

and then

$$Q^* = e_1\rangle\langle x_1 + \cdots + e_n\rangle\langle x_n,$$

that is, Q^* is the matrix whose rows are $x_1, \cdots, x_n$. Let z be the vector whose components are $\zeta_1, \zeta_2, \cdots, \zeta_n$. Put

$$z = Qy;$$

then

$$2T = \langle \dot{z}, A\dot{z}\rangle = \langle Q\dot{y}, AQ\dot{y}\rangle = \langle \dot{y}, Q^*AQ\dot{y}\rangle$$

and

$$2V = \langle z, Bz\rangle = \langle y, Q^*BQy\rangle.$$

From (2.50) and (2.51) we get

$$Q^*AQ = \sum_k\sum_j e_k\rangle\langle x_k, Ax_j\rangle\langle f_j = \sum_k e_k\rangle\langle f_k = I$$

and

$$Q^*BQ = \sum_k\sum_j e_k\rangle\langle x_k, Bx_j\rangle\langle f_j = \sum_k \lambda_k e_k\rangle\langle f_k.$$

This shows that

$$2T = \dot{\eta}_1^2 + \cdots + \dot{\eta}_n^2$$
$$2V = \lambda_1\eta_1^2 + \cdots \lambda_n\eta_n^2.$$

We state our conclusions in

Theorem 2.22. *If A and B are real self-adjoint matrices, and, if A is positive-definite, there exists a real matrix Q such that*

$$Q^*AQ = I, \quad Q^*BQ = D,$$

where D is a diagonal matrix.

PROBLEM

2.38. Consider the following set of first-order differential equations:

$$A\frac{dz}{dt} = Bz.$$

Here A and B are real symmetric matrices, and z is a vector with components $\xi_1, \xi_2, \cdots, \xi_n$. Solve these equations by making the substitution

$$z = e^{\lambda t}y,$$

where y is a constant vector, and then by introducing normal coordinates.

Spectral Representation

We have seen that if the eigenvectors of a real self-adjoint operator L span the space, they form an O.N. basis for the space. In terms of this basis the operator L takes a particularly simple form which is very important for later applications.

Let $x_1, x_2, \cdots$ be the O.N. eigenvectors of L corresponding to the eigenvalues $\lambda_1, \lambda_2, \cdots$ respectively. Then any vector x can be written as follows:

$$x = \xi_1 x_1 + \xi_2 x_2 + \cdots, \tag{2.52}$$

and we have

$$Lx = \xi_1\lambda_1 x_1 + \xi_2\lambda_2 x_2 + \cdots. \tag{2.53}$$

Since the eigenvectors are O.N., we find that

$$\xi_k = \langle x_k, x\rangle.$$

The representation of x and Lx in (2.52) and (2.53) is called the *spectral representation* of the operator L. It is apparent that problems involving L are simplified when this representation is used. For example, given a, to find x such that

$$Lx = a,$$

we use (2.52), (2.53), and the corresponding representation for a, namely,

$$a = \alpha_1 x_1 + \alpha_2 x_2 + \cdots$$

to get

$$\xi_1\lambda_1 x_1 + \xi_2\lambda_2 x_2 + \cdots = \alpha_1 x_1 + \alpha_2 x_2 + \cdots.$$

Since the eigenvectors form a basis for the space, we have

$$\xi_k = \alpha_k/\lambda_k;$$

consequently,

$$x = \Sigma\alpha_k x_k/\lambda_k,$$

where

$$\alpha_k = \langle x_k, a \rangle.$$

The ideas presented here are often applicable also when L is not an operator in a finite-dimensional space and hence is not a matrix. Then the number of vectors in a basis is infinite, and there are difficulties as to whether the infinite series in (2.52) and (2.53) converge. These difficulties may be partly avoided by the following reasoning.

If a is a vector in the space, the coefficients $\alpha_k = \langle x_k, a \rangle$ always exist even though a series such as (2.52) does not converge. If we *assume* there exists a solution of $Lx = a$, then

$$\langle x_k, a \rangle = \langle x_k, Lx \rangle = \langle Lx_k, x \rangle = \lambda \langle x_k, x \rangle$$

so that again $\xi_k = \alpha_k/\lambda_k$. Now that we know the coefficients ξ_k, the vector may usually be determined either by using the series (2.52) if it converges or by some type of summability method. We shall not discuss summability methods but we shall assume that, if the coefficients ξ_k are known, the vector x can be determined.

If L is not a self-adjoint operator, the spectral representation becomes more complicated because, as we have already observed, the eigenvectors of L may not span the space. However, we shall assume that the generalized eigenvectors of L span the space and thus give a spectral representation.

Let $x_{11}, x_{12}, \cdots, x_{1k_1}$ be a chain of generalized eigenvectors of L corresponding to the eigenvalue λ_1 so that

$$(L - \lambda_1)x_{1j} = x_{1,\,j-1}, \quad (j = 1, 2, \cdots, k_1),$$

where $x_{10} = 0$. Similarly, let $x_{21}, x_{22}, \cdots, x_{2k_2}$ be generalized eigenvectors of L corresponding to λ_2, and so on. Now if we assume that the generalized eigenvectors x_{ij} $(i = 1, 2, \cdots, j = 1, 2, \cdots, k_i)$ span the space, any vector x may be written as follows:

$$\begin{aligned} x = \xi_{11}x_{11} + \cdots + \xi_{1k_1}x_{1k_1} \\ + \xi_{21}x_{21} + \cdots + \xi_{2k_2}x_{2k_2} \cdots \end{aligned}$$

and we have

$$\begin{aligned} Lx = (\lambda_1\xi_{11} + \xi_{12})x_{11} + (\lambda_1\xi_{12} + \xi_{13})x_{12} + \cdots + \lambda_1\xi_{1k_1}x_{1k_1} \\ + (\lambda_2\xi_{21} + \xi_{22})x_{21} + (\lambda_2\xi_{22} + \xi_{23})x_{22} + \cdots \\ + (\lambda_2\xi_{2k_2-1} + \xi_{2k_2})x_{2k_2-1} + \cdots. \end{aligned}$$

These formulas are the spectral representation for the operator L. The last formula may be more readily understood if we remember that, whenever L is a matrix, its representation is as follows:

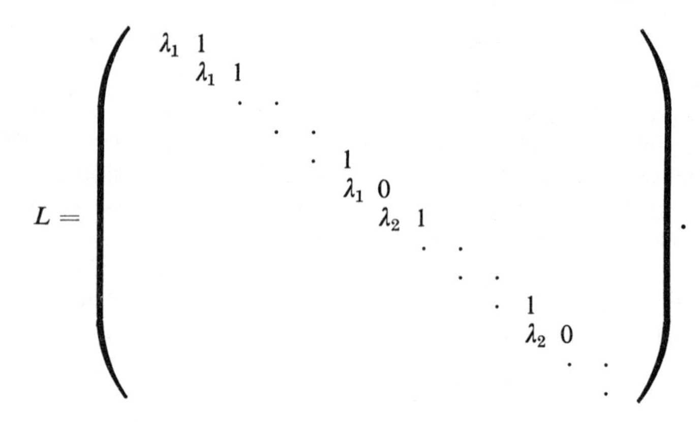

There still remains the problem of determining the coefficients a_{ij} of x. Since L is not self-adjoint, its eigenvectors are no longer mutually orthogonal. To obtain a simple formula for the coefficients, we must use the eigenvectors for the adjoint operator L^*. Theorem 2.11 states that under certain conditions λ_1 is also an eigenvalue of L^* and that there exists a generalized eigenvector of L^* of rank k_1 corresponding to λ_1. At present we shall restrict ourselves to operators L such that every eigenvalue of L is also an eigenvalue of L^*, all eigenvectors of both L and L^* are of rank one, and the eigenvectors of either L or L^* span the space. We call such operators *simple.*

In this case, the spectral representation for L takes the following form:

$$x = \xi_1 x_1 + \xi_2 x_2 \cdots$$

$$Lx = \lambda_1 \xi_1 x_1 + \lambda_2 \xi_2 x_2 \cdots.$$

Let $y_1, y_2, \cdots$ be the eigenvectors of L^* corresponding to the eigenvalues $\lambda_1, \lambda_2, \cdots$ respectively; then, if y is any vector in the space, we may write

$$y = \eta_1 y_1 + \eta_2 y_2 + \cdots$$

$$L^* y = \eta_1 \lambda_1 y_1 + \eta_2 \lambda_2 y_2 + \cdots.$$

In order to find the coefficients in the expansion of x and y, we shall use the fact which was proved in Theorem 2.11 that the eigenvectors of L form a *bi-orthogonal* set to the eigenvectors of L^*; that is,

$$\langle y_i, x_j \rangle = \delta_{ij}. \tag{2.54}$$

With the help of this bi-orthogonality relation, the coefficients of x or of y are determined by the following formulas:

$$\xi_k = \langle y_k, x\rangle,$$
$$\eta_j = \langle x_j, y\rangle.$$

PROBLEM

2.39. Suppose that L and L^* satisfy the conditions of Theorem 2.11. Show how to determine the coefficients ξ_{ij} in the expansion of an arbitrary vector x in terms of the generalized eigenvectors of L.

Functions of an Operator

Let us return to the spectral representation of a simple operator L. If x is an arbitrary vector, we may write

$$x = \xi_1 x_1 + \xi_2 x_2 + \cdots, \tag{2.55}$$
$$Lx = \xi_1 \lambda_1 x_1 + \xi_2 \lambda_2 x_2 + \cdots,$$

where

$$\xi_k = \langle x_k, x\rangle \tag{2.56}$$

if L is self-adjoint, or where

$$\xi_k = \langle y_k, x\rangle$$

if L is not self-adjoint. Hereafter, in order to avoid convergence difficulties, we shall consider relations such as (2.55) only as short methods of indicating that the coefficients of x and Lx in terms of the eigenvectors of L are ξ_k and $\lambda_k \xi_k$, respectively. Now

$$L^2 x = L(Lx) = \xi_1 \lambda_1^2 x_1 + \xi_2 \lambda_2^2 x_2 + \cdots,$$

and, similarly,

$$L^n = \xi_1 \lambda_1^n x_1 + \xi_2 \lambda_2^n x_2 + \cdots;$$

hence, if $q(\lambda)$ is any polynomial in λ, we have

$$q(L)x = \Sigma \xi_k q(\lambda_k) x_k. \tag{2.57}$$

We may generalize (2.57) to continuous functions of λ by the following definition:

$$f(L)x = \Sigma \xi_k f(\lambda_k) x_k. \tag{2.58}$$

This formula has many important applications. For example, the inverse operator $(L - \lambda)^{-1}$ is given by

$$(L - \lambda)^{-1} x = \Sigma (\lambda_k - \lambda)^{-1} \xi_k x_k. \tag{2.59}$$

Of course, if λ equals some eigenvalue λ_k, (2.59) is meaningless unless the corresponding coefficient ξ_k is equal to 0.

For another illustration of the use of (2.58), let us suppose that the space is finite-dimensional. Suppose that x depends on a parameter t and is a solution of the following system of differential equations:

$$\frac{dx}{dt} = Lx.$$

The general solution of this system will be

$$x = e^{Lt}x_0 \tag{2.60}$$

where x_0 is a vector independent of t. If we represent L in a basis where it becomes a diagonal matrix, then, by (2.58), (2.60) becomes

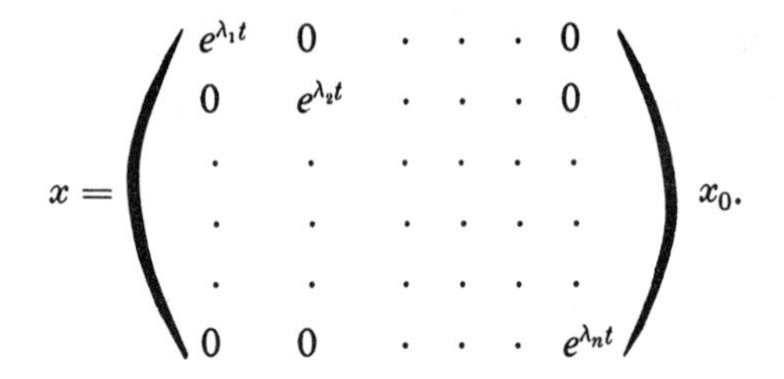

$$x = \begin{pmatrix} e^{\lambda_1 t} & 0 & \cdot & \cdot & \cdot & 0 \\ 0 & e^{\lambda_2 t} & \cdot & \cdot & \cdot & 0 \\ \cdot & \cdot & \cdot & \cdot & \cdot & \cdot \\ \cdot & \cdot & \cdot & \cdot & \cdot & \cdot \\ \cdot & \cdot & \cdot & \cdot & \cdot & \cdot \\ 0 & 0 & \cdot & \cdot & \cdot & e^{\lambda_n t} \end{pmatrix} x_0.$$

If L is represented by a matrix A which is not diagonal, then, by Theorem 2.11, there exists a matrix P such that $P^{-1}AP = D$, D being a diagonal matrix. If we solve for A, we find that $A = PDP^{-1}$, and since

$$A^k = (PDP^{-1})(PDP^{-1}) \cdots (PDP^{-1}) = PD^kP^{-1},$$

we have finally

$$q(A) = \Sigma\alpha_k A^k = \Sigma\alpha_k PD^kP^{-1} = Pq(D)P^{-1}. \tag{2.61}$$

This result can be shown to extend to analytic functions $f(\lambda)$ by using the power series for $f(\lambda)$.

Formula (2.58) is not valid unless L can be represented as a diagonal matrix. From Theorem 2.11 we know that any operator L in a finite-dimensional space can be represented in Jordan canonical form as follows:

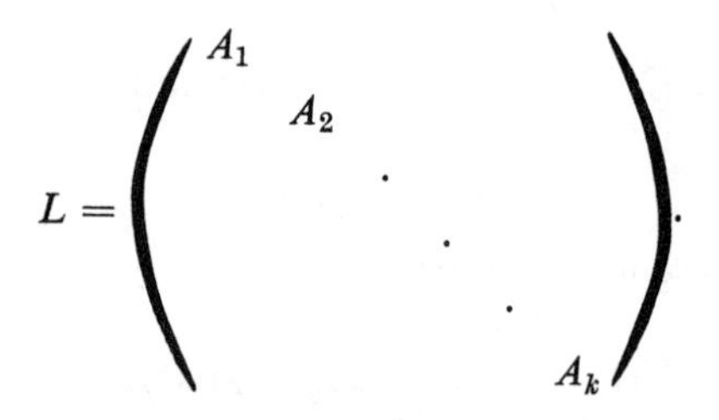

$$L = \begin{pmatrix} A_1 & & & & \\ & A_2 & & & \\ & & \cdot & & \\ & & & \cdot & \\ & & & & \cdot & \\ & & & & & A_k \end{pmatrix}.$$

Here A_j $(1 \leq j \leq k)$ is a matrix such that the elements on the main diagonal are all λ_j, the elements on the diagonal above the main diagonal are either zero or one, and all other elements of the matrix are zero.

We may write

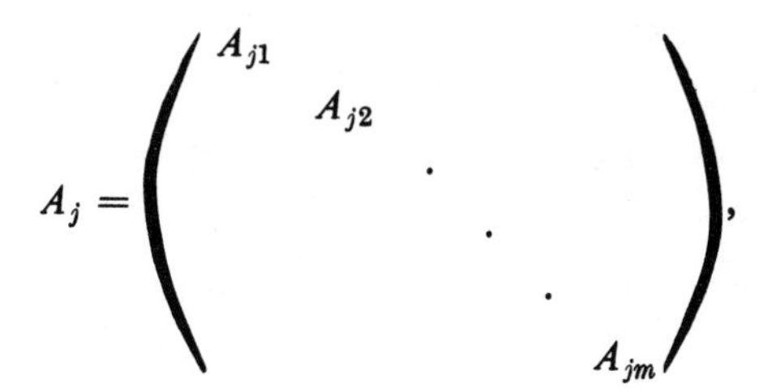

$$A_j = \begin{pmatrix} A_{j1} & & & & \\ & A_{j2} & & & \\ & & \cdot & & \\ & & & \cdot & \\ & & & & \cdot \\ & & & & & A_{jm} \end{pmatrix},$$

where $m = m(j)$, and where

$$A_{ji} = \begin{pmatrix} \lambda_j & 1 & \cdot & \cdot & \cdot & 0 & 0 \\ 0 & \lambda_j & \cdot & \cdot & \cdot & 0 & 0 \\ \cdot & \cdot & \cdot & \cdot & \cdot & \cdot & \cdot \\ \cdot & \cdot & \cdot & \cdot & \cdot & \cdot & \cdot \\ \cdot & \cdot & \cdot & \cdot & \cdot & \cdot & \cdot \\ 0 & 0 & \cdot & \cdot & \cdot & \lambda_j & 1 \\ 0 & 0 & \cdot & \cdot & \cdot & 0 & \lambda_j \end{pmatrix}.$$

Here all the elements on the diagonal above the main diagonal are ones.

By actual matrix multiplication, we see that

$$A_{ji}^2 = \begin{pmatrix} \lambda_j^2 & 2\lambda_j & 1 & \cdot & \cdot & \cdot & 0 \\ 0 & \lambda_j^2 & 2\lambda_j & \cdot & \cdot & \cdot & 0 \\ 0 & 0 & \lambda_j^2 & \cdot & \cdot & \cdot & \cdot \\ \cdot & \cdot & \cdot & \cdot & \cdot & \cdot & \cdot \\ \cdot & \cdot & \cdot & \cdot & \cdot & \cdot & 1 \\ \cdot & \cdot & \cdot & \cdot & \cdot & \cdot & 2\lambda_j \\ 0 & 0 & 0 & \cdot & \cdot & \cdot & \lambda_j^2 \end{pmatrix}$$

and that

$$A_{ji}^3 = \begin{pmatrix} \lambda_j^3 & 3\lambda_j^2 & 3\lambda_j & 1 & \cdot & \cdot & \cdot & 0 \\ 0 & \lambda_j^3 & 3\lambda_j^2 & 3\lambda_j & \cdot & \cdot & \cdot & \cdot \\ 0 & 0 & \lambda_j^3 & 3\lambda_j^2 & \cdot & \cdot & \cdot & \cdot \\ 0 & 0 & 0 & \lambda_j^3 & \cdot & \cdot & \cdot & 1 \\ \cdot & \cdot & \cdot & \cdot & \cdot & \cdot & \cdot & 3\lambda_j \\ \cdot & \cdot & \cdot & \cdot & \cdot & \cdot & \cdot & 3\lambda_j^2 \\ 0 & 0 & 0 & 0 & \cdot & \cdot & \cdot & \lambda_j^3 \end{pmatrix}.$$

These results suggest

Theorem 2.23. *If $q(\lambda)$ is a polynomial,*

$$q(A_{ji}) = \begin{pmatrix} q(\lambda_j) & q'(\lambda_j) & q''(\lambda_j)/2! & q'''(\lambda_j)/3! & \cdots \\ 0 & q(\lambda_j) & q'(\lambda_j) & q''(\lambda_j)/2! & \cdots \\ 0 & 0 & q(\lambda_j) & q'(\lambda_j) & \cdots \\ \cdot & \cdot & \cdot & \cdot & \cdots \\ \cdot & \cdot & \cdot & \cdot & \cdots \\ \cdot & \cdot & \cdot & \cdot & \cdots \end{pmatrix}.$$

Proof. From the preceding formulas, it is clear that the theorem is true for $q(\lambda) = \lambda$, λ^2, or λ^3. By mathematical induction on the integer n, we can show that the theorem is true for $q(\lambda) = \lambda^n$. Because the derivative of a sum of terms is the sum of the derivatives of each term, the theorem is true when $q(\lambda)$ is any polynomial.

Now, since by matrix multiplication

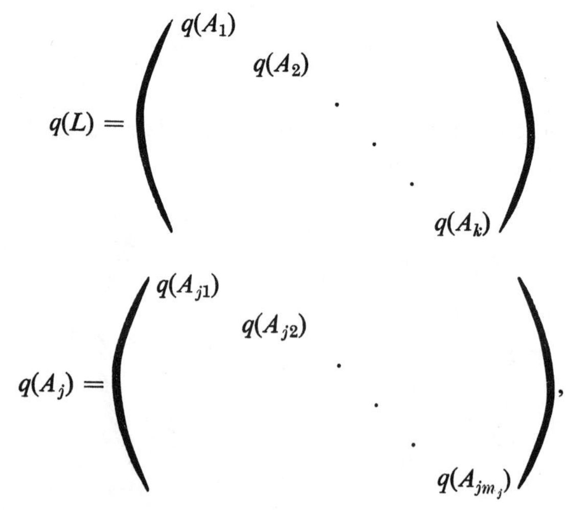

$$q(L) = \begin{pmatrix} q(A_1) & & & & \\ & q(A_2) & & & \\ & & \cdot & & \\ & & & \cdot & \\ & & & & \cdot \\ & & & & & q(A_k) \end{pmatrix}$$

and

$$q(A_j) = \begin{pmatrix} q(A_{j1}) & & & & \\ & q(A_{j2}) & & & \\ & & \cdot & & \\ & & & \cdot & \\ & & & & \cdot \\ & & & & & q(A_{jm_j}) \end{pmatrix},$$

we have found a matrix representation for $q(L)$.

The results above can be extended to analytic functions $f(\lambda)$ by using the power series which converge to $f(\lambda)$. We give one illustration:

$$\exp(tA_{ji}) = \begin{pmatrix} e^{t\lambda_j} & te^{t\lambda_j} & t^2e^{t\lambda_j}/2 & \cdots \\ 0 & e^{t\lambda_j} & te^{t\lambda_j} & \cdots \\ 0 & 0 & e^{t\lambda_j} & \cdots \\ \cdot & \cdot & \cdot & \cdots \\ \cdot & \cdot & \cdot & \cdots \end{pmatrix}.$$

PROBLEMS

2.40. Find the solution of

$$\frac{dx}{dt} = \begin{pmatrix} 3 & 2 \\ 2 & 3 \end{pmatrix} x$$

such that at $t = 0$, $x = \begin{pmatrix} 1 \\ 1 \end{pmatrix}$. (*Hint.* Use (2.61).)

2.41. Find the general solution of

$$\frac{dx}{dt} = \begin{pmatrix} \lambda & 1 & 0 \\ 0 & \lambda & 0 \\ 0 & 0 & \lambda \end{pmatrix} x.$$

2.42. Find the formula for the nth power of the following matrix:

$$\begin{pmatrix} 1 & 3 \\ 3 & 1 \end{pmatrix}.$$

2.43. Suppose that A is a simple operator with eigenvalues $0 \leq \lambda_1 \leq \lambda_2 \leq \cdots \leq \lambda_n$, and suppose that $f(\lambda)$ is an analytic function of λ regular in the circle $|\lambda| < a$. Show that the power series for $f(A)$ converges if and only if $a > \lambda_n$.

Spectral Representation and Complex Integration

Suppose that L is a simple operator with eigenvectors $x_1, x_2, \cdots, x_n$ corresponding to the eigenvalues $\lambda_1, \lambda_2, \cdots, \lambda_n$. The theory of the previous sections shows that, if x is an arbitrary vector, we may write

$$x = \xi_1 x_1 + \cdots + \xi_n x_n, \tag{2.62}$$
$$Lx = \xi_1 \lambda_1 x_1 + \cdots + \xi_n \lambda_n x_n,$$

and then

$$f(L)x = \xi_1 f(\lambda_1) x_1 + \cdots + \xi_n f(\lambda_n) x_n. \tag{2.63}$$

Now, consider the following integral in the complex λ-plane:

$$\frac{1}{2\pi i} \oint d\lambda (\lambda - L)^{-1} x.$$

Because of (2.63), this integral becomes

$$\frac{1}{2\pi i} \oint d\lambda \sum_1^n (\lambda - \lambda_k)^{-1} \xi_k x_k = \sum_1^n \frac{\xi_k x_k}{2\pi i} \oint \frac{d\lambda}{\lambda - \lambda_k} \tag{2.64}$$
$$= \Sigma \xi_k x_k = x$$

if the contour over which the integral is taken encloses the spectrum of L, that is, encloses all the eigenvalues of L.

Let us illustrate this result by assuming

$$L = \begin{pmatrix} 6 & 3 \\ -1 & 2 \end{pmatrix} \quad \text{and} \quad x = \begin{pmatrix} \xi_1 \\ \xi_2 \end{pmatrix}.$$

We find that

$$(\lambda - L)^{-1}x = \begin{pmatrix} \lambda - 6 & -3 \\ 1 & \lambda - 2 \end{pmatrix}^{-1} \begin{pmatrix} \xi_1 \\ \xi_2 \end{pmatrix} = \frac{1}{\lambda^2 - 8\lambda + 15} \begin{pmatrix} \lambda - 2 & 3 \\ -1 & \lambda - 6 \end{pmatrix} \begin{pmatrix} \xi_1 \\ \xi_2 \end{pmatrix}$$

$$= \frac{1}{\lambda^2 - 8\lambda + 15} \begin{pmatrix} \xi_1(\lambda - 2) + 3\xi_2 \\ -\xi_1 + \xi_2(\lambda - 6) \end{pmatrix}.$$

The integral

$$\frac{1}{2\pi i} \oint (\lambda - L)^{-1} x \, d\lambda$$

is a vector whose first component is

$$\frac{1}{2\pi i} \oint \frac{\xi_1(\lambda - 2) + 3\xi_2}{\lambda^2 - 8\lambda + 15} \, d\lambda$$

and whose second component is

$$\frac{1}{2\pi i} \oint \frac{-\xi_1 + \xi_2(\lambda - 6)}{\lambda^2 - 8\lambda + 15} \, d\lambda.$$

Assume that the contour is a curve enclosing the eigenvalues of L, which are $\lambda = 3$ and $\lambda = 5$. Then, evaluating the residues of the integrals at the poles $\lambda = 3$ and $\lambda = 5$, we have

$$(2.65) \quad \frac{1}{2\pi i} \oint (\lambda - L)^{-1} x \, d\lambda = -\frac{1}{2} \begin{pmatrix} \xi_1 + 3\xi_2 \\ -\xi_1 - 3\xi_2 \end{pmatrix} + \frac{1}{2} \begin{pmatrix} 3\xi_1 + 3\xi_2 \\ -\xi_1 - \xi_2 \end{pmatrix}$$

$$= \frac{\xi_1 + 3\xi_2}{2} \begin{pmatrix} -1 \\ 1 \end{pmatrix} + \frac{\xi_1 + \xi_2}{2} \begin{pmatrix} 3 \\ -1 \end{pmatrix} = \begin{pmatrix} \xi_1 \\ \xi_2 \end{pmatrix} = x.$$

This verifies (2.64).

The evaluation in (2.65) illustrates another important point. The vectors $\begin{pmatrix} -1 \\ 1 \end{pmatrix}$ and $\begin{pmatrix} 3 \\ -1 \end{pmatrix}$, which appeared in (2.65), are the eigenvectors of L corresponding to the eigenvalues 3 and 5, respectively. Suppose, then, that we do not know the spectral representation of L. We need only evaluate the complex integral

$$\frac{1}{2\pi i} \oint (\lambda - L)^{-1} x \, d\lambda,$$

and the calculation of residues would automatically give the spectral representation for the vector x. Of course, in practice, this is an awkward and inconvenient way of finding the eigenvalues and eigenvectors of a matrix. However, we shall find that for more general operators, such as differential operators, this method of evaluating the complex integral of the inverse of $\lambda - L$ will be a convenient way to obtain the spectral representation for L.

We now formulate a result somewhat more general than (2.65).

Theorem 2.24. *Let L be a simple finite-dimensional operator and let $g(\lambda)$ be a function analytic on the spectrum of L; then*

$$\frac{1}{2\pi i}\oint g(\lambda)(\lambda - L)^{-1}\, d\lambda = g(L),$$

where the integration is over any curve which encloses the spectrum of L. In particular,

$$\frac{1}{2\pi i}\oint (\lambda - L)^{-1}\, d\lambda = I,$$

I being the identity operator.

The proof of this theorem can be obtained by using (2.63), with $f(L) = g(L)(\lambda - L)^{-1}$ and evaluating the resulting integrals by the calculus of residues.

The Characteristic Equation and Functions of Matrices

The methods suggested in the preceding section for finding the function of a matrix A require a knowledge of the transformation which diagonalizes A. In practical applications of matrices, the determination of the diagonalizing matrix is quite difficult since it requires determining the eigenvalues by solving an nth order polynomial equation and also determining the eigenvectors by solving n sets of n simultaneous linear equations. We shall show that a knowledge of the diagonalizing matrix is not necessary since a function of the matrix can be found by the use of the characteristic equation satisfied by the matrix. For example, suppose that

$$A = \begin{pmatrix} 1 & -3 \\ 3 & 1 \end{pmatrix}$$

and we wish to find $A^4 + 6A$. The characteristic equation of A is found by expanding the determinant of $A - \lambda I$ and substituting the matrix A for λ in the resulting polynomial. This gives the following expression for the characteristic equation (see Theorem 2.13):

$$g(A) = A^2 - 2A + 10 = 0.$$

Now, when $A^4 + 6A$ is divided by $g(A)$, we find that

$$A^4 + 6A = (A^2 - 2A + 10)(A^2 + 2A - 6) - 26A + 60I.$$

Since $A^2 - 2A + 10$ is identically zero,

$$A^4 + 6A = -26A + 60I = \begin{pmatrix} 34 & 78 \\ -78 & 34 \end{pmatrix}.$$

This example suggests the following

Method. *If $q(\lambda)$ is a polynomial and if A is an arbitrary n-dimensional matrix whose characteristic equation is $g(A) = 0$, then $q(A)$ is equal to $r(A)$, where $r(\lambda)$ is the remainder when $q(\lambda)$ is divided by $g(\lambda)$.*

The proof of this method is as follows. Suppose that when $q(\lambda)$ is divided by $g(\lambda)$, the quotient is $q_1(\lambda)$ and the remainder $r(\lambda)$. Then we have

$$q(\lambda) = g(\lambda)q_1(\lambda) + r(\lambda),$$

which, since it is an identity, will also be valid if we replace λ by the matrix A. Thus

$$q(A) = g(A)q_1(A) + r(A).$$

By Theorem 2.13, $g(A)$ is identically zero; this justifies the method.

The above method applies only to polynomial functions of A. Suppose that we wish to find $f(A)$ when $f(\lambda)$ is an analytic function of λ. We shall try to do it by following the procedure of the above method. We look for a polynomial $r(\lambda)$ of degree $n - 1$ such that

$$f(\lambda) = g(\lambda)h(\lambda) + r(\lambda), \tag{2.66}$$

where $h(\lambda)$ is an analytic function of λ. In (2.66) we substitute the eigenvalues of A, namely, $\lambda_1, \lambda_2, \cdots, \lambda_n$, for λ. Since $g(\lambda_i) = 0 \ (1 \leq i \leq n)$, we get the following n linear equations for the n coefficients of the polynomial $r(\lambda)$:

$$\begin{aligned} f(\lambda_1) &= r(\lambda_1), \\ &\cdots \\ f(\lambda_n) &= r(\lambda_n). \end{aligned} \tag{2.67}$$

Determine $r(\lambda)$ to satisfy the equations (2.67); then

$$\frac{f(\lambda) - r(\lambda)}{g(\lambda)}$$

will be an analytic function of λ since the zeroes of the denominator are also zeroes of the numerator. Call this function $h(\lambda)$. Since (2.66) holds for all values of λ in a circle around the origin, we may substitute A for λ and get the following identity:

$$f(A) = g(A)h(A) + r(A).$$

Again, using the fact that $g(A)$ is identically zero, we find that

$$f(A) = r(A).$$

We formulate this result as

Theorem 2.25. *If A is a matrix whose eigenvalues, arranged in order of increasing absolute value, are $\lambda_1, \lambda_2, \cdots, \lambda_n$ and if $f(\lambda)$ is an analytic function of λ in a circle around the origin with radius greater than $|\lambda_n|$, then $f(A)$ equals $r(A)$, the polynomial of degree $n - 1$ for which*

$$f(\lambda_k) = r(\lambda_k), \quad k = 1, 2, \cdots, n.$$

As an illustration of this theorem, we evaluate e^{tA}, where

$$A = \begin{pmatrix} 6 & -1 \\ 3 & 2 \end{pmatrix}.$$

Put

$$e^{tA} = \alpha I + \beta A,$$

where α and β must be determined. The eigenvalues of A are $\lambda_1 = 3$, $\lambda_2 = 5$. After substitution, we find that

$$e^{3t} = \alpha + 3\beta, \quad e^{5t} = \alpha + 5\beta.$$

The solution of these equations is

$$\beta = (e^{5t} - e^{3t})/2,$$
$$\alpha = (5e^{3t} - 3e^{5t})/2;$$

therefore

$$e^{tA} = \frac{5e^{3t} - 3e^{5t}}{2}\begin{pmatrix} 1 & 0 \\ 0 & 1 \end{pmatrix} + \frac{e^{5t} - e^{3t}}{2}\begin{pmatrix} 6 & -1 \\ 3 & 2 \end{pmatrix}$$
$$= \frac{1}{2}\begin{pmatrix} 3e^{5t} - e^{3t}, & e^{3t} - e^{5t} \\ 3e^{5t} - 3e^{3t}, & 3e^{3t} - e^{5t} \end{pmatrix}.$$

PROBLEMS

2.44. Show that, when $\lambda_1 = \lambda_2 = \cdots = \lambda_\nu$ so that λ_ν is an eigenvalue of multiplicity ν for A, Theorem 2.25 should be modified as follows: $r(A)$ is that polynomial of degree $n - 1$ for which $f(\lambda_k) = r(\lambda_k)$, $(k = \nu + 1, \nu + 2, \cdots, n)$ and also $f^{(j)}(\lambda) = r^{(j)}(\lambda)$, where $j = 0, 1, 2, \cdots, \nu - 1$.

2.45. Find a formula for the nth power of the following matrices:

(*a*)
$$\begin{pmatrix} a & b \\ c & d \end{pmatrix}.$$

(*b*)
$$\begin{pmatrix} \cos\theta & \sin\theta \\ -\sin\theta & \cos\theta \end{pmatrix}.$$

(*c*)
$$\begin{pmatrix} 1 & 2 & 3 \\ -1 & 3 & 1 \\ 1 & 0 & 2 \end{pmatrix}.$$

2.46. Find e^{tA}, where A is each of the matrices in Problem 2.45.

Systems of Difference or Differential Equations

This section will show how the previously developed methods for dealing with functions of matrices can be used to solve systems both of first-order difference equations and of first-order differential equations. Consider the following recurrence relation:

$$p_{n+1} = 3p_n + 4q_n,$$
$$q_{n+1} = p_n + 3q_n.$$

We shall show how to find a formula for p_n and q_n in terms of the initial values p_0 and q_0. Consider p_n and q_n as the components of a vector x_n in two-dimensional space. Then the recurrence relation may be written as follows:

$$x_{n+1} = Ax_n,$$

where A is the matrix

$$\begin{pmatrix} 3 & 4 \\ 1 & 3 \end{pmatrix}.$$

It is now clear that $x_n = A^n x_0$. Since the eigenvalues of A are 5 and 1, if we put

$$A^n = \alpha I + \beta A,$$

we must have

$$5^n = \alpha + 5\beta, \quad 1^n = \alpha + \beta.$$

The solution of these equations is $\beta = (5^n - 1)/4$, $\alpha = (5 - 5^n)/4$. Consequently,

$$A^n = \frac{1}{4}\begin{pmatrix} 2 \cdot 5^n + 2 & 4(5^n - 1) \\ 5^n - 1 & 2 \cdot 5^n + 2 \end{pmatrix};$$

and then

$$p_n = \tfrac{1}{2}(5^n + 1)p_0 + (5^n - 1)q_0,$$
$$q_n = \tfrac{1}{4}(5^n - 1)p_0 + \tfrac{1}{2}(5^n + 1)q_0.$$

The method illustrated here is applicable to systems of recurrence relations in k unknowns. We formulate it as follows:

Method. *To solve a system of k recurrence relations in k unknowns, consider the k unknowns as the components of a k-dimensional vector* x_n; *then the recurrence relations may be written as* $x_{n+1} = Ax_n$, *where A is the matrix of the system. Use Theorem 2.25 to express* A^n *in terms of the eigenvalues of A, and then* $x_n = A^n x_0$ *will be a formula for the nth set of unknowns in terms of the initial set of unknowns.*

Consider now the second-order difference equation

$$p_{n+1} = 5p_n - 6p_{n-1}.$$

To find a formula for p_n in terms of the initial values p_0 and p_1, we write the difference equation as a set of two recurrence relations, as follows:

$$p_{n+1} = 5p_n - 6q_n$$
$$q_{n+1} = p_n.$$

Applying the method described above, we find that

$$p_{n+1} = (3^{n+1} - 2^{n+1})p_1 - 6(3^n - 2^n)p_0.$$

Method. *To solve the difference equation of kth order with constant coefficients*

$$p_{k+n} + a_1 p_{n+k-1} + \cdots + a_k p_n = 0,$$

consider $p_{n+k}, p_{n+k-1}, \cdots, p_{n+1}$ as the components of a k-dimensional vector x_{n+1}. The difference equation is then equivalent to the following set of recurrence relations:

$$x_{n+1} = Ax_n,$$

where

$$A = \begin{pmatrix} -a_1 & -a_2 & \cdot & \cdot & \cdot & -a_{k-1} & -a_k \\ 1 & 0 & \cdot & \cdot & \cdot & 0 & 0 \\ \cdot & 1 & \cdot & \cdot & \cdot & \cdot & \cdot \\ \cdot & \cdot & \cdot & \cdot & \cdot & \cdot & \cdot \\ \cdot & \cdot & \cdot & \cdot & \cdot & \cdot & \cdot \\ 0 & 0 & \cdot & \cdot & \cdot & 0 & 0 \\ 0 & 0 & \cdot & \cdot & \cdot & 1 & 0 \end{pmatrix}.$$

This recurrence relation can now be solved by the method given on page 122.

As a final application of the theory, we shall discuss the solution of a system of first-order differential equations. Let $x(t)$ be a k-dimensional vector depending on a parameter t, and suppose that

$$\frac{dx}{dt} = Ax, \tag{2.68}$$

where A is a constant matrix; then

$$x(t) = e^{At}x_0, \tag{2.69}$$

where x_0 is the value of $x(t)$ for $t = 0$. The proof of this result is as follows: The matrix e^{At} may be defined either as the infinite series of matrices $1 + At + \frac{(At)^2}{2!} + \cdots$ or as the matrix whose eigenvalues are

$e^{\lambda_1 t}, \cdots, e^{\lambda_n t}$, where $\lambda_1, \cdots, \lambda_n$ are the eigenvalues of A. In either case, we find that

$$\frac{de^{At}}{dt} = Ae^{At}.$$

Therefore,

$$\frac{dx}{dt} = \frac{de^{At}}{dt} x_0 = Ae^{At}x_0 = Ax,$$

and hence $x(t)$ as given by (2.69) is a solution of the system in (2.68). We state this result as

Theorem 2.26. *The solution of the system*

$$\frac{dx}{dt} = Ax,$$

where A is a constant matrix, is

$$x(t) = e^{At}x_0,$$

where x_0 is the value of $x(t)$ for $x = 0$.

PROBLEMS

2.47. Show that if

$$A(t) = \begin{pmatrix} t & 1 \\ 0 & 1 \end{pmatrix}$$

and we define

$$e^A = 1 + A + \frac{A^2}{2!} + \frac{A^3}{3!} + \cdots,$$

then

$$\frac{de^A}{dt} \neq \frac{dA}{dt} e^A.$$

2.48. Suppose that $A(t)$ is a matrix whose elements depend on a parameter t and suppose that

$$A(t_1)A(t_2) = A(t_2)A(t_1)$$

for all values of t_1 and t_2. Define

$$\frac{df(A)}{dt} = \lim_{h \to 0} \frac{f(A(t+h)) - f(A(t))}{h}.$$

Show that

$$\frac{dA^n}{dt} = nA^{n-1}\frac{dA}{dt}, \quad \frac{de^A}{dt} = \frac{dA}{dt} e^A.$$

2.49. Find a formula for p_n and q_n in terms of p_0 and q_0 if

$$p_{n+1} = 2p_n + q_n,$$
$$q_{n+1} = 2p_n + 3q_n.$$

2.50. Find a formula for p_n in terms of p_0 and p_1 if

$$p_{n+1} = p_n + p_{n-1}.$$

Find the limit of p_{n+1}/p_n as $n \to \infty$.

2.51. Find a formula for p_n, q_n, and r_n in terms of p_0, q_0, and r_0 if

$$p_{n+1} = 2p_n + q_n + r_n,$$
$$q_{n+1} = 2q_n + r_n,$$
$$r_{n+1} = 2r_n.$$

2.52. Solve

$$\frac{df(t)}{dt} = 3f(t) + 4g(t),$$

$$\frac{dg(t)}{dt} = f(t) + 3g(t),$$

given that $f(0) = 1$, $g(1) = 1$.

Operators in General Spaces

There are two reasons why the spectral theory for operators in general spaces is more complicated than the theory thus far presented in this chapter. First, the general operator may have eigenvalues of infinite index; and, second, the eigenvectors and generalized eigenvectors of the operator may not span the space. We shall not discuss how these specific difficulties may be treated but, instead, we shall present some of the concepts which are needed in the general theory.

The spectral representation for an operator L depends on the study of the *inverse* of the operator $L - \lambda$ for all complex values of λ. The operator $L - \lambda$ is said to have an inverse if for any vector a in the range of $L - \lambda$ there exists a unique vector x such that

$$(L - \lambda)x = a. \tag{2.70}$$

By Theorem 1.4, a necessary and sufficient condition for the existence of an inverse is that the homogeneous equation

$$(L - \lambda)x = 0 \tag{2.71}$$

have only the trivial solution $x = 0$.

If equation (2.71) has a non-trivial solution, then λ is an eigenvalue and the solution x is an eigenvector of L. In such a case, λ is said to belong to the *point spectrum* or the *discrete spectrum* of L.

Suppose that λ is such that the equation (2.71) has only the trivial solution $x = 0$; then the operator $L - \lambda$ has an inverse and we may distinguish three mutually exclusive possibilities. First, the closure of the range of $L - \lambda$ may be the whole space $\mathcal{S}$ and the inverse is a bounded operator. This means that for every vector a in $\mathcal{S}$ there exists a unique vector x satisfying (2.70) and such that the ratio $\langle x, x\rangle/\langle a, a\rangle$ is bounded. In this case, λ is said to belong to the *resolvent set* of the operator L. Second, the range or the closure† of the range of $L - \lambda$ is the whole space, but the inverse is an unbounded operator. In this case, λ is said to be in the *continuous spectrum* of L. In the third case, the range or the closure of the range of $L - \lambda$ may be a proper subset of $\mathcal{S}$. In this case, λ is said to be in the *residual spectrum* of L. The *spectrum* of L consists of all values of λ which belong to either the discrete, the continuous, or the residual spectrum.

A few illustrations will clarify these concepts. In E_∞ put

$$x = (\xi_1, \xi_2, \cdots)$$

and consider the following operators:

$$Cx = (0, \xi_1, \xi_2, \cdots),$$

$$Dx = (\xi_2, \xi_3, \xi_4, \cdots),$$

$$Ex = (\xi_1, \xi_2/2, \xi_3/3, \cdots).$$

The operator E has the values $\lambda = 1/k$, $(k = 1, 2, \cdots)$ in the discrete spectrum because

$$(E - 1/k)e_k = 0,$$

where e_k is the vector whose kth component is one and the other components are zero. Since the eigenvectors e_k form a basis for E_∞, we find that the solution of

$$(E - \lambda)x = a$$

is given by $x = \Sigma\xi_k e_k$, where

$$\xi_k = -(\lambda - 1/k)^{-1}\alpha_k$$

if $a = \Sigma\alpha_k e_k$. Clearly, the inverse of $E - \lambda$ exists for all $\lambda \neq 1/k$; however, when $\lambda = 0$ the inverse is unbounded since $E^{-1}e_k = ke_k$ and $\langle E^{-1}e_k, E^{-1}e_k\rangle/\langle e_k, e_k\rangle = k^2$. The point $\lambda = 0$, which is a limit point in the discrete spectrum, is therefore in the continuous spectrum but all other values of $\lambda \neq 1/k$ $(k = 1, 2, \cdots)$ are in the resolvent set for L. Note that even though the eigenvectors of E form a basis for the space, the continuous spectrum of E is not empty.

The operator C provides an illustration of the residual spectrum. The point $\lambda = 0$ is clearly not an eigenvalue of C. The closure of the range

† The closure of a set is the set together with all its limit points.

of C is not the whole space E_∞ since all vectors whose first component is not zero are lacking; therefore, we conclude that zero is in the residual spectrum for the operator C. The adjoint of C is D, and this latter operator has zero in the discrete spectrum because $De_1 = 0$ where e_1 is the vector whose first component is one and all the other components are zero. This is an illustration of the following fact:

If the value λ_1 is in the residual spectrum of L, then λ_1 is in the discrete spectrum of L^*.

The proof of this result follows from the fact that Theorem 1.5 implies that the null space of $L^* - \lambda_1$ is the orthogonal complement of the range of $L - \lambda_1$. Since by hypothesis the closure of the range of $L - \lambda_1$ is not the whole space, we conclude that the null space of $L^* - \lambda_1$ is not empty, and consequently λ_1 is an eigenvalue of L^*.

The definition of the continuous spectrum is not easy to apply in practice. It is easier to use the concept of the *approximate spectrum.* A value λ is in the approximate spectrum of L if there exists a sequence of vectors x_n such that $|x_n| = 1$ and such that

$$|(L - \lambda)x_n| < 1/n. \tag{2.72}$$

To illustrate this concept, consider the operator E. Put $x_n = e_n$; then $|e_n| = 1$ and $|Ee_n| = 1/n$; therefore, zero is in the approximate spectrum of E.

If we put $(L - \lambda)x_n = a_n$, we have

$$\frac{|x_n|}{|a_n|} > n;$$

consequently, the inverse of $L - \lambda$ cannot be bounded. This shows that *any value of λ in the approximate spectrum of an operator cannot be in the resolvent set.*

Any eigenvalue λ_0 belongs to the approximate spectrum because if we put $x_n = x_0$ $(n = 1, 2, \cdots)$, where x_0 is the eigenvector corresponding to λ_0, then (2.72) will be satisfied. Any value λ in the continuous spectrum of L belongs also to the approximate spectrum of L. By the definition of the continuous spectrum, the inverse of $L - \lambda$ is unbounded; consequently, there exist vectors x_n such that if we put $(L - \lambda)x_n = a_n$, then

$$\frac{\langle x_n, x_n\rangle}{\langle a_n, a_n\rangle} > n.$$

Normalize the vectors x_n so that $|x_n| = 1$; then equation (2.72) is satisfied. This proves that λ is in the approximate spectrum. We have therefore proved

Theorem 2.27. *The approximate spectrum contains all λ which are in the discrete and continuous spectrums, but it does not contain any λ which is in the resolvent set.*

A value of λ in the residual spectrum need not belong to the approximate spectrum. For example, zero is in the residual spectrum of C, but since $|Cx| = |x|$ for all x, it is clear that zero cannot be in the approximate spectrum of C.

PROBLEMS

2.53. Show that a self-adjoint operator has no residual spectrum.

2.54. Show that if λ is in the continuous spectrum of L, then $L - \lambda$ does not have a closed range.

2.55. If λ is in the residual spectrum of L and if the range of $L^* - \lambda$ is closed, show that λ is an eigenvalue of infinite index for L^*.

Illustration—the Continuous Spectrum

In this section we shall discuss the spectral representation for a typical operator whose spectrum contains the continuous spectrum only.

Let the vector x in E_∞ have as its nth component ξ_n $(n = 1, 2, \cdots)$. Consider the operator Fx which has as its nth component $\xi_{n-1} + \xi_{n+1}$ $(n = 1, 2, \cdots)$. Here we have assumed $\xi_0 = 0$.

The operator F is clearly self-adjoint. To find its spectral representation, we first investigate whether F has any eigenvalues. Consider the equation

$$(F - \lambda)x = 0.$$

The nth component of this equation is

$$\xi_{n-1} + \xi_{n+1} - \lambda\xi_n = 0, \quad (n = 1, 2, 3, \cdots). \tag{2.73}$$

This is a recurrence relation which may be treated by the methods of the section on difference and differential equations or by the following equivalent method.

Assume that $\xi_n = \rho^n$; then (2.73) becomes

$$\rho^2 - \lambda\rho + 1 = 0. \tag{2.74}$$

If we put $\lambda = 2\cos\theta$, where θ may be real or complex, the roots of (2.74) are $\rho = e^{i\theta}$ or $\rho = e^{-i\theta}$; consequently, for arbitrary values of α and β, the formula

$$\xi_n = \alpha e^{in\theta} + \beta e^{-in\theta}$$

will satisfy (2.73) when $n = 1, 2, 3, \cdots$. The extra condition that $\xi_0 = 0$ will be satisfied if $\alpha + \beta = 0$. We see then that

$$\xi_n = \alpha(e^{in\theta} - e^{-in\theta}) = 2i\alpha \sin n\theta.$$

Since α is an arbitrary constant, we may put $2i\alpha = 1$ and then we have $\xi_n = \sin n\theta$ as a solution of (2.73).

However, the vector whose nth component is $\sin n\theta$ has infinite length if $\theta \neq 0$, for the infinite series

$$\sin^2 \theta + \sin^2 2\theta + \sin^2 3\theta + \cdots$$

does not converge. This follows from the relations

$$\sigma_n^2 = \sum_{k=1}^{n} \sin^2 k\theta = \frac{n}{2} - \frac{1}{2}\sum_{k=1}^{n} \cos 2k\theta = \frac{n}{2} + \frac{1}{4} - \frac{1}{4}\sum_{-n}^{n} e^{2ik\theta}$$
$$= \frac{2n+1}{4} - \frac{1}{4}\frac{\sin(2n+1)\theta}{\sin\theta}.$$

Since the only solutions of (2.73) are vectors which do not belong to the space, we conclude that the operator F has no eigenvectors [when $\theta = 0$, the solution of (2.73) is the zero vector] and therefore F has no discrete spectrum. However, we can show that every value of λ between -2 and 2, excluding the extremes, belongs to the approximate spectrum and therefore also to the continuous spectrum. To see this, let v_n be the vector in E such that its kth component is $\sin k\theta$ for $1 \leq k \leq n$ and such that the remainder of its components are zero. Put $(F - \lambda)v_n = b_n$; then the nth component of b_n is $-\sin(n+1)\theta$, the $(n+1)$th component is $\sin n\theta$, and all other components are zero. We have $|v_n| = \sigma_n$ and

$$\langle b_n, b_n \rangle = \sin^2 n\theta + \sin^2(n+1)\theta.$$

Put $u_n = v_n/\sigma_n$ and $a_n = b_n/\sigma_n$; then, if $-2 < \lambda < 2$, θ is real and we find that $|u_n| = 1$ and

$$\langle a_n, a_n \rangle < \frac{2}{\sigma_n^2}.$$

Since this converges to zero, we conclude that the approximate spectrum contains all λ between -2 and 2. Note that if θ is not real, $\sin^2(n+1)\theta$ becomes infinite and the length of a_n does not approach zero. It can be shown that all values of θ outside the interval $(-2, 2)$ belong to the resolvent set of F.

Thus we have found the spectrum of F, but for our purposes the important thing is the spectral representation. We may obtain it by the following heuristic reasoning.

If F had a purely discrete spectrum with the eigenvalues $\lambda_1, \lambda_2, \cdots$ and the eigenvectors $x_1, x_2, \cdots$, we would have the spectral representation

$$x = \Sigma \xi_k x_k,$$
$$Fx = \Sigma \lambda_k \xi_k x_k.$$

However, F actually has a purely continuous spectrum. Instead of a discrete set of eigenvalues, $\lambda_1, \lambda_2, \cdots$, we have a continuous set λ_θ, where $0 < \theta < \pi$. The usual method in going from a discrete set of elements to a continuous set is to replace sums by integrals. We expect then that the spectral representation will become

$$x = \int_0^\pi \xi_\theta x_\theta \, d\theta, \tag{2.75}$$

$$Fx = \int_0^\pi \lambda_\theta \xi_\theta x_\theta \, d\theta,$$

where

$$\xi_\theta = \langle x, x_\theta \rangle. \tag{2.76}$$

These formulas will be correct if they are interpreted properly. The symbol x_θ will represent the vector, not in E_∞, whose nth component is $(2/\pi)^{1/2} \sin n\theta$ $(n = 1, 2, \cdots)$. The scalar product in (2.76) is formally

$$(2/\pi)^{1/2}[\alpha_1 \sin \theta + \alpha_2 \sin 2\theta + \cdots]$$

if we assume

$$x = (\alpha_1, \alpha_2, \cdots).$$

This infinite series does not converge in the general case, but it can be shown† that, if $\Sigma \alpha_k^2$ converges, there exists a function ξ_θ in $\mathcal{L}_2$ such that its nth Fourier sine coefficient is α_n. With this function ξ_θ, the formulas in (2.75) are true in the sense that the nth components of each side are equal.

The vectors x_θ are the continuous analogue of the eigenvectors. The O.N. property of the eigenvectors of a self-adjoint operator would require $\langle x_\theta, x_{\theta'} \rangle = \delta_{\theta\theta'}$. However, this scalar product does not exist even in the sense by which ξ_θ was defined. To get (2.76) from (2.75) we need the formula

$$\int_0^\pi \xi_\theta \langle x_\theta, x_{\theta'} \rangle \, d\theta = \xi_{\theta'},$$

or

$$\int_0^\pi \xi_\theta \delta_{\theta\theta'} \, d\theta \qquad = \xi_{\theta'}.$$

The symbol $\delta_{\theta\theta'}$ in this equation is known to physicists as Dirac's δ-function. The next chapter will show exactly in what sense this symbol (it is not a function) must be understood.

† Titchmarsh, *Theory of Functions*, Chapter XIII, Clarendon Press, Oxford, 1939.

APPENDIX

PROOF OF THEOREM 2.11

Theorem 2.11. *Let λ be an eigenvalue of finite index for the operator L and let $L - \lambda$ and $L^* - \lambda$ both have closed ranges; then λ is an eigenvalue of L^* also. Moreover, to any chain of length k for L, namely, vectors $e_1, e_2, \cdots, e_k$, such that*

$$e_j = (L - \lambda)^{j-1} e_1 \quad (j = 1, 2, \cdots, k),$$

there corresponds a chain of length k for L^*, namely, vectors $f_1, f_2, \cdots, f_k$, such that

$$f_j = (L^* - \lambda)^{k-j} f_k \quad (j = 1, 2, \cdots, k)$$

and such that

$$\langle f_j, e_i \rangle = \delta_{ij} \quad (i, j = 1, 2, \cdots, k).$$

We prove this theorem first in the case where the generalized null space of $L - \lambda$ contains only the chain $e_1, \cdots, e_k$. Consider the equation

$$(L - \lambda)x = e_1. \tag{2A.1}$$

If this equation had a solution $x = e_0$, the vectors $e_0, e_1, \cdots, e_k$ would form a chain of length $k + 1$ contrary to our hypothesis; consequently, (2A.1) has no solution. However, if the equation

$$(L^* - \lambda)y = 0 \tag{2A.2}$$

has only the trivial solution $y = 0$, then, since $L - \lambda$ has a closed range, equation (2A.1) must have a solution, by Theorem 1.5 This contradiction shows that there must exist a non-trivial solution f_1 of (2A.1) such that $\langle f_1, e_1 \rangle \neq 0$; therefore λ is an eigenvalue and f_1 is an eigenvector of L^*. Normalize f_1 so that $\langle f_1, e_1 \rangle = 1$. Note that for $2 \leq j \leq k$,

$$\langle f_1, e_j \rangle = \langle f_1, (L - \lambda)^{j-1} e_1 \rangle = \langle (L^* - \lambda)^{j-1} f_1, e_1 \rangle = 0$$

since $(L^* - \lambda) f_1 = 0$.

Now, consider the equation

$$(L^* - \lambda)y = f_1. \tag{2A.3}$$

Again, by Theorem 1.5, since $L^* - \lambda$ has a closed range and since $\langle f_1, e_k \rangle = 0$, equation (2A.3) must have a solution $y = g$. The solution of (2A.3) is not unique since $g - \alpha f_1$ (α any scalar) is also a solution. Choose α so that

(2A.4) $$\langle g - \alpha f_1, e_1\rangle = \langle g, e_1\rangle - \alpha\langle f_1, e_1\rangle = 0.$$

Put $f_2 = g - \alpha f_1$, then

$$(L^* - \lambda)^2 f_2 = (L^* - \lambda)f_1 = 0,$$

and

$$\langle f_2, e_2\rangle = \langle f_2, (L - \lambda)e_1\rangle = \langle (L^* - \lambda)f_2, e_1\rangle = \langle f_1, e_1\rangle = 1.$$

Also, for $3 \leq j \leq k$, we have

$$\langle f_2, e_j\rangle = \langle f_2, (L - \lambda)^{j-1}e_1\rangle = \langle (L^* - \lambda)^{j-1}f_2, e_1\rangle = 0.$$

Since, by equation (2A.4), $\langle f_2, e_1\rangle = 0$, we conclude that

$$\langle f_2, e_j\rangle = \delta_{2j} \quad (j = 1, 2, \cdots, k).$$

In the same way, by considering the equation

$$(L^* - \lambda)y = f_2,$$

we obtain a solution f_3 such that $\langle f_3, e_1\rangle = 0$; then, as before, we find that

$$\langle f_3, e_j\rangle = \delta_{3j} \quad (j = 1, 2, \cdots, k).$$

It is clear that the process continues until we have obtained the chain $f_1, f_2, \cdots, f_k$ described in Theorem 2.11, where

$$f_j = (L^* - \lambda)^{k-j}f_k \quad (j = 1, 2, \cdots, k)$$

and where

$$\langle f_i, e_j\rangle = \delta_{ij}.$$

If the generalized null space contains chains other than the one generated by e_1, the above proof must be modified. The proof of the existence of the eigenvector f_1 is the same as before, but now, in addition to the eigenvector f_1, there must exist an eigenvector of L^* for each additional chain in the generalized null space. To see this, suppose that there is a chain generated by e_1'; then we can find a scalar β such that

$$\langle f_1, e_1' - \beta e_1\rangle = 0.$$

If f_1 were the only eigenvector of L^*, this would imply that the equation

$$(L - \lambda)x = e_1' - \beta e_1$$

has a solution $x = e_0'$. This is impossible since e_0' would generate a chain whose length is greater than the lengths of either the e_1- or the e_1'-chain; consequently, we conclude that the operator $L^* - \lambda$ has another eigenvector f'. A similar argument shows that every chain produces an eigenvector for $L^* - \lambda$.

Since any linear combination of the eigenvectors of $L^* - \lambda$ is still an eigenvector of $L^* - \lambda$, it is possible to choose an eigenvector which is orthogonal to all the chain-generating vectors except e_1. We shall denote this eigenvector by f_1. We normalize f_1 so that $\langle f_1, e_1 \rangle = 1$; then we have, as before,

$$\langle f_1, e_j \rangle = \delta_{1j} \quad (1 \leq j \leq k) \tag{2A.5}$$

and also

$$\langle f_1, e' \rangle = 0, \tag{2A.6}$$

where e' is a vector in any of the other chains which form the basis for the null space.

From (2A.5) and (2A.6) we see that f_1 is orthogonal to all the solutions of $(L - \lambda)x = 0$; consequently, the equation

$$(L^* - \lambda)y = f_1 \tag{2A.7}$$

has a solution. The solution is not unique since the homogeneous equation has σ independent solutions, where σ is the number of chains in the basis for $\mathcal{N}$. Just as in the case of one chain, it is possible to so choose a solution f_2 of (2A.7) that

$$\langle f_2, e_1 \rangle = \langle f_2, e' \rangle = 0;$$

then we shall have

$$\langle f_2, e_j \rangle = \delta_{2j}.$$

Similarly, we can define $f_3, \cdots, f_k$ and thus complete the proof of Theorem 2.11.

3

GREEN'S FUNCTIONS

Introduction

The methods discussed in the previous chapters will now be applied to the solution of ordinary linear differential equations. The general theory of linear equations suggests two methods which can be used to solve the equation

$$Lu = f,$$

where L is an ordinary linear differential operator, f a known function, and u the unknown function.

One method is to find the spectral representation of L by studying the solutions of the equation

$$Lu = \lambda u,$$

where λ is an arbitrary constant. This method will be discussed in the next chapter. The other method is to find the operator inverse to L, that is, to find an operator L^{-1} such that the product $L^{-1}L$ is the identity operator. This chapter will be devoted to a complete exposition of this method. During this exposition, we shall find that, as might be expected, the inverse of a differential operator is an integral operator. The kernel of this integral operator will be called the *Green's function* of the differential operator. The techniques which we shall provide for finding the Green's function use a tool which has proved valuable in many branches of applied mathematics, namely, the Dirac δ-function.† We shall therefore begin with a discussion of the meaning and significance of the δ-function.

The Identity Operator as an Integral Operator.

Suppose L is a linear differential operator acting on a space of functions $u(x)$. The operator inverse to L is the operator L^{-1} such that $L^{-1}L = LL^{-1} = I$, where I is the identity operator. Let us assume that L^{-1} is an integral operator with kernel $g(x, t)$ so that

$$L^{-1}u = \int g(x, t)u(t)\, dt;$$

† See Dirac, *The Principles of Quantum Mechanics*, Clarendon Press, Oxford, 1947.

then, by formal (not necessarily rigorous) manipulations, we find that

$$u(x) = Iu = LL^{-1}u = L\int g(x, t)u(t)\,dt = \int Lg(x, t)u(t)\,dt. \tag{3.1}$$

Consider the kernel $Lg(x, t)$. Since L is an operator involving differentiation with respect to x, it seems that we may write

$$Lg(x, t) = \delta(x, t),$$

where $\delta(x, t)$ is a function of x and t, and then (3.1) becomes

$$u(x) = \int \delta(x, t)u(t)\,dt. \tag{3.2}$$

However, it is easy to show that, if (3.2) holds for all continuous functions $u(t)$, then $\delta(x, t)$ must be zero if $x \neq t$ so that we may put

$$\delta(x, t) = \delta(t - x),$$

and equation (3.2) becomes

$$u(x) = \int \delta(t - x)u(t)\,dt. \tag{3.3}$$

The symbol $\delta(x)$ is known as Dirac's δ-function.

It can be shown (see Problem 3.1) that if (3.3) holds for every continuous function $u(x)$, then $\delta(x) = 0$ if $x \neq 0$. This fact agrees with Dirac's definition: the δ-function is zero for every value of x except the origin, where it is infinite in such a way that

$$\int_{-\infty}^{\infty} \delta(x)\,dx = 1.$$

Mathematically, this definition is nonsense. If a function is zero everywhere except at one point, its integral, no matter what definition of the integral is used, is necessarily equal to zero. However, the δ-function has proved to be of such great utility that the physicist rightly refuses to give it up. It is the mathematician's task to find a method by which the use of the δ-function symbol can be justified.

Recently, Laurent Schwartz has proposed a method called the Theory of Distributions which justifies not only the use of the δ-function but also the use of all derivatives of the δ-function. The method is very powerful because it enables us to interchange limiting operations where such an interchange is not valid for ordinary functions and because it enables us to use series which ordinarily would be called divergent. For example, the fact that $\delta(x)$ is not a function indicates that (3.3) is not a mathematically valid consequence of (3.1). Equation (3.3) is not valid because in (3.1) we have interchanged the operations of differentiation and integration, and in this case the interchange is not justified. However, if the equations are understood in the sense of the Theory of Distributions, the interchange of operations is legitimate, and (3.3) becomes a valid consequence of (3.1).

We shall state, without proof, such results of this theory as we need. For proofs, we refer to the original sources.†

PROBLEM

3.1. Show that if $\delta(x, t)$ is a continuous function such that (3.2) is satisfied for any continuous function $u(t)$, then $\delta(x, t)$ is zero unless $t = x$. (*Hint.* Suppose that $\delta(t + \alpha, t) \neq 0$ for $a \leq t \leq b$; then take $u(t) = \delta(t + \alpha, t)$ for $a + \varepsilon \leq t \leq b - \varepsilon$, $u(t) = 0$ for t outside the interval (a, b) and let $u(t)$ be continuous in the intervals $(a, a + \varepsilon)$ and $(b - \varepsilon, b)$.)

Interpretation of the δ-function

What should be understood by the δ-function symbol? Its most important property and the one that makes it so useful is the following: for every continuous function $\phi(x)$, we have, from (3.3), that

$$\int_{-\infty}^{\infty} \delta(x)\phi(x)\, dx = \phi(0). \tag{3.4}$$

This property may also be expressed thus: given a continuous function $\phi(x)$, the δ-function picks out the value of that function at the origin.

It turns out that the δ-function can be handled algebraically as if it were an ordinary function. However, any equation involving the δ-function should be understood in the following sense: if the equation is multiplied by an arbitrary continuous function $\phi(x)$ and then integrated from $-\infty$ to ∞, with (3.4) used to evaluate integrals involving δ-functions, the result will be a correct equation involving ordinary functions. For example,

$$x\delta(x) = 0 \tag{3.5}$$

because if $\phi(x)$ is a continuous function and if we put $x\phi(x) = \psi(x)$, then

$$\int x\delta(x)\phi(x)\, dx = \int \delta(x)\psi(x)\, dx = \psi(0) = 0.$$

We shall assume that the usual techniques of integration such as substitution and integration by parts may be applied to integrals involving the δ-function.‡ As an illustration of this, we show that if $f(x)$ is a monotonic function of x which vanishes for $x = x_0$, then

$$\delta(f(x)) = \frac{\delta(x - x_0)}{|f'(x_0)|}\S \tag{3.6}$$

where by the symbol $\delta(x - x_0)$ we understand that

$$\int_{-\infty}^{\infty} \phi(x)\delta(x - x_0)\, dx = \phi(x_0).$$

† L. Schwartz, "Théorie des distributions," *Actualités scientifiques et industrielles*, No. 1091 and 1122, Hermann & Cie, Paris, 1950–1951.

‡ Proofs are in L. Schwartz, *op. cit.*

§ The absolute value of the derivative ensures that the integration will always be from $-\infty$ to $+\infty$.

In the integral

$$\int_{-\infty}^{\infty} \delta(f(x))\phi(x)\, dx$$

put $y = f(x)$ and $\psi(y) = \dfrac{\phi(x)}{|f'(x)|}$;

then it becomes

$$\int_{-\infty}^{\infty} \delta(y)\psi(y)\, dy = \psi(0) = \frac{\phi(x_0)}{|f'(x_0)|}.$$

As special cases of (3.6) we have

$$\int_{-\infty}^{\infty} \delta(ax - b)\phi(x)\, dx = |a|^{-1}\phi(ba^{-1})$$

and

$$\delta(x) = \delta(-x).$$

We notice that the function $\delta(x)$ is treated exactly as if it were an ordinary function except that we shall never talk about the "values" of $\delta(x)$. We talk about the values of integrals involving $\delta(x)$.

PROBLEM

3.2. Prove that $f(x)\delta(x) = f(0)\delta(x)$ if $f(x)$ is a continuous function.

Testing Functions and Symbolic Functions

The function $\phi(x)$ that we have used in the preceding section to test the validity of (3.5) and (3.6) is an example of what we shall call *testing functions.* For work with differential equations it is convenient to *restrict the term testing functions to functions $\phi(x)$ which are continuous, have continuous derivatives of all orders, and vanish identically outside some finite interval.* Note that this restriction of the testing function does not change the validity of (3.5) and (3.6).

The set of testing functions forms a linear vector space. It is not convenient to introduce a scalar product in this space; instead, we shall introduce the notion of convergence. We say a sequence of testing functions $\phi_n(x)$ converges to zero if the functions $\phi_n(x)$ and all their derivatives converge uniformly to zero and if all the functions $\phi_n(x)$ vanish identically outside the same finite interval.

Just as in Chapter 1, we may define linear functionals $F(\phi)$ on the space of testing functions. We repeat the definition of linear functional: $F(\phi)$ is a linear functional if to every testing function $\phi(x)$ a real or complex number $F(\phi)$ is assigned such that

$$F(\phi_1 + \phi_2) = F(\phi_1) + F(\phi_2),$$
$$F(\lambda\phi) = \lambda\, F(\phi),$$

where λ is any scalar. A functional $F(\phi)$ is said to be *continuous* if the sequence of numbers $F(\phi_n)$ converges to zero whenever the sequence of testing functions $\phi_n(x)$ converges to zero in the sense defined previously. The following are examples of continuous linear functionals:

$$F_1(\phi) = \phi'(0)$$

$$F_2(\phi) = \int_0^1 \phi(x)\, dx.$$

Laurent Schwartz† calls any continuous linear functional on the space of testing functions a *distribution.*

We have seen in Chapter 1 that a continuous linear functional on a vector space having a scalar product can be expressed as a scalar product. For the space of testing functions this result does not apply, since no scalar product was defined. It seems natural to introduce the following definition of scalar product:

$$\langle \phi, \psi \rangle = \int_{-\infty}^{\infty} \phi\psi\, dx.$$

However, with this definition we find that the space of testing functions is not complete and the results of Chapter 1 still do not apply. Nevertheless, we should like to express any continuous linear functional $F(\phi)$ as follows:

$$F(\phi) = \int_{-\infty}^{\infty} s(x)\phi(x)\, dx.$$

Sometimes this is possible; for example, when

$$F(\phi) = \int_0^1 \phi(x)\, dx,$$

$s(x) = 1$, $0 < x < 1$, and $s(x) = 0$ otherwise. At other times this is not possible; for example, when $F(\phi) = \phi(0)$ or $F(\phi) = \phi'(0)$.

We have seen, however, that it is possible to introduce a symbol $\delta(x)$ such that

$$F(\phi) = \phi(0) = \int_{-\infty}^{\infty} \delta(x)\phi(x)\, dx.$$

The δ-function is an example of a *symbolic function.* Given any continuous linear functional $F(\phi)$ on the space of testing functions, we shall introduce a functional symbol, say $s(x)$, and put

$$\int_{-\infty}^{\infty} s(x)\phi(x)\, dx = F(\phi).$$

We shall say $s(x)$ is a symbolic function. Note that a symbolic function need not have values. It produces values only when multiplied by a testing function and then integrated.

If $f(x)$ is an integrable function,

$$\int_{-\infty}^{\infty} f(x)\phi(x)\, dx$$

† L. Schwartz, *op. cit.*

is clearly a continuous linear functional on the space of testing functions (Problem 3.4); consequently, every integrable function is a symbolic function. However, there are some symbolic functions, such as the δ-function, which are neither integrable nor functions. Another example of such a symbolic function is the symbol $\delta^{(n)}(x)$ which we define as:

$$\int_{-\infty}^{\infty} \delta^{(n)}(x)\phi(x)\,dx = (-)^n \frac{d^n\phi}{dx^n}\bigg|_{x=0}.$$

Since the right-hand side is a continuous linear functional on the space of testing functions, the symbol $\delta^{(n)}(x)$ is a symbolic function.

Often the functional may be extended to a wider class of functions than to the testing functions; for example, the δ-function produces a functional which is applicable to the space of continuous functions. In such cases we shall use the functional on the larger space without explicit statement.

It can be shown† that symbolic functions can be combined algebraically as if they were ordinary functions, except that the product or quotient of two symbolic functions may not have a meaning. We shall assume that in the integrals involving symbolic functions the usual rules of integration are valid.

To summarize, symbolic functions are used as if they were ordinary functions. Any equation involving symbolic functions is to be understood in the following sense: if the equation is multiplied by an arbitrary testing function and integrated from $-\infty$ to ∞, with the functional property of the symbolic functions used to evaluate the integrals, the result should be a correct equation involving ordinary functions.

PROBLEMS

3.3. Show that the function

$$\begin{aligned} \phi(x) &= \exp(-x^{-2}) \exp[-(x-a)^{-2}], \quad && 0 \le x \le a \\ &= 0, && x \le 0 \text{ or } a \le x \end{aligned}$$

is a testing function.

3.4. Prove the statement in the text that an integrable function $f(x)$ defines a symbolic function. (*Hint.* Show that $\int f(x)\phi(x)\,dx$ is a linear continuous functional.)

Derivatives of Symbolic Functions

We wish to define the concept of the derivative of a symbolic function in such a way that it will be valid even if the symbolic function is an

† L. Schwartz, *op. cit.*

ordinary function. Since the symbolic function is defined by the functionals it produces, we must use an integration property of the derivatives of ordinary functions. Such a property is that of integration by parts. If $f(x)$ is a function with a continuous derivative, then

$$\int_{-\infty}^{\infty} f'(x)\phi(x)\,dx = -\int_{-\infty}^{\infty} f(x)\phi'(x)\,dx. \tag{3.7}$$

Note that because $\phi(x)$ vanishes identically outside a finite interval, the integrals converge and the boundary terms are zero. We shall use (3.7) to define the derivative of a symbolic function $s(x)$ and we shall say $s'(x)$ is the *derivative* of $s(x)$ if

$$\int_{-\infty}^{\infty} s'(x)\phi(x)\,dx = -\int_{-\infty}^{\infty} s(x)\phi'(x)\,dx \tag{3.8}$$

for every testing function $\phi(x)$. Since the right-hand side of (3.8) always exists, the left-hand side can be used to define the derivatives of $s(x)$. For example, $\delta'(x)$ is defined by the relation:

$$\int_{-\infty}^{\infty} \delta'(x)\phi(x)\,dx = -\int_{-\infty}^{\infty} \delta(x)\phi'(x)\,dx = -\phi'(0);$$

consequently, $\delta'(x)$ produces the functional which assigns the value $-\phi'(0)$ to a testing function $\phi(x)$.

In the same way, we can define the second derivative $\delta''(x)$ of the δ-function as the derivative of $\delta'(x)$. We have:

$$\int_{-\infty}^{\infty} \delta''(x)\phi(x)\,dx = -\int_{-\infty}^{\infty} \delta'(x)\phi'(x)\,dx = \phi''(0).$$

We see then that the symbolic function $\delta^{(n)}(x)$ defined in the previous section is actually the nth derivative of $\delta(x)$, that is,

$$\delta^{(n)}(x) = \frac{d^n\delta(x)}{dx^n}.$$

We can show that the δ-function itself is the derivative of the function $H(x)$, which is defined by the relation

$$\int_{-\infty}^{\infty} H(x)\phi(x)\,dx = \int_{0}^{\infty} \phi(x)\,dx. \tag{3.9}$$

To see this, use (3.8). We have

$$\int_{-\infty}^{\infty} H'(x)\phi(x)\,dx = -\int_{-\infty}^{\infty} H(x)\phi'(x)\,dx = -\int_{0}^{\infty} \phi'(x)\,dx = \phi(0)$$

since $\phi(\infty) = 0$. Consequently,

$$H'(x) = \delta(x). \tag{3.10}$$

The symbolic function $H(x)$ defined in (3.9) is equal to the ordinary function having the following values:

$$\begin{aligned} H(x) &= 1, \quad x > 0 \\ &= 0, \quad x < 0. \end{aligned}$$

This function is well known to physicists and engineers as the *Heaviside unit function.* Note that the derivative of the function $H(x)$ is zero for $x < 0$, zero for $x > 0$, and undefined for $x = 0$.

It can be shown that the derivative as defined in (3.8) has the usual properties of a derivative: it is a linear, homogeneous operation, and the usual product rule for differentiation holds if the products involved have a meaning (Problem 3.7).

PROBLEMS

3.5. Prove that $f(x)\delta'(x) = f(0)\delta'(x) - f'(0)\delta(x)$.

3.6. Prove that $x^m\delta^{(n)}(x) = 0$ if $m \geq n + 1$.

3.7. If $f(x)$ is a continuous differentiable function and if $s(x)$ is a symbolic function,

$$[f(x)s(x)]' = f(x)s'(x) + f'(x)s(x).$$

(*Hint.* $\int fs\phi' \, dx = \int sf\phi' \, dx = -\int sf'\phi \, dx + \int s[f\phi' + f'\phi] \, dx$
$= -\int sf'\phi \, dx - \int s'f\phi \, dx$.)

Symbolic Derivatives of Ordinary Functions

If a continuous function $f(x)$ has a piecewise continuous derivative, the derivative of $f(x)$ as defined by (3.8) will be the customary derivative $f'(x)$. Suppose, however, that $f(x)$ has a jump of magnitude a_1 at $x = x_1$ whereas for all other values of x it has a piecewise continuous derivative. The derivative of the function $f(x)$ is $f'(x)$ for $x < x_1$ and for $x > x_1$ but is undefined for $x = x_1$. However, we shall see that the symbolic function which is defined by (3.8) to be the derivative of $f(x)$ is not the ordinary derivative. We shall call this symbolic function the *symbolic derivative* of $f(x)$, and we shall denote it by $f'_s(x)$. Put

$$g(x) = f(x) - a_1 H(x - x_1),$$

where

$$\begin{aligned} H(x - x_1) &= 1, \quad x > x_1 \\ &= 0, \quad x < x_1. \end{aligned}$$

We have

$$\begin{aligned} \int_{-\infty}^{\infty} f(x)\phi'(x) \, dx &= \int_{-\infty}^{\infty} g(x)\phi'(x) \, dx + a_1 \int_{-\infty}^{\infty} H(x - x_1)\phi'(x) \, dx \\ &= -\int_{-\infty}^{\infty} g'(x)\phi(x) \, dx - a_1\phi(x_1) \end{aligned}$$

since $g'(x)$ is continuous and piecewise differentiable. From (3.8) we see that

$$\int_{-\infty}^{\infty} f'_s(x)\phi(x) \, dx = \int_{-\infty}^{\infty} [g'(x) + a_1\delta(x - x_1)]\phi(x) \, dx;$$

therefore

$$f'_s(x) = g'(x) + a_1\delta(x - x_1).$$

Since $g'(x) = f'(x)$ wherever it exists, we find that

$$f_s'(x) = f'(x) + a_1\delta(x - x_1).$$

This relation is easily generalized to functions $f(x)$ having jumps of magnitude $a_1, \cdots, a_n$ at the points $\alpha_1, \cdots, \alpha_n$. We get

$$f_s'(x) = f'(x) + a_1\delta(x - \alpha_1) + \cdots + a_n\delta(x - \alpha_n).$$

We state this result as

Theorem 3.1. *The symbolic derivative of a piecewise differentiable function with jumps is the ordinary derivative, where it exists, plus the sum* of *δ-functions at the jumps multiplied by the magnitude of the jumps.*

In a similar way we see that

$$\begin{aligned} f_s''(x) = f''(x) &+ a_1\delta'(x - \alpha_1) + \cdots + a_n\delta'(x - \alpha_n) \\ &+ b_1\delta(x - \beta_1) + \cdots + b_m\delta(x - \beta_m) \end{aligned}$$

where $\beta_1, \cdots, \beta_m$ are the points of discontinuity in $f'(x)$ and $b_1, \cdots, b_m$ are the corresponding jumps in the value of $f'(x)$.

For our purposes it will be more convenient to use symbolic derivatives. Henceforth, we shall drop the subscript s and use the symbols $f'(x)$, $f''(x)$, etc., to denote the symbolic derivatives.

Examples of Symbolic Derivatives

The symbolic derivative of the function $|x|$ is the function which is -1 for x negative and $+1$ for x positive. We call this function *signum* x and denote it by sgn x. We have from Theorem 3.1

$$\frac{d}{dx} \operatorname{sgn} x = 2\delta(x) \tag{3.11}$$

because the sgn function is constant except for a jump of magnitude two at the origin.

As an application of this result, consider the integral

$$I = \int_{-1}^{1} |x|\psi''(x)\, dx.$$

Using repeated integration by parts, we find that

$$\begin{aligned} I &= |x|\psi'(x)\Big|_{-1}^{1} - \int_{-1}^{1} |x|'\psi'(x)\, dx \\ &= \Big[|x|\psi'(x) - |x|'\psi(x)\Big]_{-1}^{1} + \int_{-1}^{1} |x|''\psi(x)\, dx \\ &= \psi'(1) - \psi'(-1) - \psi(1) - \psi(-1) + 2\psi(0). \end{aligned}$$

This method for evaluating I is simpler than the usual one in which I is written as

$$-\int_{1}^{0} x\psi''(x)\, dx + \int_{0}^{1} x\psi''(x)\, dx$$

and integration by parts is applied to each integral separately.

As another illustration, consider the problem of maximizing the integral

$$K = \int_{-1}^{1} f(x)(x - x_1)(x - x_2)\, dx,$$

where $f(x)$ is an integrable real function such that $|f(x)| \leq 1$, and where x_1, x_2 are arbitrary numbers between -1 and 1. For given x_1, x_2, the integral K will be a maximum if $f(x) = +1$ when $(x - x_1)(x - x_2)$ is positive and if $f(x) = -1$ when $(x - x_1)(x - x_2)$ is negative; therefore

$$K \leq K^* = \int_{-1}^{1} (x - x_1)(x - x_2) \operatorname{sgn}(x - x_1)(x - x_2)\, dx.$$

Now K^* will be a maximum for those values of x_1 and x_2 which make

$$\frac{\partial K^*}{\partial x_1} = \frac{\partial K^*}{\partial x_2} = 0.$$

We have

$$\frac{\partial K^*}{\partial x_1} = -\int_{-1}^{1} (x - x_2) \operatorname{sgn}[(x - x_1)(x - x_2)]\, dx + \int_{-1}^{1} (x - x_1)(x - x_2) 2\delta(x - x_1)\, dx.$$

The second integral is zero since $(x - x_1)(x - x_2)$ vanishes when $x = x_1$. Finally,

$$\frac{\partial K^*}{\partial x_1} = -\int_{-1}^{x_1} (x - x_2)\, dx + \int_{x_1}^{x_2} (x - x_2)\, dx - \int_{x_2}^{1} (x - x_2)\, dx.$$

There is a similar formula for $\dfrac{\partial K^*}{\partial x_2}$. The remainder of the solution may be left to the reader.

PROBLEMS

3.8. If $f(x)$ is absolutely continuous, show that the derivative of the distribution defined by $f(x)H(x)$ is $f'(x)H(x) + f(0)\,\delta(x)$.

3.9. If $f(x)$ is absolutely continuous with simple zeroes at $x_1, \cdots, x_n$, show that the derivative of sgn $(f(x))$ is

$$2\delta(x - x_1) \operatorname{sgn} f'(x_1) + \cdots + 2\delta(x - x_n) \operatorname{sgn} f'(x_n).$$

3.10. Show that the integral

$$u(x) = \int |x - \zeta| \phi(\zeta)\, d\zeta$$

satisfies the differential equation

$$u'' = 2\phi(x).$$

(*Hint.* Use (3.11).)

3.11. Show that the integral

$$u(x) = \int e^{ik|x-\zeta|} \phi(\zeta)\, d\zeta$$

satisfies the equation

$$u'' + k^2 u = 2ik\phi(x).$$

(*Hint.* Use (3.11) and the fact that $(\operatorname{sgn} x)^2 = 1$.)

3.12. If $p(x)^{-1}$ is an integrable function, show that

$$u = H(x)\int_0^x \frac{d\zeta}{p(\zeta)}$$

is a solution of

$$(pu')' = \delta(x).$$

The Inverse of a Differential Operator—Example

We now return to the basic problem of this chapter, that of inverting a differential operator L. Suppose that ψ and ϕ are testing functions and consider the equation

$$L\psi = \phi.$$

As in a previous section, we assume that the inverse operator L^{-1} is an integral operator with some kernel $g(x, t)$ such that

$$L^{-1}\phi = \int g(x, t)\phi(t)\, dt,$$

but now we permit $g(x, t)$ to be a symbolic function. Applying the differential operator L to both sides of this equation, we get

$$LL^{-1}\phi = \phi = \int Lg\phi\, dt.$$

This equation will be satisfied if we find g such that

$$Lg = \delta(x - t), \tag{3.12}$$

where the differentiation is to be understood as symbolic differentiation.

To illustrate the method for inverting an operator, consider the special case in which $L = \dfrac{d^2}{dx^2}$; then (3.12) becomes

$$\frac{d^2}{dx^2} g(x, t) = \delta(x - t). \tag{3.13}$$

This equation may be solved by straightforward integration. We know that the δ-function is the derivative of the Heaviside unit function; therefore,

$$\frac{d}{dx} g(x, t) = H(x - t) + \alpha(t),$$

where $\alpha(t)$ is an arbitrary function. Integrating again, we get

$$\begin{aligned} g(x, t) &= \int H(x - t)\, dx + x\alpha(t) + \beta(t) \\ &= (x - t)H(x - t) + x\alpha(t) + \beta(t), \end{aligned} \tag{3.14}$$

where $\beta(t)$ is another arbitrary function. It can be proved that any symbolic function which is a solution of (3.13) may be written in the form (3.14).† Note that $g(x, t)$ is a continuous, piecewise, differentiable

† L. Schwartz, *op. cit.*

function, and note also that if $f(x)$ is any integrable function which vanishes outside a finite interval, then it is easy to show that the function

$$u(x) = \int g(x, t) f(t)\, dt \tag{3.15}$$

satisfies the differential equation

$$u'' = f(x). \tag{3.16}$$

We leave the verification to the reader.

By a suitable choice of the functions $\alpha(t)$ and $\beta(t)$ we can in general find a solution of (3.16) which satisfies two conditions. Thus, to find a solution of (3.16) which satisfies the conditions $u(0) = u(1) = 0$, we proceed as follows:

From (3.15) we have

$$u(x) = \int_{-\infty}^{x} (x - t) f(t)\, dt + x \int_{-\infty}^{\infty} \alpha(t) f(t)\, dt + \int_{-\infty}^{\infty} \beta(t) f(t)\, dt.$$

Substitute $x = 0$ and $x = 1$ in this equation. We get

$$0 = -\int_{-\infty}^{0} t f(t)\, dt + 0 + \int_{-\infty}^{\infty} \beta(t) f(t)\, dt$$

$$0 = \int_{-\infty}^{1} (1 - t) f(t)\, dt + \int_{-\infty}^{\infty} \alpha(t) f(t)\, dt + \int_{-\infty}^{\infty} \beta(t) f(t)\, dt.$$

From the first of these equations we see that $\beta(t) = tH(-t)$, and then from the second we see that $\alpha(t) = -1 + tH(t)$, $-\infty \leq t \leq 1$, and $\alpha(t) = 0$ for all other values of t. When these values of $\alpha(t)$ and $\beta(t)$ are substituted in (3.15), the result is

$$u(x) = \int_{0}^{x} (x - t) f(t)\, dt - x \int_{0}^{1} (1 - t) f(t)\, dt.$$

In this case the kernel

$$g(x, t) = (x - t) H(x - t) - x(1 - t), \quad 0 \leq x, \quad t \leq 1 \tag{3.17}$$

also satisfies the boundary conditions

$$g(0, t) = g(1, t) = 0.$$

In later sections we shall discuss methods for finding the kernel of the inverse operator in more general cases, and we shall find that, just as in this case, the kernel as a function of x satisfies the same conditions as those satisfied by the solution of the differential equation.

PROBLEMS

3.13. Find α and β such that the solution of (3.16) will satisfy the following sets of conditions:

(1) $u(0) = u'(0) = 0$;

(2) $u(0) = 0$ and $\int_0^1 u(x)\, dx = 0$.

3.14. Can α and β be found so that the solution of (3.16) will satisfy the conditions $u(0) = u(1)$ and $u'(0) = u'(1)$? Explain. (*Hint.* Consider the homogeneous equation corresponding to (3.16).)

The Domain of a Linear Differential Operator

Before we can continue with the problem of inverting a differential operator L, it is necessary to give a complete definition of L. In this chapter we shall consider primarily second-order differential operators. We put

$$L = a(x)\frac{d^2}{dx^2} + b(x)\frac{d}{dx} + c(x), \tag{3.18}$$

where the coefficients $a(x)$, $b(x)$, and $c(x)$ are continuous functions. In later sections we shall discuss the meaning of L when the coefficients are discontinuous functions.

First, we specify the linear vector space $\mathcal{S}$ of functions on which the differential operator acts. Since for the most part we shall consider differential equations over a finite interval only, which for convenience we take to be the interval (0, 1), we shall take for $\mathcal{S}$ the space of all real-valued functions which are Lebesgue square integrable† over (0, 1), that is, $\mathcal{S}$ contains all real functions $u(x)$ defined for $0 \leq x \leq 1$ such that

$$\int_0^1 u(x)^2\,dx < \infty.$$

If u and v belong to $\mathcal{S}$, we define

$$\langle u, v\rangle = \int_0^1 u(x)v(x)\,dx.$$

For some applications, particularly to the theory of Bessel functions, it is convenient to modify the definitions of the space $\mathcal{S}$ and of the scalar product by introducing a non-negative weight function $w(x)$. $\mathcal{S}$ is now the space of all functions $u(x)$ such that

$$\int_0^1 u(x)^2 w(x)\,dx < \infty,$$

and the scalar product becomes

$$\langle u, v\rangle = \int_0^1 u(x)v(x)w(x)\,dx.$$

Since the operator L is a differential operator, it cannot be applied to every function in $\mathcal{S}$ because the function may not be differentiable. Moreover, even if the function is differentiable, the result of applying L

† The reader may consider all integrals to be Riemann integrals. For physical applications the distinction between Riemann and Lebesgue integrals is unimportant. The mathematical reason for Lebesgue integrals is that as a consequence $\mathcal{S}$ will be a complete space. See Stone, *Linear Transformations in Hilbert Space and Their Applications to Analysis*, American Mathematical Society, New York, 1932.

to it may be a function not in $\mathcal{S}$. For example, $u(x) = x \sin(x^{-1})$ is in $\mathcal{S}$ and is differentiable, but its derivative $u'(x) = \sin(x^{-1}) - x^{-1}\cos(x^{-1})$ is not in $\mathcal{S}$. As a consequence of these facts we shall consider L as acting only on those functions u for which Lu exists and belongs to $\mathcal{S}$.

There is a second requirement for a complete definition of L. We have seen in the preceding section that putting $Lu = u''$ is not enough because then the equation

$$u'' = f(x) \tag{3.16}$$

does not have a unique solution. To make the solution unique we must require the solution u to satisfy a set of two conditions such as, for example, the sets of conditions specified in Problem 3.13.

Since different sets of conditions will give different solutions to (3.16), we see that for the complete definition of L it is necessary to specify not only that $Lu = u''$ but also the conditions satisfied by u. For precise notation, we should use a different symbol for the operator each time the conditions are changed. However, for convenience, we shall use the same letter for the differential operator under all conditions but always specify in addition the conditions that the solutions of $Lu = f$ satisfy.

We take the conditions in the following form:

$$\begin{aligned} B_1(u) &= \alpha_{10}u(0) + \alpha_{11}u'(0) + \beta_{10}u(1) + \beta_{11}u'(1) = 0, \\ B_2(u) &= \alpha_{20}u(0) + \alpha_{21}u'(0) + \beta_{20}u(1) + \beta_{21}u'(1) = 0, \end{aligned} \tag{3.19}$$

where α's and β's are given constants. These conditions include both initial and boundary conditions. For example, if all α's and β's except α_{10} and α_{21} are zero, the conditions (3.18) are initial conditions; if all α's and β's except α_{10} and β_{20} are zero, the conditions are boundary conditions. For our purposes the distinction is immaterial. We must assume, however, in any case that the conditions (3.19) are independent, that is, there do not exist constants c_1 and c_2 such that

$$c_1B_1(u) + c_2B_2(u) = 0$$

for all functions $u(x)$. Notice that in all cases the conditions (3.19) are linear and homogeneous; consequently, if u_1 and u_2 satisfy (3.19), then so will $\alpha u_1 + \beta u_2$, where α and β are arbitrary constants.

We may now define the *domain* of the operator L. It is the set of *all functions* $u(x)$ *in* $\mathcal{S}$ *which have a piecewise continuous second derivative, which satisfy* (3.19) *and are such that* Lu *belongs to* $\mathcal{S}$.

It is easy to see that the domain is a linear manifold in $\mathcal{S}$. It is not a subspace because it is not closed, that is, there exists a sequence of functions $u_n(x)$ in the domain which converges to a limit $u(x)$ in $\mathcal{S}$ but the limit $u(x)$ is not in the domain. For an example of this see Problem 3.15.

PROBLEM

3.15. Suppose that the conditions (3.19) reduce to $u(0) = u(1) = 0$. Put $u_n(x) = \sin(n\pi x/2)$ for $0 \leq x \leq (n)^{-1}$, $u_n(x) = 1$ for $(n)^{-1} \leq x \leq 1 - (n)^{-1}$, $u_n(x) = \sin[n\pi(1-x)/2]$ for $1 - (n)^{-1} \leq x \leq 1$. Show that the sequence $u_n(x)$ converges to a limit $u(x)$ in $\mathcal{S}$ but that $u(0) \neq 0$, $u(1) \neq 0$.

The Adjoint Differential Operator, Hermitian Operators

In order to be able to apply the concepts of Chapters 1 and 2 to a differential operator L, it is necessary to define its adjoint L^*. The adjoint was defined previously by considering the equation

$$\langle v, Lu \rangle = \langle w, u \rangle$$

and putting $w = L^*v$. For differential operators a similar method will be used. To illustrate this, suppose $L = \dfrac{d}{dx}$ on the manifold $\mathcal{M}$ defined by the conditions $u(0) = 2u(1)$; then by integrating by parts we get

$$\begin{aligned}\langle v, Lu \rangle &= \int_0^1 v\frac{du}{dx}\,dx = vu\Big|_0^1 - \int_0^1 u\frac{dv}{dx}\,dx \\ &= u(1)[v(1) - 2v(0)] - \int_0^1 u\frac{dv}{dx}\,dx = \langle w, u \rangle.\end{aligned}$$

We see that L^* consists of two parts; a differential operator $-\dfrac{d}{dx}$, and some boundary terms. The differential operator $-\dfrac{d}{dx}$ is called the *formal adjoint* to the differential operator $\dfrac{d}{dx}$. *The adjoint to L on the manifold* $\mathcal{M}$ will be the formal adjoint $-\dfrac{d}{dx}$ on the manifold defined by the condition $v(1) = 2v(0)$. Now

$$\langle v, Lu \rangle = \langle L^*v, u \rangle$$

where

$$L^*v = -\frac{dv}{dx}$$

and where v satisfies the condition $v(1) = 2v(0)$.

It is to be noted that in this example L acts on the manifold of square integrable functions $u(x)$ such that $u(0) = 2u(1)$, but L^* acts on the manifold of square integrable functions $v(x)$ such that $v(0) = v(1)/2$. This is an illustration of the general situation where the manifold on which L^* acts may be different from the manifold on which L acts. We call one such manifold the *dual* of the other.

If $L = L^*$, the differential operator is said to be *formally self-adjoint.*

If, in addition, the boundary conditions for L and L^* are equivalent, that is, if they define the same linear manifold, the operator is said to be *self-adjoint* or *Hermitian.* As an illustration, suppose that

$$L = e^x \frac{d^2}{dx^2} + e^x \frac{d}{dx}$$

on the manifold defined by the conditions

$$u'(0) = 0, \quad u(1) = 0.$$

Then

$$\begin{aligned}\langle v, Lu\rangle &= \int_0^1 v(e^x u'' + e^x u')\, dx = \int_0^1 v(e^x u')'\, dx \\ &= \Big[ve^x u'\Big]_0^1 - \Big[v'e^x u\Big]_0^1 + \int_0^1 u(e^x v')'\, dx \\ &= u'(1)v(1)e^1 + v'(0)u(0) + \int_0^1 u(e^x v'' + e^x v')\, dx.\end{aligned}$$

This shows that

$$L^* = e^x \frac{d^2}{dx^2} + e^x \frac{d}{dx}.$$

In order that

$$\langle v, Lu\rangle = \langle L^* v, u\rangle$$

for all u, the boundary conditions that v satisfies must be

$$v'(0) = 0, \quad v(1) = 0.$$

Since $L = L^*$, the differential operator is *formally self-adjoint*, and, since the boundary conditions on $u(x)$ and $v(x)$ are the same, the differential operator is self-adjoint.

The discussion of this section may be summarized in the following

Method. *To find the adjoint of a differential operator L in a space $\mathcal{S}$, consider the scalar product $\langle v, Lu\rangle$. With the help of integration by parts, consider it as the scalar product of u with some vector w, which depends on v. The transformation from v to w defines the adjoint operator L^*. The boundary conditions on v are determined by the requirement that the terms resulting from the integration by parts vanish.*

It is to be noted that the form of the boundary conditions on v is not unique; for, if $B_1^*(v) = B_2^*(v) = 0$ are any two boundary conditions on v, then the conditions

$$\alpha_1 B_1^*(v) + \alpha_2 B_2^*(v) = \gamma_1 B_1^*(v) + \gamma_2 B_2^*(v) = 0$$

are completely equivalent to the original ones as long as $\alpha_1\gamma_2 - \alpha_2\gamma_1 \neq 0$. Nevertheless, despite the fact that the boundary conditions are not unique, it is easy to show that the manifold defined by any set of boundary conditions is always the same.

Note that, if L is defined by (3.18), then its formal adjoint is

$$L^*v = \frac{d^2}{dx^2}(av) - \frac{d}{dx}(bv) + cv.$$

We have

$$\int_\alpha^\beta [vLu - uL^*v]\,dx = J(v, u)\Big|_\alpha^\beta \tag{3.20}$$

where

$$J(v, u) = avu' - u(av)' + buv.$$

We shall call $J(v, u)$ the *conjunct* of the functions v and u. It is clear that $J(v, u)$ is a linear homogeneous function of v and u and their derivatives.

PROBLEMS

3.16. Find the adjoint differential operator L^* and the manifold on which it acts if

(*a*) $Lu = u'' + a(x)u' + b(x)u$, where $u(0) = u'(1)$ and $u(1) = u'(0)$;

(*b*) $Lu = -(p(x)u')' + q(x)u$, where $u(0) = u(1)$ and $u'(0) = u'(1)$.

Assume that the scalar product is

$$\langle u, v\rangle = \int_0^1 uv\,dx.$$

3.17. Suppose that $Lu = u''$, where $a_1u(0) + b_1u'(0) + c_1u(1) + d_1u'(1) = 0$ and $a_2u(0) + b_2u'(0) + c_2u(1) + d_2u'(1) = 0$. Find L^* and the manifold on which it acts if we use the scalar product of Problem 3.16. For what values of the constants $a_1, b_1, \cdots, c_2, d_2$ is the operator L self-adjoint?

3.18. Show that the differential operator

$$Lu = -\frac{1}{w(x)}(p(x)u')' + q(x)u$$

is formally self-adjoint if the scalar product is

$$\langle u, v\rangle = \int_0^1 u(x)v(x)w(x)\,dx.$$

3.19. Show that the differential operator

$$Lu = a(x)u'' + b(x)u' + c(x)u$$

can be identified with the operator considered in Problem 3.18 by putting $q = c$, $p = \exp \int b/a\,dx$, $w^{-1} = -a \exp[-\int b/a\,dx]$. It follows that the general second-order differential operator is formally self-adjoint if the appropriate scalar product is used.

3.20. Consider the complex-type scalar product

$$\langle v, u\rangle = \int_0^1 \overline{v(x)}u(x)w(x)\,dx,$$

where $w(x)$ is a non-negative real function. Is $Lu = -\frac{1}{w}(pu')' + qu$ formally self-adjoint in a space with the above scalar product? Distinguish between the case where p and q are both real functions and the case where p and q are complex functions.

Self-adjoint Second-Order Differential Operators

Since a large number of the differential equations that occur in mathematical physics are of the second order, we shall give special consideration to second-order differential operators. From Problems 3.18 and 3.19 we see that any such operator is self-adjoint if the appropriate scalar product is used; therefore, we shall write the operator as follows:

$$Lu = -\frac{1}{w}(pu')' + qu, \tag{3.21}$$

and we shall assume that the scalar product has $w(x)$ as its weight function.

The reason for the minus sign in the definition of L is that in certain cases it permits us to consider L as a *positive-definite* operator. A linear operator L is called positive-definite if $\langle u, Lu\rangle > 0$ for all values of u except $u = 0$. Now, when L is defined by (3.21), we have

$$\begin{aligned}\langle u, Lu\rangle &= \int_0^1 u\Big[-\frac{1}{w}(pu')' + qu\Big]w\,dx \\ &= \int_0^1 (pu'^2 + qwu^2)\,dx - puu'\Big|_0^1.\end{aligned}$$

Consequently, if $p > 0$, $q > 0$, $w > 0$, and if the boundary conditions on u are such as to make the last term on the right vanish, then L will be a positive-definite operator.

So far, L is only formally self-adjoint. Under what conditions will L be actually self-adjoint? To decide this, consider the difference $\langle v, Lu\rangle - \langle Lv, u\rangle$. We have from (3.20)

$$\langle v, Lu\rangle - \langle Lv, u\rangle = J(v, u)\Big|_0^1 = -p(x)(vu' - uv')\Big|_0^1. \tag{3.22}$$

L will be self-adjoint if the conjunct J vanishes identically when u and v are in the same manifold. We shall not discuss the most general conditions for which this happens (see Problem 3.17) but instead shall consider two cases which are important for applications:

(1) A boundary condition is *unmixed* if it involves the values of u and its derivatives either at $x = 0$ or at $x = 1$ but not at both. The typical unmixed boundary condition is

$$\alpha_0 u(0) + \beta_0 u'(0) = 0.$$

It is easy to show that if u satisfies an unmixed boundary condition at $x = 0$ and an unmixed boundary condition at $x = 1$, then L is self-adjoint.

(2) The boundary conditions are *periodic* if they have the form:

$$u(0) = u(1),\ u'(0) = u'(1).$$

In this case again it is easy to show that L is self-adjoint.

We now prove

Theorem 3.2. *If u is any solution of the equation $Lu = 0$ and if v is any solution of the equation $L^*v = 0$, the conjunct of u and v is a constant whose value depends on u and v.*

Here L and L^* are formally self-adjoint operators in the full space $\mathcal{S}$. An immediate consequence of the definition of the conjunct in (3.20) is the following equation:

$$\int_\alpha^\beta (vLu - uL^*v)\, dx = J(v, u)\Big|_\alpha^\beta.$$

From the hypothesis on u and v it follows that

$$J(v, u)\Big|_\alpha^\beta = 0;$$

consequently, the value of $J(v, u)$ at $x = \beta$ is the same as its value at $x = \alpha$. Since α and β are arbitrary points, this proves that the value of $J(v, u)$ is constant.

Corollary. *If L is a formally self-adjoint operator and if u_1 and u_2 are any solutions of $Lu = 0$, the conjunct of u_1 and u_2 is a constant whose value depends on the functions u_1 and u_2.*

As an illustration of Theorem 3.2 consider the following differential equation:

$$u'' + u = 0.$$

Since this equation is formally self-adjoint, v will also be a solution of it. The conjunct of v and u is now

$$J(v, u) = vu' - uv'.$$

Theorem 3.2 states that this expression is constant when u and v are any two solutions of $u'' + u = 0$. For example, if $u = \sin x$ and $v = \cos x$,

$$J(v, u) = \cos x \cos x - \sin x\,(-\sin x) = \text{constant}.$$

This is obviously correct because

$$\cos^2 x + \sin^2 x = 1$$

for all values of x.

We shall prove another

Corollary. *If L is self-adjoint, if u_1 and u_2 are solutions of $Lu = 0$, and if $J(u_1, u_2)$ vanishes for some value of x such that $p(x) \neq 0$, then u_1 and u_2 are linearly dependent.*

From the previous corollary, it follows that the conjunct vanishes for all values of x; consequently,

$$u_1u_2' - u_2u_1' = 0.$$

This implies that

$$\frac{u_1u_2' - u_2u_1'}{u_2^2} = 0.$$

By integration, we find that the ratio u_1/u_2 is a constant and, therefore, u_1 and u_2 are linearly dependent.

Symbolic Operations

It may have been noticed that until now we have considered only operators with homogeneous boundary conditions. However, in addition, we wish to treat problems in which the boundary conditions are not homogeneous, for example, the following: to find a function $u(x)$ such that $u'' = f(x)$ and such that $u(0) = a$, $u(1) = b$. This problem can be treated in many ways. One of the more usual ways is to reduce the problem to the consideration of an operator with homogeneous boundary conditions by writing $u = u_1 + u_2$, where u_1 is the solution of $u_1'' = f(x)$ such that $u_1(0) = u_1(1) = 0$ and u_2 is the solution of $u_2'' = 0$ such that $u_2(0) = a$, $u_2(1) = b$.

We now present another approach which has the advantage of enabling us to extend the meaning of the operator in a similar way to that in which we extended the meaning of differentiation in previous sections. Let us review what was done when we defined symbolic differentiation. We considered a space of testing functions and then defined the symbolic derivative of a function by integrating by parts the product of the function with the ordinary derivative of a testing function. We shall proceed similarly in the case of a differential operator L.

First we consider an illustrative example. Suppose that

$$K = -\frac{d^2}{dx^2}$$

with the boundary conditions $u(0) = u(1) = 0$. It is easily seen that K is self-adjoint and that $\mathcal{M}$, the domain of K, is the set of functions $u(x)$ in $\mathcal{S}$ such that u'' exists and belongs to $\mathcal{S}$ and such that $u(0) = u(1) = 0$. If v belongs to $\mathcal{M}$ and if w belongs to $\mathcal{S}$ and has a second derivative, we shall write

$$\langle Kv, w\rangle = \langle v, Kw\rangle$$

and use the left-hand side to define the meaning of the symbolic function Kw. (It is a symbolic function because w may not belong to the domain of K.) We have

$$\langle Kv, w\rangle = -\int_0^1 v''w\,dx = -v'(1)w(1) + v'(0)w(0) - \int_0^1 vw''\,dx,$$

where we made use of the fact that v belongs to $\mathcal{M}$, the domain of K. It is easy to see that the right-hand side of this equation is a linear continuous functional for functions in $\mathcal{M}$, and consequently it may be used to define a symbolic function Kw. We see then that

$$\int_0^1 vKw\,dx = -\int_0^1 vw''\,dx - v'(1)w(1) + v'(0)w(0).$$

It is convenient to write this equation as

$$\int_0^1 vKw\,dx = \int_0^1 v[-w'' + w(1)\delta'(x-1) - w(0)\delta'(x)]\,dx,$$

and hence we may put

$$Kw = -w'' + w(1)\delta'(x-1) - w(0)\delta'(x). \tag{3.23}$$

Notice that this use of the δ-function and its derivative is not justified by the definitions given before. The previous definitions stated that

$$\begin{aligned} \int \phi(x)\delta(x-a)\,dx &= \phi(a), \\ \int \phi(x)\delta'(x-a)\,dx &= -\phi'(a), \end{aligned} \tag{3.24}$$

if $x = a$ is a point inside the interval of integration. We shall now extend the definition of the δ-function and its derivative by assuming that (3.24) holds even if $x = a$ is an endpoint of the interval of integration. With this extended definition, the formula (3.23) is justified. It should be remarked that for some applications it is more convenient to extend the definition as follows:

$$\begin{aligned} \int \phi(x)\delta(x-a)\,dx &= \tfrac{1}{2}\phi(a), \\ \int \phi(x)\delta'(x-a)\,dx &= -\tfrac{1}{2}\phi'(a), \end{aligned}$$

if $x = a$ is an endpoint of the interval of integration. Either extension is correct as long as it is used consistently. For our purposes the formulas (3.24) will be more convenient.

If w belongs to $\mathcal{M}$, the definition of Kw given in (3.23) reduces to $-w''$ as it should. However, the definition (3.23) is also applicable to functions not in $\mathcal{M}$. For example, if $w(x) = c$, a constant, for $0 \leq x \leq 1$, then

$$Kw = c\delta'(x-1) - c\delta'(x).$$

We shall say that Kw as defined by (3.23) is the result of applying the symbolic operator K to w.

The problem we mentioned previously, namely, to find a function $u(x)$ such that $u'' = f(x)$ and such that $u(0) = a$, $u(1) = b$, can now be formulated as follows:

Find a function $u(x)$ in $\mathcal{S}$ such that

$$Ku = -f(x) + b\delta'(x-1) - a\delta'(x).$$

This problem will be solved with the help of the function $g(x, t)$ defined in (3.17). We found there that

$$g(x, t) = (x-t)H(x-t) - x(1-t), \quad 0 \leq x, \quad t \leq 1$$

was a solution of

(3.25) $$Kg = -\delta(x - t).$$

Multiplying this equation by $f(t)$ and integrating from 0 to 1, we find that the function

$$u_1 = \int f(t)g(x, t)\,dt = \int_0^x (x - t)f(t)\,dx - x\int_0^1 (1 - t)f(t)\,dt$$

is a solution of

$$Ku_1 = -\int f(t)\delta(x - t)\,dt = -f(x).$$

Differentiating (3.25) with respect to t, we get

(3.26) $$K\frac{\partial g}{\partial t} = \delta'(x - t).$$

By differentiating the formula for $g(x, t)$, we see that

$$\frac{\partial g}{\partial t} = -H(x - t) + x.$$

Using (3.26) for $t = 1$ and $t = 0$, we have

$$Ku_2 = b\delta'(x - 1) - a\delta'(x),$$

where

$$u_2 = bx - a(x - 1).$$

Consequently,

$$u = u_1 + u_2 = \int_0^x (x - t)f(t)\,dt - x\int_0^1 (1 - t)f(t)\,dt + bx - a(x - 1)$$

is a solution of the equation $u'' = f$ such that $u(0) = a$, $u(1) = b$. The reader can easily verify this statement.

The method discussed for extending the set of functions to which K can be applied can be generalized to apply to an arbitrary differential operator. Let L be an arbitrary differential operator, $\mathcal{M}$ its domain, L^* the adjoint operator, and let $\mathcal{M}^*$ be the domain of L^*. We shall say that functions belonging to $\mathcal{M}^*$ are *testing functions for the operator* L. Now, consider a function $w(x)$ belonging to $\mathcal{S}$ and not necessarily to $\mathcal{M}$. We wish to define Lw. The result may not be a function, but it will be called a symbolic function.

If v is a testing function for L, that is, if v belongs to $\mathcal{M}^*$, the scalar product $\langle L^*v, w\rangle$ has a meaning. This scalar product is a continuous linear functional on the space $\mathcal{M}^*$ of testing functions. We put

(3.27) $$\langle L^*v, w\rangle = \langle v, Lw\rangle,$$

where Lw is the symbolic function defined by (3.27).

This extended definition (3.27) of the operator enables us to restrict ourselves to the consideration of operators with only homogeneous boundary conditions because, as was seen in the case of the operator K, any problem with non-homogeneous boundary conditions may be changed to a non-homogeneous problem involving an operator with homogeneous boundary conditions. Henceforth, all operators will be considered in their extended sense.

PROBLEM

3.21. Find the symbolic meaning of Lw, if $L = \dfrac{d^2}{dx^2}$, with each of the following sets of boundary conditions:

(*a*) $$u(0) = u'(0) = 0,$$

(*b*) $$u'(0) - \alpha u(0) = u'(1) - \beta u(1) = 0,$$

(*c*) $$u'(0) - u'(1) = u(0) - u(1) = 0.$$

Green's Functions and δ-Functions

We have already seen in (3.17) that the inverse of the differential operator $\dfrac{d^2}{dx^2}$ with the boundary conditions $u(0) = u(1) = 0$ is the integral operator whose kernel is

$$g(x, t) = (x - t)H(x - t) - x(1 - t), \quad 0 \leq x, t \leq 1.$$

We shall consider more general differential operators and show that a similar result holds.

Suppose that L is an ordinary differential operator in x and that $\mathcal{M}$ is its domain. Suppose that we can find a function $g(x, t)$ such that

$$Lg = \delta(x - t). \tag{3.28}$$

Note that here L is the symbolic operator applied to g as a function of x. We shall call $g(x, t)$ the *Green's function* of the operator L. We can show that this function $g(x, t)$ is the kernel of the integral operator which inverts L; for, put

$$u(x) = \int g(x, t)f(t)\, dt,$$

then

$$Lu = \int Lgf\, dt = \int \delta(x - t)f\, dt = f(x).$$

This shows that

$$u = L^{-1}f = \int g(x, t)f(t)\, dt.$$

Also, since we are using the extended definition of the operator, the boundary conditions will be automatically satisfied.

The problem of solving (3.28) will be our main concern for the rest of this chapter. Suppose that L is the following second-order self-adjoint operator:

$$L = -\frac{d}{dx}\left(p\frac{d}{dx}\right) + q(x)$$

with some homogeneous boundary conditions. Here $p(x)$ and $q(x)$ are continuous functions in (0, 1) and $p(x) \neq 0$ in (0, 1). We solve (3.28) first in the special case in which $q(x) = 0$. Denote the solution by $g_0(x, t)$; then the equation becomes

$$-\frac{d}{dx}\left(p\frac{dg_0}{dx}\right) = \delta(x - t).$$

This can be integrated immediately. We have

$$p\frac{dg_0}{dx} = -H(x - t) + \alpha(t)$$

where $\alpha(t)$ is a constant of integration, and then

$$g_0 = -H(x - t)\int_t^x \frac{d\zeta}{p(\zeta)} + \alpha(t)\int_0^x \frac{d\zeta}{p(\zeta)} + \beta(t),$$

where $\beta(t)$ is another constant of integration. Note that $g_0(x, t)$ is a continuous function of x and that its derivative also is continuous except at $x = t$, where it has a jump of magnitude $-p(t)^{-1}$. Note also that

$$\frac{-d}{dx}\left(p\frac{dg_0}{dx}\right) = 0$$

except when $x = t$.

We shall show that the solution $g(x, t)$ in the general case has properties similar to those of $g_0(x, t)$. Put $g(x, t) = g_0(x, t) + k(x, t)$ in (3.28). We find

$$Lk = -q(x)g_0(x, t). \tag{3.29}$$

In the appendix to this chapter we shall prove the following theorem about functions differentiable in the ordinary sense:

Theorem 3.A.I. *Let $p(x)$, $q(x)$, and $f(x)$ be piecewise continuous functions of x in the closed interval* (0, 1) *and let $p(x)$ be positive in that interval. Then there exists a continuous function $u(x)$ such that pu' exists and is continuous for all x, such that $u(0) = u'(0) = 0$, and such that*

$$(pu')' - qu = f$$

for all values of x for which both sides are continuous functions of x.

Since the right-hand side of (3.29) is a continuous function of x, it follows that $k(x, t)$, $p(x)\dfrac{dk}{dx}$, and $\dfrac{d}{dx}p\dfrac{dk}{dx}$ are continuous functions for all values of x; consequently, $g(x, t)$ behaves like $g_0(x, t)$, that is, $g(x, t)$ is continuous and its derivative is continuous except at $x = t$, where it has a jump of magnitude $-p(t)^{-1}$. Since

$$Lg = Lg_0 + Lk = -\frac{d}{dx}\left(p\frac{dg_0}{dx}\right) + q(x)g_0 + Lk = -\frac{d}{dx}\left(p\frac{dg_0}{dx}\right),$$

we see that

$$Lg = 0$$

for all values of x except $x = t$.

We state these results in

Theorem 3.3. *For all values of x, except $x = t$, the Green's function satisfies the homogeneous equation. When L is given by* (3.27), *the Green's function at $x = t$ is continuous, but its derivative has a jump of magnitude* $-1/p(t)$.

Because g is a solution of the differential equation $Lg = 0$ except for $x = t$, it is not a symbolic function but an ordinary function. Since we are using the extended definition of the operator, it follows that g as a function of x must satisfy the boundary conditions because, if it did not, Lg would contain symbolic functions such as $\delta(x - 1)$ or $\delta'(x - 1)$. Henceforth we treat $g(x, t)$ as an ordinary function satisfying the boundary conditions by which the operator is defined.

Once the Green's function is known, the non-homogeneous equation

$$Lu = f(x) \tag{3.30}$$

with the assigned boundary conditions can be solved. The solution is

$$u(x) = \int_0^1 f(t)g(x, t)\,dt. \tag{3.31}$$

The proof that u as defined in (3.31) satisfies (3.30) is as follows:

$$Lu = \int_0^1 f(t)Lg\,dt = \int_0^1 f(t)\delta(x - t)\,dt = f(x). \tag{3.32}$$

u also satisfies the assigned boundary conditions because $g(x, t)$ as a function of x satisfies them.

Example 1—Green's Function

Let

$$L = -\frac{d^2}{dx^2},$$

and let the boundary conditions be

$$u(0) = u'(0) = 0.$$

To find the Green's function $g(x, t)$ we must solve

$$\frac{d^2g}{dx^2} = -\delta(x - t) \tag{3.33}$$

with the conditions

$$g(0) = g_x(0) = 0. \tag{3.34}$$

As remarked previously, g is a solution of the homogeneous equation

$$\frac{d^2g}{dx^2} = 0, \tag{3.35}$$

except at $x = t$, where it is continuous and its derivative has a jump of magnitude -1. For $x < t$, (3.34) and (3.35) imply that

$$g = 0,$$

but for $x > t$ we assume that g is an arbitrary solution of (3.35), namely,

$$g = \alpha x + \beta.$$

Since the jump in the derivative of g is -1, we must have

$$\alpha = -1;$$

since g is continuous at $x = t$, we must have

$$\beta = t.$$

Finally,

$$\begin{aligned} g &= 0, \quad x < t \\ &= t - x, \quad x > t \end{aligned}$$

or

$$g(x, t) = -(x - t)H(x - t) \tag{3.36}$$

for all values of x. It is obvious that $g(x, t)$ satisfies (3.33) and (3.34).

Now the solution of

$$-\frac{d^2u}{dx^2} = f(x) \tag{3.37}$$

with the conditions

$$u(0) = u'(0) = 0 \tag{3.38}$$

may be found. Using (3.31) and (3.36), we have

$$\begin{aligned} u(x) &= -\int_0^1 f(t)(x - t)H(x - t)\,dt \\ &= -\int_0^x f(t)(x - t)\,dt, \end{aligned}$$

and this is clearly the solution of (3.37) satisfying (3.38).

Suppose that we wish to find the solution of (3.37) which satisfies the conditions $u(0) = b$, $u'(0) = a$ instead of those in (3.38). We first consider the extended definition of L. If $\mathcal{S}$, the space with which we are dealing, contains all functions of integrable square over the interval (0, 1), then

L^* is $-\dfrac{d^2}{dx^2}$ with the boundary conditions $v(1) = v'(1) = 0$. Let v be a testing function for L, that is, v is any function in the domain of L^*. The extended definition of L acting on u is obtained by putting

$$\begin{aligned}\langle Lu, v\rangle = \langle u, Lv\rangle &= -\int_0^1 uv''\,dx = -\int_0^1 vu''\,dx - (uv' - vu)'\Big|_0^1 \\ &= -\int_0^1 vu''\,dx + bv'(0) - av(0) \\ &= \int_0^1 vf\,dx - \int_0^1 v[b\delta'(x) + a\delta(x)]\,dx;\end{aligned}$$

consequently,

$$Lu = f(x) - a\delta(x) - b\delta'(x). \tag{3.39}$$

Put

$$u_1 = -\int_0^x f(t)(x - t)\,dt.$$

We know that $Lu_1 = f(x)$. If we can find a function u_2 such that $Lu_2 = -a\delta(x) - b\delta'(x)$, then $u = u_1 + u_2$ will be a solution of (3.39). Since

$$Lg(x, t) = \delta(x - t),$$

we see that

$$L\left[-ag(x, 0) + b\frac{\partial g(x, 0)}{\partial t}\right] = -a\delta(x) - b\delta'(x).$$

But $g(x, 0) = -x$ and $\dfrac{\partial g(x, 0)}{\partial t} = 1$; consequently,

$$u_2 = ax + b.$$

Using this result, we find that the solution of (3.37) satisfying the conditions $u(0) = b$, $u'(0) = a$ is

$$u = -\int_0^x f(t)(x - t)\,dt + ax + b.$$

Of course, the fact that $u_2 = ax + b$ could have been found directly. However, we have used the general method in order to illustrate the technique. The reason for using $\dfrac{\partial g}{\partial t}$ instead of $\dfrac{\partial g}{\partial x}$ will be made clear later.

Example 2—Green's Function

Let L again be $-\dfrac{d^2}{dx^2}$, but now suppose that the conditions are

$$u(0) = 0, \quad u(1) = 0. \tag{3.40}$$

To find the Green's function we must solve

$$\frac{d^2g}{dx^2} = -\delta(x - t)$$

on the manifold defined by the boundary conditions

$$g(0, t) = g(1, t) = 0.$$

We shall illustrate a very useful technique for finding the Green's function. For $x < t$, the solution of the homogeneous equation that satisfies the left boundary condition $g(0, t) = 0$ will be proportional to x. For $x > t$, the solution of the homogeneous equation that satisfies the right boundary condition, $g(1, t) = 0$, will be proportional to $1 - x$.

Write

$$\begin{aligned} g(x, t) &= x, \quad x < t \\ &= 1 - x, \quad x > t. \end{aligned}$$

This is wrong because $g(x, t)$ is not continuous for $x = t$. The value of g as x approaches t from below is t, but the value of g as x approaches t from above is $1 - t$. Multiply the first expression for g by the value of the second expression at $x = t$, and multiply the second expression for g by the value of the first expression at $x = t$. Then write

$$\begin{aligned} g(x, t) &= x(1 - t), \quad x < t \\ &= (1 - x)t, \quad x > t. \end{aligned} \tag{3.41}$$

This function is continuous at $x = t$. Because the derivative of this function for $x < t$ is $(1 - t)$ and the derivative for $x > t$ is $-t$, the jump in the derivative at $x = t$ is

$$-t - (1 - t) = -1.$$

Equation (3.41) is therefore the Green's function for the operator $-\dfrac{d^2}{dx^2}$ in the manifold defined by the conditions (3.40). We may write the expression for $g(x, t)$ as follows:

$$\begin{aligned} g(x, t) &= x(1 - t)H(t - x) + (1 - x)tH(x - t) \\ &= \frac{x + t}{2} - xt - \frac{|x - t|}{2}. \end{aligned}$$

Note that $g(x, t)$ is symmetric in x and t. We shall see later that this is generally true for the Green's function of a self-adjoint operator.

By differentiation, we may check that (3.41) is actually the solution of the differential equation. We have

$$\frac{dg}{dx} = \frac{1}{2} - t - \frac{1}{2}\operatorname{sgn}(x - t)$$

$$\frac{d^2g}{dx^2} = -\delta(x - t).$$

As in the preceding example, we find that the solution of

$$-\frac{d^2}{dx^2}u = f(x)$$

with the boundary conditions

$$u(0) = u(1) = 0$$

is

$$u = \int_0^1 f(t)\,[x(1-t)H(t-x) + (1-x)tH(x-t)]\,dt$$
$$= (1-x)\int_0^x tf(t)\,dt + x\int_x^1 (1-t)f(t)\,dt.$$

Example 3—Green's Function

Consider the operator

$$L = -\frac{d^2}{dx^2} - 4$$

in the manifold defined by

$$u(0) = u'(1) = 0.$$

The Green's function is found by solving the equation

$$\frac{d^2g}{dx^2} + 4g = -\,\delta(x-t), \tag{3.42}$$

where

$$g(0, t) = g_x\,(1, t) = 0. \tag{3.43}$$

We use the same technique as in Example 2. The solutions of the homogeneous equation corresponding to (3.42) are any linear combination of $\sin 2x$ and $\cos 2x$. We pick two solutions of the homogeneous equation, one to satisfy the boundary condition at zero, the other to satisfy the boundary condition at unity. We may then write

$$g(x, t) = \sin 2x, \quad x < t$$
$$= \cos 2(1-x), \quad x > t.$$

Since $g(x, t)$ is not continuous for $x = t$, we multiply the first expression by the value of the second expression at $x = t$ and also multiply the second expression by the value of the first at $x = t$. We get

$$g(x, t) = \sin 2x \cos 2(1-t), \quad x < t$$
$$= \cos 2(1-x) \sin 2t, \quad x > t.$$

Now $g(x, t)$ is continuous, but the jump in the magnitude of the derivative at $x = t$ is

$$2 \sin 2(1-t) \sin 2t - 2 \cos 2t \cos 2(1-t)$$
$$= -\,2 \cos 2(t + 1 - t) = -\,2 \cos 2.$$

Since it follows from (3.42) that the jump in the derivative should be -1, we see that this last formula for $g(x, t)$ is still wrong. To correct it, we must divide it by $2\cos 2$. We then have the final and correct results:

$$g(x, t) = \sin 2x \cos 2(1 - t)/2\cos 2, \quad x < t$$
$$= \cos 2(1 - x) \sin 2t/2\cos 2, \quad x > t$$

or

$$(3.44) \qquad g(x, t) = \frac{\sin 2x \cos 2(1 - t)}{2\cos 2} H(t - x) + \frac{\cos 2(1 - x) \sin 2t}{2\cos 2} H(x - t).$$

Note again that $g(x, t)$ is symmetric in x and t.

It is easy to show by straightforward differentiation that $g(x, t)$ as given by (3.44) satisfies (3.42). We have

$$\frac{dg}{dx} = \frac{2\cos 2x \cos 2(1 - t)}{2\cos 2} H(t - x) + \frac{2\sin 2(1 - x) \sin 2t}{2\cos 2} H(x - t) - \frac{\sin 2x \cos 2(1 - t)}{2\cos 2} \delta(t - x) + \frac{\cos 2(1 - x) \sin 2t}{2\cos 2} \delta(x - t).$$

The sum of the last two terms is zero because

$$f(x)\delta(x - t) = f(t)\delta(x - t)$$

and

$$\left.\frac{\sin 2x \cos 2(1 - t)}{2\cos 2}\right|_{x=t} = \left.\frac{\cos 2(1 - x) \sin 2t}{2\cos 2}\right|_{x=t}.$$

Differentiating the formula for $\frac{dg}{dx}$, we find that

$$\frac{d^2g}{dx^2} = -4\frac{\sin 2x \cos 2(1 - t)}{2\cos 2} H(t - x) - \frac{4\cos 2(1 - x) \sin 2t}{2\cos 2} H(x - t) - \frac{\cos 2x \cos 2(1 - t)}{\cos 2} \delta(t - x) + \frac{\sin 2(1 - x) \sin 2t}{\cos 2} \delta(x - t)$$

$$= -4g + \frac{\delta(x - t)}{\cos 2}[\sin 2(1 - x) \sin 2t - \cos 2x \cos 2(1 - t)]$$

$$= -4g - \delta(x - t),$$

which proves that $g(x, t)$ satisfies (3.42).

Suppose that we wish to solve the following non-homogeneous problem: Find a solution $u(x)$ of the equation

$$u'' + 4u = -f(x)$$

such that

$$u(0) = a, \; u'(1) = b.$$

Let us first consider the extended definition of L. Clearly, L is self-adjoint, so that $L = L^*$. Let v be any function in the domain of L; then by the extended definition of L we have

$$\begin{aligned}\langle Lu, v\rangle = \langle u, Lv\rangle &= -\int_0^1 u(v'' + 4v)\,dx \\ &= -\int_0^1 v(u'' + 4u)\,dx - (uv' - u'v)\Big|_0^1 \\ &= \int_0^1 vf(x)\,dx + av'(0) + bv(1).\end{aligned}$$

Consequently,

$$Lu = f(x) + b\delta(x - 1) - a\delta'(x).$$

Put

$$u_1 = \int_0^1 f(t)g(x, t)\,dt,$$

and we find that $Lu_1 = f$. Now we need to find a function u_2 such that

$$Lu_2 = b\delta(x - 1) - a\delta'(x),$$

and then $u = u_1 + u_2$ will be the desired solution.

Since

$$Lg = \delta(x - t),$$

we see that

$$L\left[bg(x, 1) + a\frac{\partial g\,(x, 0)}{\partial t}\right] = b\,\delta(x - 1) - a\delta'(x).$$

From (3.44) we find that

$$g(x, 1) = \frac{\sin 2x}{2\cos 2}$$

and that

$$\frac{\partial g\,(x, 0)}{\partial t} = \frac{\cos 2(1 - x)}{\cos 2};$$

consequently,

$$u_2 = \frac{b\sin 2x}{2\cos 2} + \frac{a\cos 2(1 - x)}{\cos 2}.$$

Finally,

$$u = \int_0^1 f(t)g(x, t)\,dt + \frac{b\sin 2x}{2\cos 2} + \frac{a\cos 2(1 - x)}{\cos 2}.$$

Green's Function if Boundary Conditions are Unmixed

The technique that has been used in the two previous examples can be applied to any second-order differential operator if the boundary conditions are not *mixed*, that is, if each boundary condition involves just one point of the boundary but not both. A condition such as $u'(0) = u(0)$ is not mixed, but a condition such as $u(1) = u(0)$ is mixed.

We shall obtain the Green's function for the general second-order self-adjoint operator

$$L = -\frac{d}{dx}\left(p\frac{d}{dx}\right) + q$$

when the domain is defined by the general unmixed boundary conditions

$$B_1(u) = B_2(u) = 0. \tag{3.45}$$

We shall assume that $p(x) \neq 0$ in the interval (0, 1) and that $B_1(u)$ involves values of u and its derivatives at $x = 0$ only, whereas $B_2(u)$ involves values of u and its derivatives at $x = 1$ only.

The Green's function is the solution of

$$(pg_x)_x - qg(x, t) = -\delta(x - t) \tag{3.46}$$

with the boundary conditions

$$B_1(g) = B_2(g) = 0. \tag{3.47}$$

Consider the homogeneous equation

$$(pu_x)_x - qu = 0 \tag{3.48}$$

without any boundary conditions. Suppose that $v_1(x)$ and $v_2(x)$ are any two independent solutions of (3.48). Let $w_1(x)$ be a linear combination of $v_1(x)$ and $v_2(x)$ which satisfies the condition

$$B_1(w_1) = 0.$$

Similarly, let $w_2(x)$ be a linear combination of $v_1(x)$ and $v_2(x)$ which satisfies the condition

$$B_2(w_2) = 0.$$

Then we start the construction of the Green's function by writing

$$\begin{aligned} g(x, t) &= w_1(x), \quad x < t \\ &= w_2(x), \quad x > t. \end{aligned}$$

This expression for $g(x, t)$ will satisfy the differential equation (3.46) for $x \neq t$ and also the boundary conditions (3.47), but it will not satisfy the continuity or the jump conditions at $x = t$. To make $g(x, t)$ continuous, multiply the first expression by the value of the second at $x = t$ and multiply the second expression by the value of the first at $x = t$ so that we have

$$\begin{aligned} g(x, t) &= w_1(x)w_2(t), \quad x < t \\ &= w_2(x)w_1(t), \quad x > t. \end{aligned} \tag{3.49}$$

This expression still does not satisfy the jump condition in the derivative at $x = t$. The jump in the derivative at $x = t$ is given by

$$w_2'(t)w_1(t) - w_1'(t)w_2(t) = -\frac{J(w_2, w_1)}{p(t)}$$

while it should be $-1/p(t)$. This can be remedied by dividing (3.49) by $J(w_2, w_1)$, the conjunct of w_2, w_1. The final and correct formula is then

(3.50)
$$\begin{aligned} g(x, t) &= w_1(x)w_2(t)/J(w_2, w_1), \quad x < t \\ &= w_2(x)w_1(t)/J(w_2, w_1), \quad x > t, \end{aligned}$$

or, written in another way,

(3.51)
$$g(x, t) = \frac{w_1(x)w_2(t)H(t - x) + w_2(x)w_1(t)H(x - t)}{J(w_2, w_1)}.$$

The conjunct in the denominator of these expressions is to be evaluated at $x = t$. However, as we have shown in Theorem 3.2,

$$J(w_2, w_1) = \text{constant};$$

consequently, instead of evaluating the conjunct at $x = t$, we may evaluate it at any convenient point. For a simple illustration of the usefulness of this fact, consider Example 3 of the Green's function. There, the conjunct of $\cos 2(1 - x)$ and $\sin 2x$ had to be found. Now, since $p(x) = 1$, we have

$$\begin{aligned} J(\cos 2(1 - x), \sin 2x) &= \text{constant} \\ &= 2 \cos 2(1 - x) \cos 2x - 2 \sin 2(1 - x) \sin 2x \\ &= 2 \cos 2 \end{aligned}$$

when we put $x = 0$. Before, this result was obtained by using the addition theorem for the cosine, but now it is an immediate consequence of the properties of the conjunct.

Formulas (3.50) and (3.51) break down if $J(w_2, w_1) = 0$. By the second corollary to Theorem 3.2 the fact that the conjunct is zero implies that one function, say $w_2(x)$, is a multiple of the other, $w_1(x)$. Since $B_1(w_1) = 0$ and $B_2(w_2) = 0$, it follows that $B_1(w_2) = B_2(w_2) = 0$. This means that $w_2(x)$ is a non-trivial solution of the homogeneous equation

$$(pu')' - qu = 0$$

with the boundary conditions

(3.45)
$$B_1(u) = B_2(u) = 0.$$

Consequently, $w_2(x)$ is an eigenfunction of the operator L with the boundary conditions (3.45), and the corresponding eigenvalue is $\lambda = 0$.

The preceding discussion may be summarized in the following theorem:

Theorem 3.4. *Let $w_1(x)$ be the solution of*

$$Lw = -(pw')' + qw = 0,$$

satisfying the unmixed boundary conditions $B_1(w) = 0$, and let $w_2(x)$ be the solution of the same differential equation satisfying the unmixed condition $B_2(w) = 0$. Then the Green's function for L with the boundary conditions B_1 and B_2 is given by the following formula:

$$g(x, t) = \frac{w_1(x)w_2(t)H(t - x) + w_2(x)w_1(t)H(x - t)}{J(w_2, w_1)},$$

where $J(w_2, w_1)$ is the conjunct of w_2 and w_1 evaluated at any point.

PROBLEMS

3.22. Find the Green's function for $L = -\frac{d^2}{dx^2}$ in the manifold defined by the boundary conditions $u'(0) = 0$, $u(1) = 0$.

3.23. Find the Green's function for $L = -\frac{d^2}{dx^2} - k^2$, k constant, in the manifold defined by the boundary conditions $u(0) = 0$, $u(1) = 0$. For what values of k does the formula break down?

3.24. Find the Green's function for $L = -\frac{d}{dx}\left(x\frac{d}{dx}\right)$ in the manifold defined by the boundary conditions $u(0) = 0$, $u'(1) = 0$.

3.25. Use the Green's function to find a function $u(x)$ such that $u'' + k^2u = f(x)$ and such that $u'(0) = u'(1) = 0$.

3.26. Show that formula (3.51) satisfies the conditions of Theorem 3.3.

Non-homogeneous Boundary Conditions

Consider the problem of finding a solution of $Lu = f$ such that $B_1(u) = a$, $B_2(u) = b$. As we know, the function

$$u_1 = \int_0^1 f(t)g(x, t)\, dt$$

will be a solution of $Lu_1 = f$ such that $B_1(u_1) = 0$, $B_2(u_1) = 0$. Just as before, in order to take into account the non-homogeneous boundary conditions, we must extend the definition of L.

Suppose that the boundary conditions are of the form:

$$B_1(u) = u(0) \cos \alpha + u'(0) \sin \alpha,$$
$$B_2(u) = u(1) \cos \beta + u'(1) \sin \beta.$$

Since L is self-adjoint, any function v in the domain of L is a testing function for L. The extended definition of L acting on u is obtained by putting

$$
\begin{aligned}
(3.52) \qquad \langle Lu, v\rangle &= \langle u, Lv\rangle \\
&= \int_0^1 u[-(pv')' + qv]\, dx \\
&= \int_0^1 v[-(pu')' + qu]\, dx - p(uv' - u'v)\Big|_0^1.
\end{aligned}
$$

Now

$$
\begin{aligned}
uv' - u'v = {} & (u \sin \alpha - u' \cos \alpha)(v \cos \alpha + v' \sin \alpha) \\
& - (u \cos \alpha + u' \sin \alpha)(v \sin \alpha - v' \cos \alpha).
\end{aligned}
$$

A similar result holds with α replaced by β. We have then that

$$
\begin{aligned}
(3.53) \qquad p(uv' - u'v)\Big|_0^1 = {} & - p(1)b[v(1) \sin \beta - v'(1) \cos \beta] \\
& + p(0)a[v(0) \sin \alpha - v'(0) \cos \alpha].
\end{aligned}
$$

From (3.52) and (3.53) we see that the extended definition of Lu is this:

$$
\begin{aligned}
Lu = {} & f(x) + bp(1)[\delta(x-1) \sin \beta + \delta'(x-1) \cos \beta] \\
& - ap(0)[\delta(x) \sin \alpha + \delta'(x) \cos \alpha].
\end{aligned}
$$

If we now find a function u_2 such that

$$
\begin{aligned}
(3.54) \qquad Lu_2 = {} & bp(1)[\delta(x-1) \sin \beta + \delta'(x-1) \cos \beta] \\
& - ap(0)[\delta(x) \sin \alpha + \delta'(x) \cos \alpha],
\end{aligned}
$$

then $u = u_1 + u_2$ will be a solution of the problem considered.

Since

$$Lg(x, t) = \delta(x - t),$$

we see that, if we put

$$
\begin{aligned}
u_2 = {} & bp(1)\left[g(x, 1) \sin \beta - \frac{\partial g\,(x, 1)}{\partial t} \cos \beta\right] \\
& - ap(0)\left[g(x, 0) \sin \alpha - \frac{\partial g\,(x, 0)}{\partial t} \cos \alpha\right],
\end{aligned}
$$

it will be a solution of (3.54). Note that we cannot use $\dfrac{\partial g}{\partial x}$ in the above formula even though

$$\frac{\partial}{\partial x} Lg = \delta'(x - t)$$

because

$$L\frac{\partial g}{\partial x} \neq \frac{\partial}{\partial x} Lg.$$

However, we do have

$$L\frac{\partial g}{\partial t} = \frac{\partial}{\partial t} Lg = -\delta'(x - t).$$

From Theorem 3.4 we find that

$$g(x, 1) = \frac{w_1(x)w_2(1)}{J(w_2, w_1)}, \quad g(x, 0) = \frac{w_2(x)w_1(0)}{J(w_2, w_1)},$$

$$\frac{\partial g(x, 1)}{\partial t} = \frac{w_1(x)w_2'(1)}{J(w_2, w_1)}, \quad \frac{\partial g(x, 0)}{\partial t} = \frac{w_2(x)w_1'(0)}{J(w_2, w_1)};$$

consequently,

$$u_2 = \frac{bp(1)w_1(x)}{J(w_2, w_1)}[w_2(1) \sin \beta - w_2'(1) \cos \beta]$$
$$- \frac{ap(0)w_2(x)}{J(w_2, w_1)}[w_1(0) \sin \alpha - w_1'(0) \cos \alpha].$$

Finally, we have

Theorem 3.5. *If w_1 and w_2 are the solutions of $Lw = 0$ defined in Theorem* 3.4, *then the solution of*

$$-(pu')' + qu = f$$

which satisfies the conditions

$$B_1(u) = a, \quad B_2(u) = b$$

is

$$u = \frac{w_2(x)}{J}\int_0^x w_1(t)f(t)\,dt + \frac{w_1(x)}{J}\int_x^1 w_2(t)f(t)\,dt$$
$$+ \frac{bp(1)w_1(x)}{J}[w_2(1) \sin \beta - w_2'(1) \cos \beta]$$
$$- \frac{ap(0)w_2(x)}{J}[w_1(0) \sin \alpha - w_1'(0) \cos \alpha].$$

PROBLEM

3.27. Solve the following equations:

(*a*) $u'' = f(x), \quad u'(0) = a, \quad u(1) = b,$

(*b*) $u'' - k^2u = f, \quad u(0) - u'(0) = a, \quad u(1) = b,$

(*c*) $-(xu')' = f, \quad u(0) = 0, \quad u'(1) = b.$

If the Homogeneous Equation Has a Non-trivial Solution

Suppose that there exists a non-zero solution $w(x)$ of the homogeneous equation

$$-Lw = (pw_x)_x - qw = 0$$

such that

$$B_1(w) = B_2(w) = 0.$$

Consider now the problem of finding $u(x)$ such that

$$(pu_x)_x - qu = f(x)$$

with the conditions

$$B_1(u) = B_2(u) = 0.$$

In Chapter 1 it was proved that if L is self-adjoint and has a closed range, and if the homogeneous equation

$$Lu = 0$$

has a single non-trivial solution w, the non-homogeneous equation

$$Lu = f$$

has a solution if and only if

$$\langle f, w \rangle = 0.$$

We shall show that a similar theorem is valid for differential operators.

Theorem 3.6. *Let $w(x)$ be the solution, unique except for a constant factor, of the self-adjoint homogeneous differential equation*

$$Lw = -(pw')' + qw = 0$$

such that $B_1(w) = B_2(w) = 0$; then the non-homogeneous equation

$$Lu = f \tag{3.55}$$

has a solution satisfying the boundary conditions $B_1(u) = B_2(u) = 0$ if and only if

$$\int_0^1 f(x)w(x)\,dx = 0. \tag{3.56}$$

We shall here assume that the boundary conditions are unmixed. The theorem is valid for arbitrary boundary conditions.

If a function u exists satisfying (3.55) and the boundary conditions, then, since L is self-adjoint, we have

$$0 = \langle u, Lw \rangle - \langle Lu, w \rangle = -\langle f, w \rangle,$$

and this shows that (3.56) is satisfied. Conversely, if we suppose that (3.56) is satisfied, we may solve (3.55). Let $v(x)$ be a solution of

$$(pu_x)_x - qu = 0,$$

and let $v(x)$ be independent of $w(x)$. Then, by using the Green's function technique as illustrated in Example 1, we find that a solution of (3.55) which satisfies the conditions $u(0) = u'(0) = 0$, instead of the conditions $B_1(u) = B_2(u) = 0$, is

$$u(x) = \frac{w(x)}{J(w, v)}\int_0^x fv\,d\tau - \frac{v(x)}{J(w, v)}\int_0^x fw\,d\tau. \tag{3.57}$$

Note that, as shown previously, $J(w, v)$ is constant for all values of x.

We can now show that the function defined by (3.57) will be a solution of the original problem. It is clear that (3.57) satisfies the differential equation (3.55). We now investigate the boundary conditions. Since the boundary conditions are linear, homogeneous, and unmixed, we may write them as follows:

$$B_1(u) = \alpha_1 u(0) + \beta_1 u'(0),$$
$$B_2(u) = \alpha_2 u(1) + \beta_2 u'(1),$$

where α_1, β_1, α_2, β_2 are given constants. Since $u(x)$ in (3.57) was constructed so that $u(0) = u'(0) = 0$, it is obvious that $B_1(u) = 0$. From (3.57) we have

$$u(1) = \frac{w(1)}{J(w, v)}\int_0^1 fv\, d\tau, \tag{3.58}$$

because from (3.56)

$$\int_0^1 fw\, d\tau = 0.$$

If we differentiate (3.57), using condition (3.56), and put x equal to unity, we obtain

$$u'(1) = \frac{w'(1)}{J(w, v)}\int_0^1 fv\, d\tau. \tag{3.59}$$

Formulas (3.58) and (3.59) together show that

$$B_2(u) = \frac{B_2(w)}{J(w, v)}\int_0^1 fv\, d\tau = 0$$

because of the assumption on w. We have therefore proved Theorem 3.6, and we have also shown that if (3.56) is satisfied then the solution of (3.55) which satisfies the conditions

$$u(0) = u'(0) = 0$$

will also satisfy the boundary conditions

$$B_1(u) = B_2(u) = 0.$$

We could not use the Green's function technique in this case because the equation

$$(pu_x)_x - qu = -\,\delta(x - t),$$

with the boundary conditions $B_1(u) = B_2(u) = 0$, has no solution since

$$\int \delta(x - t)w(x)\, dx \neq 0,$$

that is, (3.56) is not satisfied. If the right side of the above equation is modified so that it is orthogonal to $w(x)$, we obtain a problem which can be solved. This may be done by writing

$$(pu_x)_x - qu = -\,\delta(x - t) + w(t)\,\frac{\beta_1\delta(x) + \alpha_1\delta'(x)}{\beta_1 w(0) - \alpha_1 w'(0)}.$$

This equation can be solved in the same way as (3.55) was. The details will be left to the reader.

PROBLEM

3.28. In each of the following cases find the appropriate orthogonality relation that $f(x)$ must satisfy in order that the equation may have a solution, and then find a solution:

(*a*) $$u'' = f(x), \quad u'(0) = u'(1) = 0;$$

(*b*) $$u'' + \pi^2 u = f(x), \quad u(0) = u(1) = 0;$$

(*c*) $$(xu')' = f(x), \quad xu'(x)|_{x=0} = 0, \quad u'(1) = 0;$$

(*d*) $$u'' = f(x), \quad u(0) = u(1), \quad u'(0) = u'(1).$$

Green's Function for General Boundary Conditions

In the general case where the boundary conditions are mixed, the Green's function may still be found in a straightforward way. Suppose that the equation is

$$(pg_x)_x - qg = -\delta(x - t), \tag{3.60}$$

with the boundary conditions

$$B_1(g) = B_2(g) = 0. \tag{3.61}$$

Let $v(x)$ and $w(x)$ be any two linearly independent solutions of

$$(pu_x)_x - qu = 0.$$

Then write

$$g(x, t) = \alpha v(x) + \beta w(x) + \frac{v(x)w(t)H(t - x) + w(x)v(t)H(x - t)}{J(w, v)}, \tag{3.62}$$

where α and β are constants which will be determined so that $g(x, t)$ satisfies the boundary conditions (3.61). The denominator $J(w, v)$ is the conjunct of w and v at the value t. Note that it cannot be zero for any value of t because, if it were zero, the functions $v(x)$ and $w(x)$ would not be linearly independent; this will be shown in Problem 3.31.

We show that $g(x, t)$ as defined in (3.62) satisfies the appropriate conditions at $x = t$, namely, that $g(x, t)$ is continuous and its derivative has a jump of magnitude $-1/p(t)$. The continuity is obvious. From the definition of the conjunct it follows that the jump in the derivative at $x = t$ is

$$\frac{w'(t)v(t) - v'(t)w(t)}{J(w, v)} = -\frac{1}{p(t)}.$$

Now consider the boundary conditions. Since they are linear and homogeneous, we have

$$B_1(g) = \alpha B_1(v) + \beta B_1(w) + B_1(r),$$
$$B_2(g) = \alpha B_2(v) + \beta B_2(w) + B_2(r),$$

where we have used $r(x)$ to represent the function

$$\frac{v(x)w(t)H(t-x)+w(x)v(t)H(x-t)}{J(w, v)}.$$

The boundary conditions (3.61) will consequently give us two linear equations to determine the values of α and β. These equations will have a solution if the determinant

$$\begin{vmatrix} B_1(v) & B_1(w) \\ B_2(v) & B_2(w) \end{vmatrix}$$

does not vanish.

If the determinant vanishes, either $B_1(v) = B_2(v) = 0$, which implies that v is an eigenfunction of the operator corresponding to the eigenvalue zero, or there exists a constant c such that

$$B_1(w) + cB_1(v) = 0,$$
$$B_2(w) + cB_2(v) = 0.$$

These equations imply that $B_1(w + cv) = B_2(w + cv) = 0$; hence $w + cv$ is an eigenfunction of the operator corresponding to the eigenvalue zero.

We see then that *if there is no eigenfunction corresponding to the eigenvalue zero, the Green's function exists and is given by* (3.62) *where* α *and* β *are found by solving* (3.61).

PROBLEMS

3.29. Find the Green's function for the operator $L = -\dfrac{d^2}{dx^2} - k^2$, with the periodic boundary conditions

$$u(0) = u(1) \text{ and } u'(0) = u'(1).$$

3.30. Explain why it is impossible to find a Green's function for the operator $L = -\dfrac{d^2}{dx^2} - k^2$ with the boundary conditions $u(0) = u(1)$ and $u'(0) = -u'(1)$.

3.31. Show that if $J(w, v) = 0$ and if $p(x) \neq 0$, and if $w(x)$ and $v(x)$ do not both vanish for the same value of x, then w and v are linearly dependent. (*Hint.* If $v(x) \neq 0$, $w(x) \neq 0$, then $v'/v = w'/w$ implies $\log v = \log Cw$, where C is a constant. If $v(x) \neq 0$, $w(x) = 0$, then $J(w, v) = 0$ implies that $w' = 0$, and by differentiation all derivatives of w are zero.)

Green's Function of the Adjoint Equation

We prove

Theorem 3.7. *Let $g(x, t)$ be the Green's function for an operator L on a manifold defined by certain boundary conditions, and let $h(x, t)$ be the Green's function for the adjoint operator L^* on the manifold defined by the adjoint boundary conditions; then*

$$g(x, t) = h(t, x).$$

In particular, if L is self-adjoint, then $g(x, t)$ is a symmetric function of x and t.

To prove the theorem, we start with the defining relations for the Green's functions, namely,

$$Lg(x, t) = \delta(x - t), \qquad (3.63)$$
$$L^*h(x, \tau) = \delta(x - \tau).$$

From the properties of the adjoint operator, we get

$$\int h(x, \tau)Lg(x, t)\, dx = \int L^*h(x, \tau)g(x, t)\, dx.$$

If we use (3.63), this equation becomes

$$\int h(x, \tau)\delta(x - t)\, dx = \int \delta(x - \tau)g(x, t)\, dx,$$

or

$$h(t, \tau) = g(\tau, t).$$

This last result, with a slight change of notation, is the conclusion of Theorem 3.7.

PROBLEM

3.32. Verify Theorem 3.7 if $L = \dfrac{d^2}{dx^2} + x\dfrac{d}{dx}$, with the conditions $u(0) = u(1)$ and $u'(0) + u'(1) = 0$.

Discontinuity Conditions

In applications it is important to consider the solution of linear differential equations in which the coefficients may be discontinuous. This situation occurs when the properties of the medium vary discontinuously, for example, the propagation of sound waves through media of different densities or the transmission of heat through materials of different thermal conductivity. The differential equation describes the behavior of the solution in each medium separately, but it is still necessary to determine how the solution behaves across the interface of the two media. Usually the behavior of the solution across the interface, from which we derive the so-called *discontinuity conditions*, is determined by physical considerations. For example, in sound propagation the pressure and velocity of the sound wave must be continuous across the interface; otherwise there would be an infinite acceleration there. Similarly, in electromagnetic theory we can show that because of Gauss' theorem the tangential components of the electric field must be continuous across the interface. Clearly, these discontinuity conditions are as much a part of the physical description of the problem as are the differential equations themselves, and consequently the question of discontinuity conditions is a physical and not a mathematical question. However, it is useful to investigate what kind of discontinuity conditions are mathematically appropriate for given differential equations.

Consider the self-adjoint differential equation

$$-Lu = (pu')' - qu = f(x), \tag{3.64}$$

where $p(x)$, $q(x)$ and $f(x)$ are continuous except for possible jumps at the points $x = x_1, x_2, \cdots, x_k$. Theorem 3.A.I in the appendix to this chapter tells us that, despite the jumps, there are solutions of (3.64) such that the functions $u(x)$ and $p(x)u'(x)$ are continuous throughout the entire interval; consequently, we may write the discontinuity conditions for this equation as follows:

$$u(x_i + 0) = u(x_i - 0),$$
$$p(x_i + 0)u'(x_i + 0) = p(x_i - 0)u'(x_i - 0), \quad i = 1, 2, \cdots, k.$$

Note that, if $p(x)$ is continuous at $x = x_i$, the last condition states that $u'(x)$ must be continuous at $x = x_i$.

Instead of using the results of a theorem such as Theorem 3.A.I about the existence of a solution of the differential equation, we may obtain the discontinuity conditions by the following heuristic argument. Integrate (3.64) across a discontinuity, say from $x = x_i - \varepsilon$ to $x = x_i + \varepsilon$, where ε is a small positive quantity. We have

$$pu'\Big|_{x_i - \varepsilon}^{x_i + \varepsilon} = \int_{x_i - \varepsilon}^{x_i + \varepsilon} (qu + f)\, dx.$$

Since q, u, and f are integrable functions, the right-hand side of this equation approaches zero as ε approaches zero; and therefore pu' is continuous across x_i. Now write (3.64) in an integral form:

$$u' = \frac{p(0)u'(0)}{p(x)} + \frac{1}{p(x)} \int_0^x (qu + f)\, dt.$$

This is valid if $p(x) \neq 0$. If we integrate again across x_i we get

$$u\Big|_{x_i - \varepsilon}^{x_i + \varepsilon} = \int_{x_i - \varepsilon}^{x_i + \varepsilon} \frac{p(0)u'(0)}{p(x)}\, dx + \int_{x_i - \varepsilon}^{x_i + \varepsilon} \frac{dx}{p(x)} \int_0^x (qu + f)\, dt.$$

The integrals again approach zero as ε approaches zero, and therefore $u(x)$ also is continuous at $x = x_i$.

The same conclusions about the continuity of $u(x)$ and pu' may be obtained by a different argument. Suppose that $u(x)$ has a jump of magnitude α at $x = x_i$; then we can write

$$u(x) = \alpha H(x - x_i) + v(x),$$

where $v(x)$ is continuous at $x = x_i$. We would now find that Lu is not a piecewise continuous function but instead contains the derivative of a delta function. Similarly, if pu' has a jump discontinuity, Lu contains a delta function. Since Lu is given as a piecewise continuous function, it follows that both u and pu' must be continuous. We state our conclusions in the following

Rule. *The discontinuity conditions for the operator* $Lu = -(pu') + qu$ *are that both* u *and* pu' *be continuous.*

This rule has an important consequence which is based on the concept of the *impedance* of a solution. We define the impedance of a solution $u(x)$ of the equation $Lu = -f(x)$ to be the ratio pu'/u. Then the above rule implies that *the impedance is continuous across a discontinuity.*

It should be noted that caution is needed in the application of the above rule for discontinuity conditions. After all, the discontinuity conditions are part of the physical problem and cannot be obtained by mathematical considerations alone. However, if the mathematical equations are properly formulated, the rule will give the correct discontinuity conditions.

To understand what is meant by a proper formulation, consider the case of one-dimensional electromagnetic wave propagation. In general, two kinds of propagation are possible; they are called transverse electric (TE) and transverse magnetic (TM) propagation. In both TE and TM propagation we must solve the equation

$$u'' + \omega^2 \varepsilon \mu u = 0,$$

where ω is the frequency of the wave, μ the magnetic permeability, and ε the dielectric constant of the medium. Suppose that we are interested in the transmission of a wave across the interface of two media, for each of which ε has a constant value, but different in each medium, whereas μ has the same constant value in both media. For TE propagation the discontinuity conditions are that u and u' be continuous across the interface, but for TM propagation the discontinuity conditions are that u and $\varepsilon u'$ be continuous. This latter case seems to contradict our rule. However, if we scrutinize the physical situation we observe that for the TE case the equation

$$u'' + \omega^2 \varepsilon \mu u = 0$$

is correct even when ε varies continuously, but for the TM case the correct equation if ε varies continuously is

$$\frac{1}{\varepsilon}(\varepsilon u')' + \omega^2 \varepsilon \mu u = 0.$$

If we apply the rule to this latter case we get the correct discontinuity conditions, namely, u and $\varepsilon u'$ continuous.

From this illustration we deduce the requirement that *the mathematical equations should be so formulated that they are valid for continuous changes of the parameters specified in the problem.* If they are so formulated, the rule of this section will give the physically correct discontinuity conditions.

We shall give a final example. Consider the equation

$$u'' + \alpha\delta(x - x_0)u = 0.$$

What is to be understood by this equation? In order that the term $u(x)$ times the delta function should make sense, $u(x)$ must be continuous at $x = x_0$. Integrate the equation from $x_0 - \varepsilon$ to $x_0 + \varepsilon$. We get

$$u'\Big|_{x_0-\varepsilon}^{x_0+\varepsilon} + \alpha u(x_0) = 0;$$

therefore the jump in $u'(x)$ at $x = x_0$ must equal $-\alpha u(x_0)$. For values of $x \neq x_0$, the value of the delta function is zero, and the equation reduces to

$$u'' = 0.$$

We conclude that the equation is to be interpreted as follows: The function u is a solution of $u'' = 0$ for all values of x different from x_0. At $x = x_0$, the function u is continuous, but its derivative has a jump of magnitude $-\alpha u(x_0)$. With these specifications the solution of the equation is completely determined as soon as two boundary conditions are given.

PROBLEMS

3.33. Find the Green's function for $Lu = k^2u - u''(-\infty < x < \infty)$, when k is a constant k_1 for $x < 0$, another constant k_2 for $x > 0$, and when the boundary conditions are that u vanish at both $\pm\infty$.

3.34. Find the general solution of the equation

$$u'' + \alpha\delta(x - x_0)u = 0.$$

Wave Propagation and Scattering

The ideas of the preceding section have important applications in both quantum mechanics and electromagnetic theory. Before discussing these applications, we shall present some of the simpler mathematical aspects of the theory of wave propagation. Consider the differential equation

$$(pu')' + qu = 0 \tag{3.65}$$

over the interval $(-\infty, \infty)$ and suppose that the functions $p(x)$ and $q(x)$ approach the constant values p_0 and q_0, respectively, as x approaches plus infinity. Put $k_0^2 = q_0p_0^{-1}$; then it is shown in Appendix II that as x approaches infinity, there exists a solution $u_1(x)$ of (3.65) which approaches e^{ik_0x} and another solution $u_2(x)$ of (3.65) which approaches e^{-ik_0x}. Since (3.65) is a second-order equation and since u_1 and u_2 are linearly independent, it follows that any solution $u(x)$ of (3.65) can be expressed as $\alpha u_1 + \beta u_2$, where α and β are constants.

If a solution $u_+^{(1)}(x)$ of (3.65) is such that $\beta = 0$, we shall say that $u_+^{(1)}(x)$ behaves like an *outgoing wave* at plus infinity. The reason for the terminology is this: If we introduce a time factor $e^{-i\omega t}$, the function $u_+^{(1)}(x)e^{-i\omega t}$ behaves at $x = \infty$ like the function $\alpha e^{\,i(k_0x-\omega t)}$, and this latter represents a wave disturbance traveling from left to right. Similarly, a solution $u_+^{(2)}(x)$ of (3.65) such that $\alpha = 0$ is said to be an *incoming wave* at plus infinity because $u_+^{(2)}(x)e^{-i\omega t}$ behaves like $\beta e^{-i(k_0x+\omega t)}$ which represents a wave disturbance going from right to left.

In an analogous way, if p and q approach constant values p_1 and q_1, respectively, as x approaches $-\infty$, we may define outgoing and incoming waves at minus infinity. Put $k_1^2 = q_1p_1^{-1}$. Let $w_1(x)$ be that solution of (3.65) which approaches e^{-ik_1x} as x approaches $-\infty$, and let $w_2(x)$ be that solution of (3.65) which approaches e^{ik_1x} as x approaches $-\infty$; then any solution $u(x)$ of (3.65) may be expressed as $\alpha_1w_1(x) + \beta_1w_2(x)$ where α_1 and β_1 are constants. If a solution $u_-^{(1)}(x)$ is such that $\beta_1 = 0$, then $u_-^{(1)}(x)e^{-i\omega t}$ behaves like $\alpha_1e^{-i(k_1x+\omega t)}$, and this represents a wave going from right to left. We shall say $u_-^{(1)}(x)$ behaves like an outgoing wave at minus infinity. Similarly, the solution $u_-^{(2)}(x)$ such that $\alpha_1 = 0$ will be said to behave like an incoming wave at minus infinity, traveling from left to right.

These concepts of incoming and outgoing waves are used to treat physical problems in which a disturbance (for example, a light wave) arrives from a very large distance (infinity), is affected by near-by conditions (for example, a pane of glass), and is then sent off to a very large distance (infinity again). The behavior of this light wave might be treated mathematically in the manner shown below.

Assume that a plane wave of light with frequency $\omega/(2\pi)$ moves along the x-axis from $x = -\infty$ to $x = +\infty$, and assume that the glass pane extends from $x = -a$ to $x = a$, with the rest of the x-axis representing air. Let $u(x)$ be the amplitude of the light wave; then $u(x)$ is a solution of the equation

$$u'' + k^2u = 0,$$

where $k^2 = \omega^2/c_a^2$ for $x < -a$ and $x > a$, and where $k^2 = \omega^2/c_g^2$ for $-a < x < a$. In these formulas, c_a and c_g are the velocities of light in air and glass, respectively.

The function $u(x)$ is not yet completely determined since we have not specified the boundary conditions that it must satisfy. The boundary conditions will be obtained from the physical situation, namely, a light wave comes in from $-\infty$, is affected by the glass pane, and the light wave then goes to ∞. It seems then that we should specify $u(x)$ as a function which starts out as an incoming wave at $-\infty$ and ends up as an outgoing wave at ∞. This specification of $u(x)$ would be appropriate if we were interested in the complete history of how an incoming wave is transformed into an outgoing wave. For many applications, however, we want to have a description of the behavior of the light wave only after a steady state has been reached, that is, we assume the light wave is steadily coming in from $-\infty$ for a time long enough for the physical situation to settle down to a state of equilibrium. In this steady state we cannot require that $u(x)$ behave like an incoming wave at $-\infty$ because the glass pane, besides transmitting light to ∞, will also reflect some light back to $-\infty$; consequently, the only boundary condition we can impose on $u(x)$ is that it behave like an outgoing wave at ∞. This boundary condition is homogeneous and in general will determine $u(x)$ uniquely except for a multiplicative constant. The value of this constant is usually obtained from the strength of the source which emits the light waves.

Problems similar to the one we have described about the light wave occur in many different branches of physics. We shall present a general formulation for these problems.

Consider again (3.65) over $(-\infty, \infty)$. Put $v = -pu'$; then we may write (3.65) as follows:

$$v' = qu. \tag{3.66}$$

This equation, together with the equation

$$u' = -p^{-1}v, \tag{3.67}$$

are examples of what Schelkunoff† has called *transmission-line equations*. We shall list a few situations in which these transmission-line equations occur.

(1) In the propagation of electric waves along transmission lines, the coordinate x will represent distance along the line, the quantity u will be the voltage difference between the transmission lines at x, and the quantity v will be the current through the line at x. In this case the coefficients p^{-1} and q are known as the distributed series impedance and the distributed shunt admittance of the line.

(2) The periodic vibration of strings under constant tension may be described by (3.66) and (3.67) if we identify u as the force on a typical point of the string at right angles to the string and v as the velocity at that point. Here, $p^{-1} = r - i\omega\rho$, $q = -i\omega T^{-1}$, where ω is 2π times the frequency of vibration, r is the resistance per unit length, ρ is the density, and T is the tension of the string. Note that in this illustration as in the others we shall assume a time factor of the form $e^{-i\omega t}$.

† Schelkunoff, *Bell System Technical Journal*, vol. XVII, p. 17, 1938.

(3) The one-dimensional motion of periodic sound waves† in a fluid will be described by (3.66) and (3.67) if we take u as the pressure in the fluid at any point and v as the velocity at that point. Here, $p^{-1} = -i\omega\rho$, $q = -(\rho c^2)^{-1}i\omega$, where ρ is the density of the fluid and c is the velocity of sound in the fluid.

(4) Transmission of periodic heat waves will also be a special case of (3.66) and (3.67) if we consider u as the temperature and v as the rate of heat flow. Here $p = K$ and $q = -i\omega c\delta$, where K is the thermal conductivity, δ is the density, and c the specific heat.

(5) Maxwell's equations reduce to (3.66) and (3.67) in the case of one-dimensional periodic plane polarized electromagnetic waves if we consider u as the electric intensity and v as the magnetic intensity. We find $p^{-1} = -i\omega\mu$, $q = \sigma - i\omega\varepsilon$, where σ, ε, μ are respectively, the conductivity, dielectric constant, and the magnetic permeability of the medium in which the waves propagate.

(6) In quantum mechanics,‡ (3.65) will be the time-independent Schrodinger's equation for a particle. We may interpret u as the probability amplitude and v as the velocity probability amplitude if we put

$$p = \hbar(im)^{-1}, \quad q = i(E - V)(2\hbar)^{-1},$$

where $\hbar$ is Planck's constant divided by 2π, m is the mass and E the total energy of the particle, and V is the potential energy of the force field in which the particle is moving.

Let us return to (3.66) and (3.67) and let us formulate the problem we wish to consider. We shall assume that $p(x)$ and $q(x)$ are piecewise constant functions of x. This assumption will enable us to discuss many interesting physical situations such as, for example, the transmission of light through a pane of glass. Even when $p(x)$ and $q(x)$ are not piecewise constant functions, they may be approximated by such functions. Suppose then that

$$\begin{aligned} p(x) &= p_0, \quad -\infty < x < x_1; & q(x) &= q_0, \quad -\infty < x < x_1 \\ &= p_1, \quad x_1 < x < x_2; & &= q_1, \quad x_1 < x < x_2 \\ &\cdots\cdots & &\cdots\cdots \\ &= p_{n-1}, \quad x_{n-1} < x < x_n; & &= q_{n-1}, \quad x_{n-1} < x < x_n \\ &= p_n, \quad x_n < x < \infty; & &= q_n, \quad x_n < x < \infty, \end{aligned}$$

where $p_0, p_1, \cdots, p_n, q_0, q_1, \cdots, q_n$ are constants. This formulation would correspond to the problem of wave propagation in $n + 1$ homogeneous

† Morse, *Vibration and Sound*, 2nd Edition, p. 222, McGraw-Hill Book Co., New York, 1948.

‡ Schiff, *Quantum Mechanics*, pp. 19–21, McGraw-Hill Book Co., New York, 1949.

media. The region from $-\infty$ to x_1 we shall call medium 0, from x_1 to x_2 medium 1, $\cdots$, and from x_n to ∞ medium n. In most physical problems, as can be seen from the illustrations given above, the p's and q's will be pure imaginary numbers.

Consider, first, (3.66) and (3.67) for $x < x_0$. They are

$$u' = -p_0^{-1}v, \quad v' = q_0 u. \tag{3.68}$$

If u is eliminated from these equations, we get

$$v'' = -(q_0/p_0)v.$$

The general solution of this equation is

$$v = A_0 e^{ik_0 x} + B_0 e^{-ik_0 x},$$

where $k_0^2 = q_0/p_0$. We shall call k_0 the *propagation constant* in medium 0. From (3.68) we find that the value of v corresponding to this expression for u is

$$u = Z_0(A_0 e^{ik_0 x} - B_0 e^{-ik_0 x}),$$

where Z_0 is that square root of $(-p_0 q_0)^{-1}$ which has a positive real part. We shall call the ratio u/v, which equals

$$Z_0 \frac{1 - B_0 A_0^{-1} e^{-2ik_0 x}}{1 + B_0 A_0^{-1} e^{-2ik_0 x}}, \quad x < x_0,$$

the *impedance of the wave motion* at the point x. The quantity Z_0 will be called the *characteristic impedance of the medium* 0.

Similarly, in medium m, $(0 \leq m \leq n)$, where $x_m < x < x_{m+1}$, the wave motion would be described by the equations

$$\begin{aligned} v &= A_m e^{ik_m x} + B_m e^{-ik_m x} \\ & \qquad\qquad\qquad\qquad x_m < x < x_{m+1} \\ u &= Z_m(A_m e^{ik_m x} - B_m e^{-ik_m x}) \end{aligned} \tag{3.69}$$

where the propagation constant is given by the formula

$$k_m^2 = q_m/p_m, \tag{3.70}$$

and the characteristic impedance is given by the formula

$$Z_m^2 = (-p_m q_m)^{-1}. \tag{3.71}$$

Formulas (3.69) do not completely specify the wave motion. As we have seen in the preceding section or, what is equivalent, from physical considerations, a solution of (3.65) must satisfy certain discontinuity conditions at the interface between the different media, that is, at the points $x = x_1, \cdots, x_n$. In this case, the discontinuity conditions are that u and v and therefore the impedance ratio u/v be continuous at those points. The condition that the impedance be continuous at $x = x_m$ gives

$$Z_{m-1} \frac{1 - B_{m-1}A_{m-1}^{-1} e^{-2ik_{m-1}x_m}}{1 + B_{m-1}A_{m-1}^{-1} e^{-2ik_{m-1}x_m}} = Z_m \frac{1 - B_m A_m^{-1} e^{-2ik_m x_m}}{1 + B_m A_m^{-1} e^{-2ik_m x_m}}.$$

Put

$$z_m = Z_m/Z_{m-1} \tag{3.72}$$
$$R_m^- = B_m A_m^{-1} e^{-2ik_m x_m},$$
$$R_m^+ = B_m A_m^{-1} e^{-2ik_m x_{m+1}};$$

then the above impedance matching equation becomes

$$\frac{1 - R_{m-1}^+}{1 + R_{m-1}^+} = z_m \frac{1 - R_m^-}{1 + R_m^-}.$$

Solving this for R_{m-1}^+, we find that

$$R_{m-1}^+ = \frac{r_m + R_m^-}{1 + r_m R_m^-} \tag{3.73}$$

where

$$r_m = \frac{1 - z_m}{1 + z_m}. \tag{3.74}$$

The quantities R_m^+, R_m^-, and r_m have an important physical significance. Consider the quantity $v(x)$ as defined in (3.69). This formula is composed of a wave $A_m e^{ik_m x}$ moving to the right and a wave $B_m e^{-ik_m x}$ moving to the left. (This is a consequence of our assumption that all functions have a time factor of the form $e^{-i\omega t}$.) We shall define the *reflection coefficient* of the wave as the ratio of the amplitude of the wave moving towards the left to the amplitude of the wave moving towards the right, namely, the ratio $B_m A_m^{-1} e^{-2ik_m x}$. We see then that R_m^- is the reflection coefficient at the left endpoint of medium m and that R_m^+ is the reflection coefficient at the right endpoint of medium m.

To understand the significance of r_m, suppose that $n = 1$; this means that there are only two media 0 and 1 to consider. Suppose that a wave $e^{ik_0 x}$ starts from $-\infty$. When it meets the discontinuity at $x = x_1$, how much will be reflected back to $-\infty$ and how much will be transmitted to medium 1? Since we have a left-going and a right-going wave in medium 0 but only a right-going wave in medium 1, the wave motion can be written as follows:

$$v = e^{ik_0 x} + re^{2ik_0 x_1} e^{-ik_0 x}, \quad x < x_1;$$
$$= Te^{+ik_1 x} e^{i(k_0 - k_1)x_1}, \quad x > x_1.$$
$$u = Z_0(e^{ik_0 x} - re^{2ik_0 x_1} e^{-ik_0 x}), \quad x < x_1;$$
$$= Z_1 T e^{ik_1 x} e^{i(k_0 - k_1)x_1}, \quad x > x_1.$$

The quantity r is the *reflection coefficient* for a wave going from medium 0 to medium 1 and the quantity T is the *transmission coefficient.*

Expressing the condition that the impedance ratio must be continuous at $x = x_1$, we see that

$$\frac{1 - r}{1 + r} = \frac{Z_1}{Z_0} = z_1.$$

Solving for r, we get

(3.75) $$r = \frac{1 - z_1}{1 + z_1}.$$

This formula expresses the reflection coefficient for a wave going from medium 0 to medium 1 in terms of the quantity z_1, which is the ratio of the characteristic impedance for medium 1 to the characteristic impedance for medium 0. Since (3.74) has the same form as (3.75), we conclude that r_m is the reflection coefficient for a wave going from medium $m - 1$ to medium m if medium m extends to infinity. This conclusion could have been obtained also from formula (3.73). If medium m extends to infinity, a wave in medium m will not be reflected and consequently $R_m^- = 0$. Substituting this in (3.73), it follows that $R_{m-1}^+ = r_m$.

Formulas (3.73) and (3.74) are consequences of one discontinuity condition, namely, the continuity of the impedance ratio. If we use another discontinuity condition such as the continuity of $v(x)$ at $x = x_m$, we get

$$A_{m-1}e^{ik_{m-1}x_m}(1 + R_{m-1}^+) = A_m e^{ik_m x}(1 + R_m^-).$$

Put

$$T_m = A_m A_{m-1}^{-1} e^{i(k_m - k_{m-1})x_m};$$

then this equation becomes

(3.76) $$T_m = \frac{1 + R_{m-1}^+}{1 + R_m^-} = \frac{1 + r_m}{1 + r_m R_m^-}.$$

The quantity T_m is called the *transmission coefficient* for a wave going from medium $m - 1$ to medium m.

We may summarize the results of our discussion in

Theorem 3.8. *The general solution of* (3.66) *and* (3.67) *when $p(x)$ and $q(x)$ are piecewise constant functions of x may be written as follows:*

(3.77)
$$\begin{aligned} v &= A_0 e^{ik_0 x}(1 + R_0^+ e^{-2ik_0(x - x_1)}), \quad x < x_1 \\ &= T_1 A_0 e^{ik_0 x_1} e^{ik_1(x - x_1)}(1 + R_1^- e^{-2ik_1(x - x_1)}), \quad x_1 < x < x_2 \\ &\quad \cdots\cdots\cdots\cdots\cdots\cdots \\ &= T_1 T_2 \cdots T_p\, A_0 \exp[ik_0 x_1 + i\sum_1^{p-1} k_m (x_{m+1} - x_m) \\ &\qquad + ik_p(x - x_p)](1 + R_p^- e^{-2ik_p(x - x_p)}),\ x_p < x < x_{p+1} \\ &\quad \cdots\cdots\cdots\cdots\cdots\cdots \\ &= T_1 T_2 \cdots T_n A_0 \exp\,[ik_0 x_1 + i\sum_1^{n-1} k_m(x_{m+1} - x_m) \\ &\qquad + ik_n(x - x_n)], \quad x_n < x. \end{aligned}$$

Here we have

$$R_{m-1}^{+} = \frac{r_m + R_m^-}{1 + r_m R_m^-}, \tag{3.78}$$

and

$$R_m^+ = R_m^- e^{-2ik_m(x_{m+1}-x_m)}, \tag{3.79}$$

with

$$r_m = \frac{1 - z_m}{1 + z_m}, \tag{3.80}$$

where $z_m = Z_m/Z_{m-1}$. *Also,*

$$T_m = \frac{1 + R_{m-1}^+}{1 + R_m^-} = \frac{1 + r_m}{1 + r_m R_m^-}. \tag{3.81}$$

The formula for u in the mth medium is obtained from the corresponding formula for v by multiplying it with Z_m and by changing R_m^- to $-R_m^-$.

Despite its complicated mathematical appearance, Theorem 3.8 has the following simple physical interpretation: In the mth region where $x_m < x < x_{m+1}$ there are two waves present, one going to the right, the other going to the left. To reach this region, the wave must have been transmitted across the $m - 1$ regions for which $x < x_m$. In going across the jth region, the wave has its amplitude multiplied by T_j and its phase increased by $k_j(x_{j+1} - x_j)$; consequently, in the mth region, there is a wave

$$v_+ = A_m \exp [ik_m(x - x_m)]$$

where

$$A_m = A_0 T_1 \cdots T_m \exp [ik_0 x_1 + ik_1(x_2 - x_1) + \cdots + ik_{m-1}(x_m - x_{m-1})]$$

going to the right. This wave goes to $x = x_{m+1}$ and is reflected back, producing the following wave going to the left:

$$v_- = R_m^- A_m \exp [-ik_m(x - x_m)].$$

The total wave in the mth region is then the sum of these two waves, and we have

$$v = v_- + v_+.$$

We shall apply Theorem 3.8 to a problem similar to the one with which we started, namely, the behavior of a light wave passing through a pane of glass. In this case, there are three media present: air, glass, and air again. We shall consider the transmission of a wave through three media bounded by interfaces at $x = -a$ and $x = a$. We know also that the first and third media are identical and that they have the propagation constant k_0 and characteristic impedance Z_0 while the medium in the middle has the propagation constant k_1 and the characteristic impedance Z_1. If a wave starts at $-\infty$ in the first medium and travels to the right, there will be no reflection in the third medium; consequently, $R_2^- = 0$. From (3.78), (3.79), and (3.80) we find that

$$R_1^+ = r_2 = \frac{1 - z_2}{1 + z_2} = \frac{Z_1 - Z_0}{Z_1 + Z_0},$$
$$R_1^- = R_1^+ e^{4ik_1a};$$

then, using these values, we get

$$R_0^+ = \frac{r_1 + R_1^-}{1 + r_1 R_1^-} = \frac{r_1 + r_2 e^{4ik_1a}}{1 + r_1 r_2 e^{4ik_1a}}.$$

Since

$$r_1 = \frac{1 - z_1}{1 + z_1} = \frac{Z_0 - Z_1}{Z_0 + Z_1} = -r_2,$$

we have

$$R_0^+ = r_1 \frac{1 - e^{4ik_1a}}{1 - r_1^2 e^{4ik_1a}}.$$

From (3.81) we find that

$$T_2 = 1 + r_2,$$
$$T_1 = \frac{1 + R_0^+}{1 + r_2 e^{4ik_1a}}.$$

Finally, we obtain the following formula:

$$\begin{aligned} v &= A_0 e^{ik_0x}\left(1 + r_1 \frac{1 - e^{4ik_1a}}{1 - r_1^2 e^{4ik_1a}} e^{-2ik_0(x+a)}\right), \quad x < -a \\ &= T_1 A_0 e^{-ik_0a} e^{ik_1(x+a)}(1 - r_1 e^{-2ik_1(x-a)}), \quad -a < x < a \\ &= T_1 T_2 A_0 e^{i(2k_1-k_0)a} e^{ik_0(x-a)}, \quad a < x. \end{aligned}$$

There will be a similar formula for u.

There is another method of solving these problems which may be more convenient to use in practice. The essence of this method is to work with trigonometrical solutions of (3.66) and (3.67) instead of exponential solutions. From (3.73) it follows that, if $R_{m-1}^+ = \pm 1$, then also $R_m^- = \pm 1$. From (3.69) we see that, if $R_m^- = 1$, then v is proportional to $\cos k_m (x - x_m)$, that is, a cosine wave centered at the interface. Similarly, if $R_m^- = -1$, then v is proportional to $\sin k_m (x - x_m)$, that is, a sine wave centered at the interface.

Using (3.76) and (3.74), we deduce the following rule for centered cosine and sine waves: *A centered cosine wave such as* $v = \cos k_{m-1}(x - x_m)$ *for* $x_{m-1} < x < x_m$ *remains centered to become* $v = \cos k_m(x - x_m)$ *for* $x_m < x < x_{m+1}$. *A centered sine wave such as* $v = \sin k_{m-1}(x - x_m)$ *for* $x_{m-1} < x < x_m$ *becomes* $v = z_m^{-1} \sin k_m(x - x_m)$ *for* $x_m < x < x_{m+1}$.

We shall use these facts to obtain again the solution for the behavior of a light wave passing through a pane of glass. Since there is no reflection in the third medium, the wave there is purely outgoing, and consequently it may be represented by $v = e^{ik_0(x-a)} = \cos k_0(x - a) + i \sin k_0(x - a)$, for $x > a$. By the above rule we find that $v = \cos k_1(x - a) + iz_1^{-1} \sin k_1(x - a)$, $-a < x < a$, where $z_1 = Z_1/Z_0$. To find v in the first medium, we replace the cosine and sine waves which have been centered at $x = a$ by cosine and sine waves centered at $x = -a$ as follows:

$$\cos k_1(x - a) = \cos [k_1(x + a) - 2k_1a] = \cos 2k_1a \cos k_1(x + a) + \sin 2k_1a \sin k_1(x + a),$$

$$\sin k_1(x - a) = -\sin 2k_1a \cos k_1(x + a) + \cos 2k_1a \sin k_1(x + a).$$

Now, applying the rule again, we find that

$$v = (+\cos 2k_1a - iz_1^{-1} \sin 2k_1a) \cos k_0(x + a) + (\sin 2k_1a + iz_1^{-1} \cos 2k_1a)z_1 \sin k_0(x + a),\ x < -a.$$

Note that we have used the fact that the relative impedance in going from the glass to the air is $Z_0/Z_1 = z_1^{-1}$.

We leave it to the reader to show that these formulas are the same as those obtained previously.

PROBLEM

3.35. Suppose that in a medium with propagation constant k_0 and characteristic impedance Z_0, a slab of thickness $2a$ is introduced. The slab has propagation constant k_1 and characteristic impedance Z_1. Suppose that at $-\infty$ there is an incoming wave $A_-e^{ik_0x}$ and also at $+\infty$ there is an incoming wave $A_+e^{-ik_0x}$ for v. Find the complete expression for v in the whole region. (*Hint.* Suppose that $v = A_-e^{ik_0x} + B_-e^{-ik_0x}$ at $x = -\infty$ and $v = A_+e^{-ik_0x} + B_+e^{ik_0x}$ at $x = +\infty$. Express the vector (B_-, B_+) in terms of the vector (A_-, A_+).) The matrix which transforms the A vector into the B vector is called the *scattering* matrix of the slab. Show that

$$|A_+|^2 + |A_-|^2 = |B_+|^2 + |B_-|^2.$$

APPENDIX I

EXISTENCE OF A SOLUTION FOR EQUATION (3.30)

We now wish to justify the statement that the solution of (3.30) is a continuously differentiable function of x. We shall prove a theorem which will contain this result as a special case.

Theorem 3.A.I. *Let $p(x)$, $q(x)$, and $f(x)$ be piecewise continuous functions of x in the closed interval* (0, 1), *and let $p(x)$ be positive in that interval; then there exists a continuous function $u(x)$ such that $u(0) = u'(0) = 0$ and such that pu' exists and is continuous for all x; furthermore*

$$(pu')' - qu = f(x) \tag{3.A.1}$$

for all values of x for which both sides are continuous functions of x.

The proof will illustrate how the Green's function can be used to transform a differential equation into an integral equation. Equation (3.A.1) may be written in the form

$$Lu = (pu')' = qu + f(x) = h(x).$$

Suppose that $h(x)$ were known; then u would be found by using the Green's function for L, just as the solution of (3.30) was given by (3.31).

Consider the equation defining the Green's function, namely,

$$(pg')' = -\delta(x - t) \tag{3.A.2}$$

with $g(0) = g'(0) = 0$. It can be solved by direct integration. We have

$$pg' = -H(x - t) + c,$$

where c, the constant of integration, must be zero, since $g'(0) = 0$. Put

$$r(x) = \int_0^x \frac{ds}{p(s)};$$

a second integration will then give

$$g(x) = [r(t) - r(x)]H(x - t).$$

Multiply (3.A.2) by $-h(t)$ and integrate from 0 to 1. We find that the solution of $Lu = h$ is

$$u = \int_0^x [r(x) - r(t)]h(t)\,dt. \tag{3.A.3}$$

Since $h(t) = q(t)u(t) + f(t)$ is not a known function of t, (3.A.1) becomes the following integral equation for $u(x)$:

$$u(x) = \int_0^x f(t)[r(x) - r(t)]\,dt + \int_0^x q(t)u(t)[r(x) - r(t)]\,dt. \tag{3.A.4}$$

We have thus shown that if (3.A.1) has a solution $u(x)$, this solution will satisfy the integral equation (3.A.4). We now show the converse. Suppose that (3.A.4) has a solution $u(x)$. Note that from its definition $r(x)$ is a continuous function; consequently, the right-hand side of (3.A.4), and therefore also $u(x)$, is a continuous function of x. More, however, is true. When we differentiate the right-hand side of (3.A.4) we find that

$$u'(x) = \int_0^x f(t)\,dt/p(x) + \int_0^x q(t)u(t)\,dt/p(x).$$

This equation shows that for all x such that $p(x)$ is continuous, $u'(x)$ exists and

$$p(x)u'(x) = \int_0^x f(t)\,dt + \int_0^x q(t)u(t)\,dt. \tag{3.A.5}$$

Since the right-hand side of this equation is a continuous function of x for all x, then at the points where $p(x)$ is discontinuous we may define $u'(x)$ so that $p(x)u'(x)$ will be continuous everywhere.

Because the integrands in (3.A.5) are piecewise continuous functions of x, we may differentiate the right-hand side of (3.A.5). We find that

$$(pu')' = f(x) + q(x)u(x)$$

whenever the right-hand side is continuous. This shows that u satisfies (3.A.1). From (3.A.5) it is clear that $u'(0) = 0$. This shows that (3.A.4) is completely equivalent to (3.A.1) and its boundary conditions.

In order to complete the proof of Theorem 3.A.I we need only show that there exists a solution of (3.A.4). At first glance, this does not seem to be a simplification since (3.A.4) seems more complicated than the original equation (3.A.1). However, (3.A.4) is actually much better suited for our purpose because it contains an integral operator, which is *bounded*, whereas (3.A.1) contains a differential operator, which is always unbounded.

The technique which we shall use to prove the existence of a solution of (3.A.4) is the same as that used in proving Theorem 1.2.

Put

$$u_1(x) = \int_0^x f(t)[r(x) - r(t)]\,dt,$$

and, for $n = 1, 2, \cdots$, put

$$u_{n+1}(x) = \int_0^x f(t)[r(x) - r(t)]\,dt + \int_0^x q(t)[r(x) - r(t)]u_n(t)\,dt.$$

We have

$$u_{n+1}(x) - u_n(x) = \int_0^x q(t)[r(x) - r(t)][u_n(t) - u_{n-1}(t)]\,dt,$$

and therefore

$$|u_{n+1}(x) - u_n(x)| \leq \int_0^x |q(t)||r(x) - r(t)||u_n(t) - u_{n-1}(t)|\,dt. \tag{3.A.6}$$

In order to estimate the right-hand side we consider the function

$$R(x) = \int_0^x [r(x) - r(t)]\, dt = \int_0^x dt \int_t^x \frac{ds}{p(s)} = \int_0^x \frac{s}{p(s)}\, ds \tag{3.A.7}$$

where we have used the previously given definition of $r(x)$, namely,

$$r(x) = \int_0^x \frac{ds}{p(s)}.$$

Since $p(s)$ is positive, it follows that both $r(x)$ and $R(x)$ are non-negative and increasing functions of x. Note that

$$\int_0^x R(t)^k[r(x) - r(t)]\, dt = \int_0^x dt\, R(t)^k \int_t^x \frac{ds}{p(s)} = \int_0^x \frac{ds}{p(s)} \int_0^s dt\, R(t)^k.$$

Since $R(t)$ is an increasing function of x,

$$\int_0^s dt\, R(t)^k \leq R(s)^k \int_0^s dt = sR(s)^k;$$

consequently, if we use the fact that from (3.A.7), $R'(x) = x/p(x)$, we have

$$\begin{aligned} \int_0^x R(t)^k[r(x) - r(t)]\, dt &\leq \int_0^x \frac{s\, ds}{p(s)} R(s)^k = \int_0^x R'(s)\, ds\, R(s)^k \\ &= \frac{R(x)^{k+1}}{k+1}. \end{aligned} \tag{3.A.8}$$

We can now get an estimate of $|u_{n+1}(x) - u_n(x)|$ from (3.A.6). First, we notice that since $p(x)$, $q(x)$, and $f(x)$ are piecewise continuous functions of x, they must be bounded in absolute value. Also, since $p(x)$ is positive in the interval (0, 1), it is greater than some fixed constant, and consequently $r(x)$ is bounded. Let m be a bound for the functions $p(x)$, $q(x)$, $f(x)$, and $r(x)$; then, from the definition, we have

$$|u_1(x)| \leq \int_0^x |f(t)|[r(x) - r(t)]\, dt \leq mR(x),$$

and from (3.A.6) and (3.A.8) we find that

$$\begin{aligned} |u_2(x) - u_1(x)| &\leq \int_0^x |q(t)|[r(x) - r(t)]mR(t)\, dt \\ &\leq m^2 \int_0^x R(t)[r(x) - r(t)]\, dt \leq \frac{m^2 R(x)^2}{2}. \end{aligned}$$

Now we use mathematical induction. Suppose that for $n = 1, 2, \cdots, k$,

$$|u_n(x) - u_{n-1}(x)| \leq \frac{m^n R(x)^n}{n!};$$

we shall prove the same result holds for $n = k + 1$. From (3.A.6) and (3.A.8) we have

$$\begin{aligned} |u_{k+1}(x) - u_k(x)| &\leq \int_0^x |q(t)|[r(x) - r(t)]m^k R(t)^k\, dt/k! \\ &\leq \frac{m^{k+1}}{k!} \int_0^x R(t)^k[r(x) - r(t)]\, dt \leq \frac{m^{k+1} R(x)^{k+1}}{(k+1)!}. \end{aligned}$$

Consider the infinite series

$$u_1(x) + [u_2(x) - u_1(x)] + [u_3(x) - u_2(x)] + \cdots.$$

We have just shown that each term of this series is less in absolute value than the corresponding term of the series:

$$mR(x) + \frac{m^2R(x)^2}{2!} + \frac{m^3R(x)^3}{3!} + \cdots = -1 + \exp[mR(x)].$$

This latter series converges absolutely and uniformly for all values of x in (0, 1); consequently, the original series converges absolutely and uniformly to some limit function $u(x)$. Since the nth partial sum of the original series is $u_n(x)$, we have

$$u(x) = \lim u_n(x)$$

as n approaches infinity.

All that remains to be shown is that $u(x)$ satisfies (3.A.4). Consider the equation defining $u_{n+1}(x)$ and let n approach infinity. We find that

$$u(x) = \int_0^x f(t)[r(x) - r(t)]\,dt + \int_0^x q(t)[r(x) - r(t)]\,dt,$$

which is (3.A.4). Consequently, we have proved Theorem 3.A.I.

APPENDIX II

SOLUTIONS FOR LARGE VALUES OF x

We shall consider the general equation

$$u'' + (k^2 + q(x))u = 0; \tag{3.A.9}$$

and we shall show that if $q(x)$ is small enough at infinity or, more precisely, if

$$\int^{\infty} |q(x)|\, dx < \infty, \tag{3.A.10}$$

then there exist solutions of (3.A.9) which behave like $e^{\pm ikx}$ at $x = \infty$. This last phrase is to be understood in the following sense:

There exists a function $u_1(x)$ which is a solution of (3.A.9) and is such that

$$\lim_{x\to\infty} |e^{-ikx}u_1(x) - 1| = 0.$$

Similarly, there exists a function $u_2(x)$ which is a solution of (3.A.9) and is such that

$$\lim_{x\to\infty} |e^{ikx}u_2(x) - 1| = 0.$$

To prove these results substitute $u = ve^{ikx}$ in (3.A.9) and we get

$$v'' + 2ikv' + qv = 0.$$

If we can now show that this equation has a solution $v_1(x)$ which approaches unity as x goes to infinity, we may put $u_1(x) = v_1(x)e^{ikx}$ and we shall have established the existence of the function $u_1(x)$. Similarly, by putting $u = ve^{-ikx}$, we can prove the existence of $u_2(x)$.

The method of proof is similar to that used in the proof of Theorem 3.A.I. We write the differential equation in the form

$$v'' + 2ikv' = -qv$$

and consider the right-hand side as a non-homogeneous term; then by using the Green's function for the left-hand side we shall obtain an integral equation for $v(x)$.

The appropriate Green's function can be found if we recall that $v(\infty) = 1$ and $v'(\infty) = 0$. We have

$$v(x) = 1 - \int_{\infty}^{x} \frac{1 - e^{-2ik(x-\xi)}}{2ik} q(\xi)v(\xi)\, d\xi. \tag{3.A.11}$$

So far, these results were obtained on the assumption that there exists a function $u_1(x)$ and a corresponding function $v(x)$. We shall now construct a function $v(x)$ which satisfies (3.A.11). Then it can be shown by differentiation that $v(x)$ satisfies the differential equation. Since it follows from (3.A.11) that $v(x)$ converges to unity as x approaches infinity, we will have established the existence of $u_1(x)$.

The existence of a solution $v(x)$ of (3.A.11) will be established by the iteration method used in proving Theorem 3.A.I. We put

$$v_{n+1}(x) = 1 - \int_\infty^x \frac{1 - e^{-2ik(x-\xi)}}{2ik} q(\xi) v_n(\xi)\, d\xi \quad (n = 1, 2, 3 \cdots),$$
$$v_1(x) = 1,$$

and we shall show that $v_n(x)$ converges to a limit function. We have

$$v_2(x) = 1 - \int_\infty^x \frac{1 - e^{-2ik(x-\xi)}}{2ik} q(\xi)\, d\xi.$$

Note that because of (3.A.10) the integral converges for all real values of k; therefore, $v_2(x)$ is bounded for all values of x. By induction it is easy to prove in the same way that $v_n(x)$ is bounded for all values of x.

Consider the difference $v_{n+1}(x) - v_n(x)$. We have

$$\text{(3.A.12)} \quad v_{n+1}(x) - v_n(x) = -\int_\infty^x \frac{1 - e^{-2ik(x-\xi)}}{2ik} q(\xi)[v_n(\xi) - v_{n-1}(\xi)]\, d\xi \quad (n = 2, 3, \cdots)$$

$$v_2(x) - v_1(x) = -\int_\infty^x \frac{1 - e^{-2ik(x-\xi)}}{2ik} q(\xi)\, d\xi.$$

Since $|\exp\{-2ik(x - \xi)\}| = 1$ for all real values of k, we conclude that

$$\text{(3.A.13)} \qquad |v_2(x) - v_1(x)| \le \frac{1}{k} \int_x^\infty |q(\xi)|\, d\xi.$$

Put

$$Q(x) = \int_x^\infty |q(\xi)|\, d\xi;$$

then (3.A.13) becomes

$$|v_2(x) - v_1(x)| \le \frac{Q(x)}{k}.$$

We now prove by induction that

$$\text{(3.A.14)} \qquad |v_{n+1}(x) - v_n(x)| \le \frac{Q(x)^n}{n!k^n}.$$

Suppose that this inequality holds for $n = 1, 2, \cdots, m - 1$; then from (3.A.12) we get

$$|v_{m+1}(x) - v_m(x)| \le \int_x^\infty |q(\xi)| \frac{Q(\xi)^{m-1}}{(m-1)!k^m}\, d\xi = \frac{1}{(m-1)!k^m} \int_x^\infty Q(\xi)^{m-1}\, dQ(\xi)$$
$$= \frac{Q(x)^m}{m!k^m},$$

which proves (3.A.14).

The infinite series

$$v_1(x) + (v_2(x) - v_1(x)) + (v_3(x) - v_2(x)) + \cdots \tag{3.A.15}$$

converges for all values of x because each term is less in absolute value than the corresponding term of the absolutely convergent series

$$1 + \frac{Q(x)}{k} + \frac{Q(x)^2}{2!k^2} + \cdots + \frac{Q(x)^n}{n!k^n} + \cdots.$$

Since the partial sums of the series (3.A.15) are $v_n(x)$, this proves that $v_n(x)$ converges to a limit, which we denote by $v(x)$. It is easy now to show that $v(x)$ satisfies the integral equation (3.A.11) and that $v(\infty) = 1$. As mentioned above, the existence of the function $u_1(x)$ is an immediate consequence of the existence of $v(x)$.

We state this result as

Theorem 3.A.II. *If k is real and if*

$$\int^{\infty} |q(x)|\, dx < \infty,$$

there exists a solution $u_1(x)$ of the differential equation

$$u'' + (k^2 + q)u = 0$$

such that

$$\lim_{x\to\infty} [u_1 e^{-ikx} - 1] = 0.$$

There also exists a solution $u_2(x)$ such that

$$\lim_{x\to\infty} [u_2 e^{+ikx} - 1] = 0.$$

By making suitable transformations we can apply Theorem 3.A.II to a much wider class of differential equations than its appearance suggests. For example, consider the differential equation

$$(xu')' + \left(k^2 x - \frac{\nu^2}{x}\right) u = 0. \tag{3.A.16}$$

If we put $u = x^{-1/2}w$ into (3.A.16) we get

$$w'' + \left(k^2 - \frac{\nu^2 - 1/4}{x^2}\right) w = 0.$$

From Theorem 3.A.II it is clear that this equation has solutions which behave at infinity like linear combinations of e^{ikx} and e^{-ikx}; consequently, the solutions of (3.A.16) behave at infinity like $x^{-1/2}$ times linear combinations of e^{ikx} and e^{-ikx}.

Finally, we state a result which is used in many applications.

Corollary. *The equation*

$$-(pu')' - \beta wu = 0$$

has solutions which have the following asymptotic behavior for large positive values of x:

$$u \sim (wp)^{-1/4} \exp [\pm i\beta^{1/2} \int^x (w/p)^{1/2} \, dx],$$

if

$$\int^\infty \left| g^{-1} \frac{d^2 g}{dy^2} \right| dy < \infty$$

where

$$g = (wp)^{1/4}, \quad y = \int^x (w/p)^{1/2} \, dx.$$

This corollary is proved in Problems 3.37 and 3.38.

PROBLEMS

3.36. Show that when the substitution $u = v \exp [-1/2 \int b \, dx]$ is made in the differential equation $u'' + b(x)u' + c(x)u = 0$, the resulting equation for v does not contain any first derivative. Apply this substitution to the equation $u'' + \frac{n}{x} u' + k^2 u = 0$, and then by the use of Theorem 3.A.II obtain the asymptotic behavior of u for large values of x.

3.37. Consider the equation $-(pu')' + \alpha qu - \beta wu = 0$, where α and β are constants. Put $v = gu$, $g = (wp)^{1/4}$, $y = \int^x (w/p)^{1/2} \, dx$, and then show that

$$-\frac{d^2 v}{dy^2} + (Q - \beta)v = 0, \text{ where } Q = \frac{1}{g}\frac{d^2 g}{dy^2} + \frac{\alpha q}{w}.$$

3.38. If $\int^\infty |Q| \, dy < \infty$, where Q is the function defined in Problem 3.37, then Theorem 3.A.II may be applied to the equation for v. Show that in this case we have the asymptotic formula $u = g^{-1} \exp [+ i\beta^{1/2} \int^x (w/p)^{1/2} \, dx]$.

3.39. Find the behavior of u for large values of x if $-u'' - \beta x^2 u + \alpha u = 0$. (*Hint.* Use Problem 3.38.)

3.40. Find the behavior of u for large values of x if $-u'' = (x + \lambda)u$. (*Hint.* Use Problem 3.39 with $q = 0$, $w = x + \lambda$.)

4

EIGENVALUE PROBLEMS OF ORDINARY DIFFERENTIAL EQUATIONS

Introduction

In the first chapter we saw that there are two fundamental methods of solving the linear equation

$$Lx = m,$$

where L is the linear operator, m a given vector, and x an unknown vector. One method, which was exemplified in Chapter 3, is to construct the inverse operator L^{-1}; then we have

$$x = L^{-1}m.$$

The other method is to use the spectral representation of the operator L. In this method, if we assume that the eigenvectors of L span the space, we have

$$m = \Sigma\alpha_n x_n,$$
$$x = \Sigma\beta_n x_n,$$

where x_n is an eigenvector of L corresponding to the eigenvalue λ_n, the values of the α_n are known, and those of β_n are unknown. From the properties of the eigenvectors we have

$$Lx = \Sigma\lambda_n\beta_n x_n = \Sigma\alpha_n x_n,$$

and therefore

$$\beta_n = \alpha_n/\lambda_n;$$

consequently

$$x = \Sigma\alpha_n x_n/\lambda_n.$$

In the next chapter we shall learn that a number of partial differential equations may be solved if we can assign a meaning to functions of an ordinary differential operator L. As we have already shown in Chapter 2, this interpretation may be obtained from the spectral representation of L as follows:

If $\phi(t)$ is an analytic function of t over the spectrum of L and if L has the spectral representation given above, then

$$\phi(L)x = \Sigma\phi(\lambda_n)\beta_n x_n.$$

We shall see later that a knowledge of this functional calculus for operators will help us also to obtain the spectral representation of L.

The present chapter will be devoted to a discussion of the spectral representation theory for ordinary differential operators. The framework of the theory will be the same as the abstract theory of spectral representation which was formulated in Chapter 2. However, in the application of this theory to differential operators, several difficulties appear which were not present in the application to matrices. The first of these difficulties is the question of the completeness of the eigenvectors, that is, does the set of eigenvectors of L span the domain of L? We shall show that in the case of a self-adjoint differential operator which is defined over a finite interval and which is such that the operator has no singularities in the interval, the answer is always yes, the set of eigenvectors do span the domain of the operator.

As we have seen in Chapter 2, a number λ belongs to the spectrum of the linear operator L if the operator $L - \lambda$ does not have a bounded inverse. If λ belongs to the spectrum of L, there are the two following possibilities: either there exists a non-zero vector x in the space over which L is defined such that $(L - \lambda)x = 0$, or no such vector exists. In the first case, λ is an eigenvalue of L and x is an eigenvector. In the second case, since we shall show later that a differential operator has no residual spectrum, it follows that λ is in the *continuous spectrum* of L.

The nature of the continuous spectrum is the second main difficulty in the theory of differential operators. It is important to know when an operator has a continuous spectrum, and if a continuous spectrum does exist, how it can be used to give a spectral representation of the operator. We shall see that these questions can be answered by a consideration of the Green's function of the operator. First, however, we shall discuss some examples of operators and the corresponding eigenfunction expansions.

Eigenfunction Expansion—Example 1

Let the operator L be $-\dfrac{d^2}{dx^2}$ and consider it acting on the domain of twice-differentiable functions $u(x)$ satisfying the boundary conditions

$$u(0) = u(1) = 0. \tag{4.1}$$

If $u(x)$ is an eigenfunction of L, then $u(x)$ is a solution of

$$-\frac{d^2u}{dx^2} = \lambda u, \tag{4.2}$$

which also satisfies the boundary conditions (4.1). A solution of (4.2) which satisfies the first condition in (4.1) is

$$u = \sin(\sqrt{\lambda}x).$$

For this function to satisfy the second condition in (4.1) we must have

$$\sin\sqrt{\lambda} = 0, \tag{4.3}$$

from which it follows that

$$\lambda = n^2\pi^2, \quad n = 1, 2, 3, \cdots.$$

Note that all the eigenvalues are positive. We shall see later that this could have been deduced from the general properties of L without solving (4.3). Note that $\lambda = 0$ is not an eigenvalue even though it satisfies (4.3) because the solution of (4.2) and (4.1) corresponding to the value $\lambda = 0$ is identically zero.

Using the values of λ derived above, we find that the eigenfunctions are $u_n = \sin nx$, $n = 1, 2, \cdots$. It is easy to see that

$$\int_0^1 u_n(x)u_m(x)\,dx = \tfrac{1}{2}\delta_{mn}.$$

This equation implies that the eigenfunctions are mutually orthogonal.

We shall show later that the eigenfunctions are complete. Consequently, any arbitrary square integral function $f(x)$ can be expanded in terms of the eigenfunctions as follows:

$$f(x) = \sum_1^\infty \alpha_m \sin m\pi x. \tag{4.4}$$

Using the orthogonality of the eigenfunctions, we find that

$$\alpha_n = 2\int_0^1 f(x)\sin n\pi x\,dx.$$

Here the 2 is needed for the normalization of the eigenfunctions since

$$\int_0^1 (\sin n\pi x)^2\,dx = 1/2.$$

Equation (4.4) is just the Fourier sine series for the function $f(x)$. It should be pointed out that the equality in (4.4) is to be understood as equality in the sense of the norm of the space, that is, (4.4) means the following:

$$\lim_{m\to\infty} \int_0^1 \left[f(x) - \sum_1^m \alpha_k \sin k\pi x\right]^2 dx = 0.$$

This result also implies that (4.4) can also be understood as an equality in the sense of distributions.

Eigenfunction Expansion—Example 2

Again let L be $-\dfrac{d^2}{dx^2}$, but now let the domain consist of twice-differentiable functions $u(x)$ such that

$$u(0) = 0, \quad u'(1) = \tfrac{1}{2}u(1). \tag{4.5}$$

The function $u(x)$ will be an eigenfunction if it is a solution of (4.2) and also satisfies (4.5). Again we start with $u = \sin \sqrt{\lambda} x$, a solution of (4.2) which satisfies the first condition of (4.5). In order that this function satisfy the second condition in (4.5) we must have

$$\lambda^{1/2} \cos \lambda^{1/2} = \tfrac{1}{2} \sin \lambda^{1/2}$$

which can be written as

$$\tan \lambda^{1/2} = 2\lambda^{1/2}. \tag{4.6}$$

From the general properties of the operator, we can show that λ is real (see the following section). We therefore consider two cases. In the first case, we assume that λ is a positive real number equal to k^2; and then (4.6) becomes

$$\tan k = 2k. \tag{4.7}$$

By considering the intersections of the curve $\eta = \tan \xi$ with the curve $\eta = 2\xi$, we see that (4.7) has an infinite number of solutions. In the second case, we assume that λ is a negative real number equal to $-k^2$; and then (4.6) becomes

$$\tanh k = 2k. \tag{4.8}$$

If we investigate the intersection of the curve $\eta = \tanh \xi$ with the curve $\eta = 2\xi$, we find that the only solution of (4.8) is $k = 0$. However, $k = 0$ implies $\lambda = 0$, and then it is easy to see that the solution of (4.2) which satisfies (4.5) is $u = 0$; consequently, $\lambda = 0$ is not an eigenvalue.

The eigenvalues are therefore given by $\lambda = k_n^2$, where k_n is a non-zero solution of (4.7). The eigenfunctions are $u_n = \sin k_n x$. It is easy to show that the eigenfunctions are mutually orthogonal. If $f(x)$ is an arbitrary square integrable function, we have

$$f(x) = \Sigma \alpha_n \sin k_n x\,, \tag{4.9}$$

where

$$\alpha_n = \frac{-2}{\cos 2k_n} \int_0^1 f(x) \sin k_n x \, dx.$$

The factor before the integral is needed to normalize the eigenfunctions.

Formula (4.9) is an example of an expansion in a non-harmonic Fourier series. In (4.4) we have an expansion in terms of trigonometric functions of πx and of integral multiples of πx. This corresponds to the vibration of a string where we have a fundamental tone (πx) and overtones (multiples of πx). Since the numbers k_n which satisfy (4.7) are not integral multiples of the smallest of them, the expansion (4.9) is non-harmonic. We shall see, however, that the general theory handles both cases equally well.

PROBLEMS

4.1 Find the eigenvalues and the normalized eigenfunctions if $L = -\dfrac{d^2}{dx^2}$ and the boundary conditions are each of the following:

(*a*) $u(0) = 0, \quad u'(1) = \alpha u(1)$. Distinguish between $\alpha < 1$ and $\alpha > 1$.

(*b*) $u'(0) = u'(1) = 0$.

(*c*) $u(0) = u(1), u'(0) = u'(1)$.

4.2. Find the eigenvalues and the normalized eigenfunctions if $Lu = -x(xu')'$ and if $u(1) = u(2) = 0$.

General Theory of Eigenfunction Representations

Let L be a differential operator defined over some domain D and let L^* be the adjoint operator defined over the adjoint domain D^*. If u is any function in the domain of L and if v is any function in the domain of L^*, then

$$\langle v, Lu\rangle - \langle L^*v, u\rangle = 0. \tag{4.10}$$

A non-zero function $u(x)$ in the domain of L is an eigenfunction of L if there exists a real or complex number λ such that

$$Lu = \lambda u. \tag{4.11}$$

Similarly a function $v(x)$ in the domain of L^* is an eigenfunction of L^* if there exists a real or complex number λ' such that

$$L^*v = \lambda' v. \tag{4.12}$$

If we subtract the scalar product of (4.11) with v from the scalar product of (4.12) with u, we find, with the help of (4.10), that

$$0 = \langle v, Lu\rangle - \langle L^*v, u\rangle = (\lambda - \lambda')\langle v, u\rangle.$$

If $\lambda \neq \lambda'$, this implies that $\langle v, u\rangle = 0$ or, in words:

An eigenfunction of L corresponding to the eigenvalue λ is orthogonal to every eigenfunction of L^* which does not correspond to the eigenvalue λ.

We now prove a theorem similar to Theorem 2.11.

Theorem 4.1. *If λ is an eigenvalue of L, then it is also an eigenvalue of L^*.*

The proof of this theorem will depend upon the following result: Either there exists for any continuous function $f(x)$ a function $v(x)$ in the domain of L^* such that

$$(L^* - \lambda)v = f$$

or there exists a non-zero solution of

$$(L^* - \lambda)v = 0,$$

and therefore an eigenfunction of L^* corresponding to the value λ. This result implies that either the range of $L^* - \lambda$ is dense in the space or the operator $L^* - \lambda$ has an eigenfunction; consequently, the differential operator does not have a residual spectrum.

We shall not prove this result, but in Problem 4.3 we indicate how a proof may be given for the special case where L^* is a second-order differential operator.

Suppose that $u(x)$ is the eigenfunction of L corresponding to the eigenvalue λ; then $(L - \lambda)u = 0$. Let v be any function in D^*; then

$$0 = \langle v, (L - \lambda)u \rangle = \langle (L^* - \lambda)v, u \rangle.$$

Consequently, u is orthogonal to every function of the form $(L^* - \lambda)v$. If $(L^* - \lambda)v$ represented every continuous function $f(x)$, we would have a contradiction because $\langle u, f \rangle = 0$ for every continuous $f(x)$ would imply that $u = 0$. Therefore, the other alternative of the result we have just quoted must hold, that is, there exists an eigenfunction of L^* corresponding to the value λ. This proves the theorem.

Suppose that the operator L is simple in the sense of Chapter 2, page 112, that is, all eigenvectors of L are of rank one and the eigenvectors span the space. The case where L is not simple can be treated, just as it was in Chapter 2, by the introduction of generalized eigenvectors in the space.

Let $\lambda_1, \lambda_2, \cdots$ be the eigenvalues of L and suppose that $u_1(x), u_2(x) \cdots$ are the corresponding eigenfunctions. By Theorem 4.1 the numbers $\lambda_1, \lambda_2, \cdots$ are also the eigenvalues of L^*. Suppose that $v_1(x), v_2(x) \cdots$ are the corresponding eigenfunctions. Just as in Chapter 2, we may so normalize the eigenfunctions that

$$\langle v_j, u_k \rangle = \delta_{jk} \tag{4.13}$$

for $j, k = 1, 2, \cdots$ Finally, we shall assume that the eigenfunctions of L or of L^* are complete, that is, we shall assume that every square integrable function $f(x)$ can be expanded in a series of eigenfunctions. We may find the coefficients in the expansions by using the orthogonality relations (4.13). Assume that

$$f(x) = \Sigma \alpha_k u_k(x); \tag{4.14}$$

then taking the scalar product of this with $v_j(x)$, we find that

$$\alpha_j = \langle v_j, f \rangle.$$

Note that

$$Lf = \Sigma\lambda_k\alpha_k u_k(x).$$

Similarly, if we assume that

$$f(x) = \Sigma\beta_j v_j(x),$$

we find that

$$\beta_k = \langle f, u_k\rangle.$$

Note that now

$$L^*f = \Sigma\lambda_j\beta_j v_j(x).$$

If L is self-adjoint, then, instead of the eigenfunctions forming a bi-orthogonal set as indicated in (4.13), they form an orthogonal set, that is,

$$\langle u_j, u_k\rangle = \delta_{jk}.$$

The coefficients in the expansion (4.14) are now given by the formula:

$$\alpha_j = \langle f, u_j\rangle.$$

Just as in Chapter 2, we can show that a real self-adjoint differential operator can have only real eigenvalues if it acts on a domain which contains $\overline{u(x)}$, whenever it contains $u(x)$. We can show also that, if L is positive-definite, that is, if $\langle u, Lu\rangle > 0$ for all non-zero u in D, the eigenvalues of L must be positive. This follows easily from the definition of eigenvalue. If λ is an eigenvalue and $u(x)$ the corresponding eigenfunction, we have $Lu = \lambda u$. The scalar product of this with $u(x)$ gives

$$\langle u, Lu\rangle = \lambda\langle u, u\rangle \tag{4.15}$$

or

$$\lambda = \frac{\langle u, Lu\rangle}{\langle u, u\rangle}. \tag{4.16}$$

Since both the numerator and the denominator are positive, this shows that λ must be positive.

For an application of these facts about eigenvalues, consider the operator

$$Lu = -\frac{1}{x}(xu')', \tag{4.17}$$

where u is in the domain of the operator defined by the boundary conditions

$$\lim_{x\to 0} xu' = 0, \quad u(1) = 0. \tag{4.18}$$

In order that L be formally self-adjoint, we use the following definition of the scalar product:

$$\langle u, v\rangle = \int_0^1 u(x)v(x)x\,dx.$$

With this definition and the boundary conditions (4.18), we see that L is self-adjoint. L is also positive-definite because

$$\langle u, Lu\rangle = -\int_0^1 u(xu')'\,dx = \int_0^1 xu'^2\,dx \geq 0$$

by the use of (4.18). We conclude, therefore, that the eigenvalues of L are positive.

To find the eigenvalues of L we must first find a function which satisfies (4.18) and is a solution of the differential equation

$$(xu')' + \lambda xu = 0. \tag{4.19}$$

A solution of this which satisfies the first condition in (4.18) is $u = J_0(\sqrt{\lambda}x)$, where $J_0(t)$ is Bessel's function of order zero. This function will not be discussed in detail. For the present all that need be known about it is that such a function exists and that the solution of (4.19) is a function of the combination $\sqrt{\lambda}x$. This can be shown by putting $t = \sqrt{\lambda}x$ in (4.19).

To satisfy the second condition in (4.18) we must have

$$J_0(\sqrt{\lambda}) = 0,$$

that is, we must find the zeroes of the Bessel function. Since λ must be positive, we have thus proved that *all the zeroes of $J_0(t)$ are real.*

PROBLEMS

4.3. Prove that if L^* is a second-order differential operator and if $f(x)$ is a continuous function, there exists either a function $v(x)$ such that $(L^* - \lambda)v = f$ or a non-zero function $u(x)$ such that $(L^* - \lambda)u = 0$. (*Hint.* Suppose that the boundary conditions are $B_1(v) = B_2(v) = 0$. Let v_1 and v_2 be linearly independent solutions of the differential equation $(L^* - \lambda)v = 0$ but not necessarily satisfying the boundary conditions. The general solution of $(L^* - \lambda)v = f$, again not necessarily satisfying the boundary conditions, is $v = v_0 + \alpha v_1 + \beta v_2$, where v_0 is a particular solution. The application of the boundary conditions to v will either determine α and β or prove the existence of the function $u(x)$. Compare Chapter 3, page 172.)

4.4 Show that the operator $L = -\dfrac{d^2}{dx^2} + q(x)$, with the boundary conditions either $u(0) = u(1) = 0$ or $u'(0) = u'(1) = 0$, is non-negative definite if $q(x)$ is a non-negative continuous function in (0, 1).

4.5 Show that if there exists a constant C and a function $w(x)$ such that either $\langle u, Lu\rangle + C\langle u, w\rangle^2 > 0$ or $\langle u, Lu\rangle + C\langle Lu, w\rangle^2 > 0$ for all u in the domain of L, then L can have at most one non-positive eigenvalue. (*Hint.* Suppose u_1 and u_2 are eigenfunctions of L corresponding to non-positive eigenfunctions; then there exists a linear combination of u_1 and u_2, which we denote by u, such that $\langle u, w\rangle = 0$ or $\langle Lu, w\rangle = 0$ and $\langle u, Lu\rangle \leq 0$.)

4.6 Show that the operator in Problem 4.1*a* has at most one non-positive eigenvalue. (*Hint.* Use the second condition in Problem 4.5 with $w(x) = x$.)

Eigenfunction Expansion—Example 3

We shall now discuss a non-self-adjoint operator which has complex eigenvalues. Let L be $-\dfrac{d^2}{dx^2}$, with the boundary conditions

$$u(0) = 0, \quad u'(0) = u(1). \tag{4.20}$$

The eigenfunctions are solutions of

$$u'' + \lambda u = 0$$

which satisfy the conditions (4.20). The only solution of the differential equation which satisfies the first condition of (4.20) is a constant multiple of $u = \sin(\sqrt{\lambda}x)$. For this function to satisfy the second condition we must have

$$\lambda^{1/2} = \sin \lambda^{1/2}. \tag{4.21}$$

Put $\lambda = k^2$ and $\lambda = -k^2$ successively in (4.21) which then becomes $k = \sin k$ or $k = \sinh k$. By considering the intersections of the line $y = k$ with the curves $y = \sin k$ or $y = \sinh k$, we see that these equations have only one solution, namely, $k = 0$. This means that (4.21) has only the one real root, $\lambda = 0$. It would appear that $\lambda = 0$ is not an eigenvalue since the corresponding solution seems to be $u = 0$. However, if we put $\lambda = 0$ in the differential equation, we find that $u = x$ satisfies both the differential equation and the boundary conditions (4.20). Consequently, $\lambda = 0$ is an eigenvalue, and the corresponding eigenfunction is $u = x$.

We can remove the difficulty that $\lambda = 0$ does not seem to give an eigenfunction if we normalize the solution which satisfies the first boundary condition in (4.20). We chose the solution $u = \sin \sqrt{\lambda}x$, but the function $u = C \sin \sqrt{\lambda}x$, where C is independent of x, is also a solution. The value of C may be uniquely determined by the requirement that u satisfy a non-homogeneous boundary condition at $x = 0$. Any boundary condition independent of λ and independent of the first condition in (4.20) may be used. For example, we may require $u'(0) = 1$; then

$$u = \frac{\sin \sqrt{\lambda}x}{\sqrt{\lambda}}$$

and the eigenvalue equation becomes

$$1 = \frac{\sin \lambda^{1/2}}{\lambda^{1/2}}.$$

This equation still has the root $\lambda = 0$, and u then reduces to x automatically without the need for again solving the differential equation.

Let us return to the eigenvalue equation (4.21). It can be shown (see

Problem 4.7) that it must have an infinite number of complex roots which are given by the asymptotic formula:

$$\lambda^{1/2} \sim \left(2m + \frac{1}{2}\right)\pi - \frac{2\log(4m+1)\pi}{(4m+1)\pi} \pm i\log(4m+1)\pi.$$

The eigenfunctions $u_0 = x$, $u_n = \sin(\lambda_n^{1/2}x)$, $n = 1, 2, \cdots$ are *not* orthogonal since L is only formally self-adjoint and not self-adjoint. If we assume that every square integrable function $f(x)$ can be expanded in a series of eigenfunctions as follows:

$$f(x) = \Sigma\alpha_k u_k(x)$$

then the value of α_k will be obtained by taking the scalar product of $f(x)$ with the eigenfunctions of L^*. Since L is formally self-adjoint, L^* is $-\dfrac{d^2}{dx^2}$, but its domain D^* is different from D.

We find D^* by an integration by parts. Suppose that u is any function in D; consequently it will satisfy (4.20). We have

$$\begin{aligned}\langle v, Lu\rangle &= -\int_0^1 vu''\,dx = \Big[-vu' + v'u\Big]_0^1 - \int_0^1 uv''\,dx \\ &= u(1)[v'(1) + v(0)] - u'(1)v(1) - \int_0^1 uv''\,dx.\end{aligned}$$

Now, in order that $\langle v, Lu\rangle = \langle u, Lv\rangle$ whenever v is in D^* and u is in D, the domain D^* must be defined by the conditions

$$v(1) = 0, \quad v'(1) + v(0) = 0.$$

The eigenfunctions of L^* are those functions in D^* which satisfy the differential equation

$$v'' + \lambda v = 0.$$

A solution of this equation which satisfies the first boundary condition of D^* is

$$v = \frac{\sin\lambda^{1/2}(1-x)}{\lambda^{1/2}}.$$

The eigenvalue equation is

$$-1 + \frac{\sin\lambda^{1/2}}{\lambda^{1/2}} = 0,$$

which is the same as (4.21). This is to be expected from Theorem 4.1. The eigenfunction $v = 1 - x$ corresponds to $\lambda = 0$, and the eigenfunction $v_n = \sin\lambda_n^{1/2}(1-x)$ to $\lambda = \lambda_n$.

The eigenfunctions will be normalized by considering the scalar product of an eigenfunction of L by an eigenfunction of L^*. We have

$$\begin{aligned}&\int_0^1 \sin\lambda_k^{1/2}x \sin\lambda_j^{1/2}(1-x)\,dx \\ &\quad= \frac{1}{2}\int_0^1\left\{\cos[\lambda_k^{1/2}x - \lambda_j^{1/2}(1-x)] - \cos[\lambda_k^{1/2}x + \lambda_j^{1/2}(1-x)]\right\}dx = 0,\end{aligned}$$

if $\lambda_k \neq \lambda_j$.

In the last equality, equation (4.21) has been used. This result could be anticipated from the general theory of the preceding section.

If $\lambda_k = \lambda_j$,

$$\int_0^1 \sin \lambda_k^{1/2}x \sin \lambda_k^{1/2}(1 - x)\, dx = \frac{\sin \lambda_k^{1/2}}{2\lambda_k^{1/2}} - \frac{1}{2}\cos \lambda_k^{1/2} = \frac{1 - \cos \lambda_k^{1/2}}{2}.$$

Note also that for $\lambda = 0$ we have

$$\int_0^1 x(1 - x)\, dx = 1/6.$$

From these formulas it follows that if the eigenfunctions are complete,

$$f(x) = \alpha_0 x + \Sigma\alpha_k \sin \lambda_k^{1/2}x,$$

where

$$\alpha_0 = 6\int_0^1 f(x)(1 - x)\, dx$$

$$\alpha_k = \frac{2}{1 - \cos \lambda_k^{1/2}}\int_0^1 f(x)v_k(x)\, dx, \quad k = 1, 2, \cdots.$$

Eigenfunction Expansion—Example 4

Consider the following eigenvalue problem: Find solutions of the differential equation

$$u'' + \lambda u = 0$$

which satisfy the conditions

$$u(0) = 0, \quad u'(1) = \lambda u(1).$$

This is not the usual type of eigenvalue problem because the eigenvalue appears in the boundary conditions; consequently, we cannot put $L = -\dfrac{d^2}{dx^2}$ and consider the problem as a special case of $Lu = \lambda u$ because D, the domain of L, depends upon λ.

In order to fit this problem into the general framework, we must enlarge our definition of L. First, we extend our space. Consider the space of two-component vectors U whose first component is a real twice-differentiable function $u(x)$ and whose second component is a real number u_1. We define the scalar product of two vectors U and V as follows:

$$\langle U, V\rangle = \int_0^1 u(x)v(x)\, dx + u_1v_1. \tag{4.22}$$

Consider now the subspace D of vectors U such that

$$u(0) = 0 \quad \text{and} \quad u(1) = u_1. \tag{4.23}$$

Second, if we put

$$LU = \begin{pmatrix} -u''(x) \\ u'(1) \end{pmatrix}, \tag{4.24}$$

the eigenvalue problem reduces to the following: find a vector U in D such that $LU = \lambda u$.

With this formulation, the problem becomes the same type as those we have considered, and therefore the results of the previous sections are applicable.

Let U and V be vectors in D; then with the help of (4.24) we find that

$$\langle V, LU\rangle = -\int_0^1 v(x)u''(x)\,dx + v_1u'(1) = \Big[-vu' + v'u\Big]_0^1 - \int_0^1 uv''\,dx + v(1)u'(1)$$

$$= -\int_0^1 uv''\,dx + u_1v'(1) = \langle LV, U\rangle.$$

This shows that L is self-adjoint, and therefore all the eigenvalues are real and the eigenfunctions are orthogonal in the sense of the scalar product (4.22). Note also that L is positive-definite for

$$\langle U, LU\rangle = -\int_0^1 uu''\,dx + u_1u'(1) = \int_0^1 u'^2\,dx > 0.$$

We may use the original formulation to find the eigenvalues and the eigenfunctions. Start with

$$u = \frac{\sin \lambda^{1/2}x}{\lambda^{1/2}};$$

then, applying the second boundary condition, we find that

$$\cos \lambda^{1/2} = \lambda^{1/2} \sin \lambda^{1/2}.$$

Let $\lambda_1, \lambda_2, \cdots$ be the roots of this equation and $u_n(x) = \sin \lambda_n^{1/2}x$ the corresponding solutions of the differential equation. To find an expansion theorem we must use our general formulation. Let F be a vector in D and let U_n be the vector in D whose first component is $u_n(x)$ and whose second component is $u_n(1)$; then we have the expansion

$$F = \Sigma\alpha_n U_n(x),$$

where

$$\alpha_n = \frac{\langle F, U_n\rangle}{(1 + \sin^2 \sqrt{\lambda_n})/2}. \tag{4.25}$$

The denominator here is the result of the evaluation of $\langle U_n, U_n\rangle$.

The expansion given for F is really the expansion of both the function $f(x)$, which is the first component of F, and of the number f_1, which is the second component of F. We have

$$f(x) = \sum_1^\infty \alpha_n \sin \sqrt{\lambda_n}x,$$

$$f_1 = \sum_1^\infty \alpha_n \sin \sqrt{\lambda_n},$$

where α_n is given by the formula

$$\alpha_n = \frac{\int_0^1 f \sin \sqrt{\lambda_n} x \, dx + f_1 \sin \sqrt{\lambda_n}}{(1 + \sin^2 \sqrt{\lambda_n})/2}.$$

Note that α_n can be defined even if f_1 does not equal $f(1)$. In fact, we may take $f(x)$ identically zero and $f_1 = 1$; then we get

$$0 = 2\sum_1^\infty \frac{\sin \sqrt{\lambda_n} \sin \sqrt{\lambda_n} x}{1 + \sin^2 \sqrt{\lambda_n}}, \quad 0 \le x < 1,$$

$$(4.26) \qquad 1 = 2\sum_1^\infty \frac{\sin^2 \sqrt{\lambda_n}}{1 + \sin^2 \sqrt{\lambda_n}}.$$

These formulas can be justified by complex integration. (See Problem 4.8.)

A careful consideration of the convergence question for this problem is given in a paper by Langer.†

PROBLEMS

4.7 Derive the asymptotic formula given in the text for the roots of $\sin \lambda^{1/2} = \lambda^{1/2}$. (*Hint.* Put $\lambda^{1/2} = \alpha + i\beta$; then $\sin \alpha \cosh \beta = \alpha$ and $\cos \alpha \sinh \beta = \beta$. If α is large, the first equation implies β large. If β is large, $\beta/\sinh \beta$ approaches zero, and therefore $\alpha = \left(n + \frac{1}{2}\right)\pi + \varepsilon_n$, where ε_n approaches zero. From the first equation, since $\cos \varepsilon_n \sim 1$, we have $\cosh \beta = \left(2m + \frac{1}{2}\right)\pi$, where $n = 2m$.)

4.8. Justify Formula 4.26 by evaluating

$$\frac{1}{2\pi i}\oint \frac{\sin kx \csc k \, dk}{k - \cot k}$$

over a large circle of radius R. (*Hint.* First, evaluate the integral by residues and then estimate the integrand on the circle $|k| = R$.)

4.9. Find the eigenvalues and the expansion theorem for the equation $u'' + \lambda u = 0$, with the conditions $u(0) = 0$ and $u'(1) = \lambda^{1/2}u(1)$. (*Hint.* Put $\lambda = k^2$. Use the space of two component vectors U whose elements are functions $u(x)$, $v(x)$; then $LU = kU$, where $LU = \begin{pmatrix} v' \\ -u' \end{pmatrix}$.)

Approximating Eigenvalues by Variational Methods

The examples of spectral representation that we have discussed in the preceding sections have shown that eigenvalues may be found by solving a transcendental equation such as $\tan k = k$. For an arbitrary second-

† R. E. Langer, *Tohoku Mathematics Journal, Japan*, Vol. 35, 260, 1932.

order differential operator L, the transcendental equation which yields the eigenvalues can be obtained if two independent solutions of the homogeneous equation $(L - \lambda)u = 0$ are known. However, it is important to have a method for finding the eigenvalues which does not depend upon a prior knowledge of the solutions of the differential equation, since in many cases the solutions of the differential equation are known only approximately. Even when the solutions of the differential equation are known, a method which does not require the solution of a transcendental equation may be useful. Such a method for calculating the eigenvalues (called the Rayleigh-Ritz method) is given by a variational principle.

Let $u(x)$ be any function in the domain D of a self-adjoint operator L and denote the ratio $\langle u, Lu\rangle/\langle u, u\rangle$ by $\rho(u)$. If $u = u_n$, an eigenfunction of L, then $\rho(u_n) = \lambda_n$, the corresponding eigenvalue. In general, we do not know the eigenfunction u_n, but if we take a function u in D which is "close" to the eigenfunction u_n, we hope that the value of $\rho(u)$ will be close to λ_n. This is to be expected because ρ is a continuous functional of u.

However, although the error in the value of most functionals is of the same order of magnitude as the error in the function, for the functional $\rho(u)$ the error in its approximation to the eigenvalue is of the second order of smallness when compared to the error in approximating to the eigenfunction. We express this fact by saying that $\rho(u)$ has a *stationary value* (in the sense of the theory of maxima and minima) whenever u is an eigenfunction u_n. To prove this fact, suppose that $u(x) = u_n(x) + \varepsilon v(x)$, where ε is assumed to be a small quantity. Using the fact that L is self-adjoint and that u_n is an eigenfunction, we find that

$$\begin{aligned}\rho(u) &= \frac{\lambda_n\langle u_n, u_n\rangle + 2\varepsilon\lambda_n\langle u_n, v\rangle + \varepsilon^2\langle v, Lv\rangle}{\langle u_n, u_n\rangle + 2\varepsilon\langle u_n, v\rangle + \varepsilon^2\langle v, v\rangle} \\ &= \lambda_n + \frac{\varepsilon^2\{\langle v, Lv\rangle - \lambda_n\langle v, v\rangle\}}{\langle u_n, u_n\rangle + 2\varepsilon\langle u_n, v\rangle + \varepsilon^2\langle v, v\rangle}.\end{aligned}$$

This equation shows that although the error in the eigenfunction, namely, $u - u_n$, is of the order of magnitude ε, the error in the eigenvalue, namely, $\rho - \lambda_n$, is of the order of magnitude ε^2.

The converse is also true. If $\rho(u)$ is stationary when $u = w$, then w is an eigenfunction of L. To see this, put $u = w + \varepsilon v$; then

$$\begin{aligned}\rho(u) &= \frac{\langle w + \varepsilon v, Lw + \varepsilon Lv\rangle}{\langle w + \varepsilon v, w + \varepsilon v\rangle} \\ &= \lambda + \frac{\langle w, Lw - \lambda w\rangle + 2\varepsilon\langle v, Lw - \lambda w\rangle + \varepsilon^2\langle v, Lv - \lambda v\rangle,}{\langle w + \varepsilon v, w + \varepsilon v\rangle}\end{aligned}$$

where λ is an arbitrary constant. The functional $\rho(u)$ will be stationary for $\varepsilon = 0$ if $\frac{\partial \rho}{\partial \varepsilon} = 0$, when $\varepsilon = 0$. The vanishing of the derivative implies that

$$\langle v, Lw - \lambda w \rangle = 0.$$

Since this should hold for any function v in the domain of L, it follows that $Lw - \lambda w = 0$, that is, w is an eigenfunction of L corresponding to the eigenvalue λ.

We state

Theorem 4.2. *If L is a self-adjoint operator with a purely discrete spectrum and if λ_1 is its smallest eigenvalue, the minimum value of $\rho(u)$ is λ_1. This minimum is attained when $u = u_1$, the eigenfunction corresponding to λ_1.*

Suppose that the eigenvalues of L arranged in order of increasing magnitude are $\lambda_1 \leq \lambda_2 \leq \lambda_3 \leq \cdots$. Suppose that the corresponding eigenfunctions are $u_1, u_2, \cdots$ and suppose that they span the space; then for any function $u(x)$ in the domain of L we have

$$u(x) = \sum_1^\infty \alpha_k u_k(x),$$

$$Lu = \sum_1^\infty \lambda_k \alpha_k u_k(x).$$

Because of the orthogonality properties of the eigenfunctions, we get

$$\langle u, Lu \rangle = \Sigma \lambda_k \alpha_k^2 = \lambda_1 \Sigma \alpha_k^2 + \Sigma(\lambda_k - \lambda_1)\alpha_k^2 \geq \lambda_1 \Sigma \alpha_k^2 = \lambda_1 \langle u, u \rangle;$$

consequently,

$$\rho(u) = \frac{\langle u, Lu \rangle}{\langle u, u \rangle} \geq \lambda_1.$$

This inequality proves the theorem. Note that if $\alpha_1 = 1$, $\alpha_2 = \alpha_3 = \cdots = 0$, then $\rho = \lambda_1$.

Some applications of this theorem are shown in the following illustrations. Consider Eigenfunction Expansion, Example 1. The operator is $L = -\frac{d^2}{dx^2}$, and the domain is defined by $u(0) = u(1) = 0$. An integration by parts shows that

$$\rho = -\frac{\int_0^1 uu'' \, dx}{\int_0^1 u^2 \, dx} = \frac{\int_0^1 u'^2 \, dx}{\int_0^1 u^2 \, dx}.$$

Now assume that $u = x(1 - x)$, a function which is obviously in the domain of the operator. We find $\rho = 10$, which is not a bad approximation to the first eigenvalue $\lambda_1 = \pi^2$. To get a better approximation, we must use a more complicated "trial" function for u. For example, assume that

$$u = \alpha x(1 - x) + x^3(1 - x) = x(1 - x)(\alpha + x^2),$$

where α is a parameter to be chosen later so that ρ will be a minimum. We find that

$$\rho = \frac{\dfrac{1}{3}\alpha^2 + \dfrac{\alpha}{5} + \dfrac{3}{35}}{\dfrac{\alpha^2}{30} + \dfrac{2\alpha}{105} + \dfrac{1}{252}}.$$

This expression is stationary when $\alpha = -1/3$ or $\alpha = -28/3$ approximately. For the latter value of α we find $\rho = 9.984$, which is a better approximation to π^2 than the previous value of ρ was.

When we substitute $\alpha = -1/3$ in the expression for ρ, we find $\rho = 39.7$. This is a good approximation to $\lambda_2 = 4\pi^2$. The reason why this value of α gives an approximation to λ_2 instead of λ_1 is that the trial function u has a zero inside the interval (0, 1) and is therefore "close" to u_2, the second eigenfunction.

Instead of applying Theorem 4.2 directly to L, we may apply it to the inverse operator L^{-1}. If $Lu = \lambda u$, then clearly $u = \lambda L^{-1}u$; consequently $1/\lambda$ is an eigenvalue for $L^{-1}u$. We have then that $\langle u, L^{-1}u\rangle/\langle u, u\rangle$ has a maximum value equal to λ^{-1}, and therefore the minimum value of $\langle u, u\rangle/\langle u, L^{-1}u\rangle$ is λ_1. Using a generalized form of the Schwarz inequality (see Chapter 2, Problem 2.34), we can show (see Problem 4.10) that, if we put $\rho' = \langle u, u\rangle/\langle u, L^{-1}u\rangle$, then $\rho' \le \rho$. Since $\lambda_1 \le \rho$, it follows that of the two approximations ρ and ρ', ρ' is closer to λ_1. For example, in the problem we have discussed above we have

$$L^{-1}u = (1 - x)\int_0^x \xi u \, d\xi + x\int_x^1 (1 - \xi)u \, d\xi.$$

Again, put $u = x(1 - x)$, and we find that

$$\rho' = 9\tfrac{15}{17} = 9.882,$$

while the correct value is $\pi^2 = 9.8696$.

Eigenfunction Expansion, Example 4, will serve as a final illustration of Theorem 4.2. We have

$$\rho = \frac{-\int_0^1 uu'' \, dx + u_1 u'(1)}{\int_0^1 u_2 \, dx + u_1^2} = \frac{\int_0^1 u'^2 \, dx}{\int_0^1 u^2 \, dx + u(1)^2}.$$

As a trial function take $u = x$; then we find that

$$\rho = 3/4$$

as an approximation to λ_1, whereas the value of λ, correct to two decimal places, is 0.74.

So far, we have obtained only an upper bound to the value of λ_1. We shall now present some methods for obtaining a lower bound for λ_1. Suppose that L is a self-adjoint operator with a purely discrete spectrum containing the eigenvalues $\lambda_1 \leq \lambda_2 \leq \lambda_3 \leq \cdots$ and suppose that the corresponding normalized eigenfunctions $u_1, u_2, \cdots$ span the space. Consider the sum

$$S(\alpha) = \sum_1^\infty (\lambda_k - \alpha)^2 \alpha_k^2,$$

where $\alpha, \alpha_1, \alpha_2, \cdots$ are arbitrary real numbers. If α is closer to the eigenvalue λ_1 than to any other eigenvalue, we have $(\lambda_k - \alpha)^2 \geq (\lambda_1 - \alpha)^2$, and consequently

$$S(\alpha) \geq (\lambda_1 - \alpha)^2 \sum_1^\infty \alpha_k^2.$$

This inequality will be used to give an estimate for λ_1, but first we see that the left-hand side will be a minimum if α is chosen such that $\Sigma(\lambda_k - \alpha)\alpha_k^2 = 0$, that is, if

$$\alpha = \frac{\Sigma \lambda_k \alpha_k^2}{\Sigma \alpha_k^2}.$$

In this case $S(\alpha)$ reduces to $\Sigma \lambda_k^2 \alpha_k^2 - \alpha^2 \Sigma \alpha_k^2$, and we obtain the estimate

$$(\lambda_1 - \alpha)^2 \leq \frac{\Sigma \lambda_k^2 \alpha_k^2}{\Sigma \alpha_k^2} - \alpha^2.$$

If we denote the right-hand side of this inequality by E^2, we get

$$\alpha - E \leq \lambda_1 \leq \alpha + E.$$

This result gives us both an upper and a lower bound for λ_1, but there remains the question of how to evaluate the sums occurring in the definition of E^2. Suppose that u is any function in the domain of L and suppose that $u = \Sigma \alpha_k u_k$; then $Lu = \Sigma \lambda_k \alpha_k u_k$. Because the eigenfunctions form an orthonormal set, we have $\langle u, u \rangle = \Sigma \alpha_k^2$, $\langle u, Lu \rangle = \Sigma \lambda_k \alpha_k^2$, and $\langle Lu, Lu \rangle = \Sigma \lambda_k^2 \alpha_k^2$. We see then that α was chosen equal to $\Sigma \lambda_k \alpha_k^2 / \Sigma \alpha_k^2 = \langle u, Lu \rangle / \langle u, u \rangle = \rho(u)$. We find also that $E^2 = -\rho^2 + \langle Lu, Lu \rangle / \langle u, u \rangle$. Substituting these results in the previously obtained inequalities we have the following bounds for λ_1:

$$\rho - \sqrt{-\rho^2 + \langle Lu, Lu \rangle / \langle u, u \rangle} \leq \lambda_1 \leq \rho + \sqrt{-\rho^2 + \langle Lu, Lu \rangle / \langle u, u \rangle}. \tag{4.27}$$

Note that if we start with α closer to λ_k than to any other eigenvalue, a procedure similar to that used above will give upper and lower bounds for λ_k.

We shall illustrate the use of (4.27) by considering the operator $L = -\dfrac{d^2}{dx^2}$, with the conditions $u(0) = u(1) = 0$. Previously, we showed that if we put $u = x(1 - x)$, then $\rho = 10$. We find that

$$\langle Lu, Lu\rangle = \int_0^1 4\,dx = 4, \quad \langle u, u\rangle = 1/30,$$

and then $E^2 = 120 - 100 = 20$; consequently,

$$10 - \sqrt{20} \leq \lambda_1 \leq 10 + \sqrt{20}.$$

This upper bound is worse than the one we had obtained previously, namely, 10, and the lower bound is quite far from the correct value. We shall present a method, essentially due to Kato,† for obtaining a better lower bound to the value of λ_1. This method makes use of a lower bound for λ_2, a lower bound which can be found by the use of an inequality for λ_2 analogous to (4.27).

Suppose that α and β are real numbers such that $\lambda_1 \leq \alpha \leq \beta \leq \lambda_2$; then

$$\sum_1^\infty (\lambda_k - \alpha)(\lambda_k - \beta)\alpha_k^2 \geq 0$$

for any real numbers $\alpha_1, \alpha_2, \cdots$. Put $\lambda_k - \alpha = \lambda_k - \rho + \rho - \alpha$ and $\lambda_k - \beta = \lambda_k - \rho + \rho - \beta$, where $\rho = \Sigma\lambda_k\alpha_k^2/\Sigma\alpha_k^2$.
We get

$$\Sigma(\lambda_k - \rho)^2\alpha_k^2 + (\rho - \alpha)(\rho - \beta)\Sigma\alpha_k^2 \geq 0$$

or

$$\rho - \frac{\Sigma\lambda_k^2\alpha_k^2 - \rho^2\Sigma\alpha_k^2}{(\beta - \rho)\Sigma\alpha_k^2} \leq \alpha.$$

Since this inequality holds for any number α larger than λ_1, it holds also for λ_1. Using the previous notation, we see that

$$\lambda_1 \geq \rho - \frac{E^2}{\beta - \rho}, \tag{4.28}$$

where β is any number less than λ_2 and larger than ρ.

Inequality (4.28) gives the desired lower bound for λ_1. We illustrate its use by considering again the first example of this section on page 209. With the assumption $u = x(1 - x)$ we found $\rho = 10$, $E^2 = 120 - 100 = 20$.

† *Journal Physical Society, Japan*, Vol. 4, 334, 1949.

We also found that 39.7 is an approximation to λ_2. Since β must be less than λ_2, put $\beta = 35$, say, then inequality (4.28) gives

$$\lambda_1 \geq 10 - \frac{20}{35 - 10} = 9.2,$$

which is quite close to the correct value of λ_1.

PROBLEMS

4.10. Prove that $\rho' = \langle u, u\rangle / \langle u, L^{-1}u\rangle$ is not greater than $\rho = \langle u, Lu\rangle / \langle u, u\rangle$. (*Hint.* Use Problem 2.34 with $x = u$, $y = L^{-1}u$, and $A = L$.)

4.11. Find an upper and a lower bound for the first zero of $J_0(t) = 0$ by estimating the lowest eigenvalue for $Lu = -\frac{1}{x}(xu')'$, with the conditions $u(0)$ regular and $u(1) = 0$.

Green's Function and Spectral Representation

Our main purpose in this chapter is not to discuss the eigenvalues and eigenfunctions for their own sakes, but to discuss how the eigenfunctions can be used to give the spectral representation. We first proceed in a purely formal manner. Let L be a simple operator and suppose that $u_1, u_2, \cdots$ are its eigenfunctions and $\lambda_1, \lambda_2, \cdots$ the corresponding eigenvalues. We shall also suppose that the eigenfunctions are complete and therefore every square integrable function $u(x)$ may be expanded as follows:

$$u(x) = \Sigma \alpha_k u_k(x),$$

where

$$\alpha_k = \langle v_k, u\rangle.$$

Here $v_k(x)$, $k = 1, 2, \cdots$ are the eigenfunctions of L^*. Now

$$Lu(x) = \Sigma \alpha_k \lambda_k u_k(x);$$

and if $f(t)$ is any function of t analytic in a region containing the eigenvalues, we define

$$f(L)u(x) = \Sigma f(\lambda_k)\alpha_k u_k(x).$$

Suppose that $f(t) = \dfrac{1}{\lambda - t}$; then

$$\frac{1}{\lambda - L}u(x) = \sum \frac{\alpha_k u_k(x)}{\lambda - \lambda_k}. \tag{4.29}$$

The left-hand side of this equation can be expressed in terms of the Green's function for the differential operator $L - \lambda$. To see this, put $w(x) = (\lambda - L)^{-1}u(x)$; then we have $(L - \lambda)w = -u$. If $G(x, \xi, \lambda)$ is the Green's function for the operator $L - \lambda$, we have shown that

$$w = -\int G(x, \xi, \lambda)u(\xi)\, d\xi;$$

consequently,

$$\frac{1}{\lambda - L} u(x) = -\int G(x, \xi, \lambda) u(\xi)\, d\xi.$$

We shall use this fact later.

Suppose that we integrate (4.29) over a large circle of radius R in the complex λ-plane. We get

$$\frac{1}{2\pi i} \oint \frac{d\lambda}{\lambda - L} u(x) = \sum \frac{1}{2\pi i} \oint \frac{d\lambda}{\lambda - \lambda_k} \alpha_k u_k(x).$$

Now as the radius of the circle approaches infinity, the right-hand side includes more and more residues. We have then

$$\lim_{R\to\infty} \frac{1}{2\pi i} \oint \frac{d\lambda}{L - \lambda} u(x) = -\Sigma \alpha_k u_k(x) = -u(x). \tag{4.30}$$

This result which connects the Green's function with the eigenfunctions was obtained by making a great many assumptions, such as that the eigenfunctions were known and that they were complete. In practice, we try to work it backwards. We start with a knowledge of the Green's function for the operator $L - \lambda$; then we consider the following integral in the complex λ-plane:

$$\frac{1}{2\pi i} \oint \frac{d\lambda}{L - \lambda} u(x) = \frac{1}{2\pi i} \oint d\lambda \int G(x, \xi, \lambda) u(\xi)\, d\xi;$$

and then, by evaluating it in terms of residues, we hope to get (4.30), that is, an expansion of $u(x)$ in terms of the eigenfunctions of L. To prove the validity of (4.30) it is necessary to prove that

$$\lim_{R\to\infty} \frac{1}{2\pi i} \oint \frac{d\lambda}{L - \lambda} u(x) = -u(x).$$

In later sections we shall prove an equivalent formula, namely,

$$\lim_{R\to\infty} \frac{1}{2\pi i} \oint G(x, \xi, \lambda)\, d\lambda = -\delta(x - \xi). \tag{4.31}$$

After (4.31) has been proved, we simplify the integral by shifting the path of integration. We can show that $G(x, \xi, \lambda)$ is an analytic function of λ except for pole and branch-point singularities; consequently, the integral in (4.31) reduces to a sum of residues at the poles plus integrals along the branch cuts.

We thus have an expansion of $\delta(x - \xi)$ as a sum of terms plus an integral. The sum will correspond to the discrete spectrum and the integral will correspond to the continuous spectrum. We can obtain a similar expansion of a continuous function $u(x)$ if we multiply the expansion of $\delta(x - \xi)$ by $u(\xi)$ and integrate with respect to ξ. A few examples will clarify the procedure.

Spectral Representation—Examples

Using the methods of the preceding section, we shall obtain the eigenfunction expansions considered previously in Examples 1 to 4. In Example 1, the Green's function is that solution of

$$-\frac{d^2G}{dx^2} - \lambda G = \delta(x - \xi)$$

which satisfies the boundary conditions

$$G(0, \xi, \lambda) = G(1, \xi, \lambda) = 0.$$

By the methods of Chapter 3 we find that

$$\begin{aligned} G &= \frac{\sin\sqrt{\lambda}x \sin\sqrt{\lambda}(1-\xi)}{\sqrt{\lambda}\sin\sqrt{\lambda}}, \quad x < \xi \\ &= \frac{\sin\sqrt{\lambda}(1-x)\sin\sqrt{\lambda}\xi}{\sqrt{\lambda}\sin\sqrt{\lambda}}, \quad x > \xi. \end{aligned} \tag{4.32}$$

It should be noticed that $\lambda = 0$ is neither a pole nor a branch point of G. This may be seen by considering the series expansion of G in the neighborhood of $\lambda = 0$. Now

$$\begin{aligned} \frac{1}{2\pi i}\oint G\,d\lambda &= \frac{1}{2\pi i}\oint d\lambda \left[\frac{\sin\sqrt{\lambda}x\sin\sqrt{\lambda}(1-\xi)}{\sqrt{\lambda}\sin\sqrt{\lambda}}H(\xi - x)\right. \\ &\qquad\qquad \left. + \frac{\sin\sqrt{\lambda}(1-x)\sin\sqrt{\lambda}\xi}{\sqrt{\lambda}\sin\sqrt{\lambda}}H(x-\xi)\right] \\ &= -2\sum_{n^2\pi^2<R}\sin n\pi x\sin n\pi\xi[H(\xi - x) + H(x - \xi)]. \end{aligned}$$

We shall prove in Problem 4.15 that the contour integral approaches $-\delta(x - \xi)$ as $R \to \infty$; consequently, we have

$$2\sum_{1}^{\infty}\sin n\pi x\sin n\pi\xi = \delta(x - \xi).$$

If we multiply this formula by $f(\xi)$ and integrate, we get

$$f(x) = 2\sum_{1}^{\infty}\sin n\pi x\int_0^1 f(\xi)\sin n\pi\xi\,d\xi,$$

which is the same as (4.4).

In Eigenfunction Expansions, Example 2, a similar treatment will yield (4.9). We leave the details to the reader.

In Eigenfunction Expansion, Example 3, the Green's function G is the solution of

$$-\frac{d^2G}{dx^2} - \lambda G = \delta(x - \xi),$$

which satisfies the boundary conditions

$$G(0, \xi, \lambda) = G_x(0, \xi, \lambda) - G(1, \xi, \lambda) = 0.$$

Using the methods of Chapter 3, we find that

$$G = \frac{\sin \sqrt{\lambda} x \sin \sqrt{\lambda}\,(1 - \xi)}{\sqrt{\lambda}\,(\sin \sqrt{\lambda} - \sqrt{\lambda})} - \frac{\sin \sqrt{\lambda}(x - \xi)}{\sqrt{\lambda}} H(x - \xi). \tag{4.33}$$

Again, $\lambda = 0$ is not a branch-point but it is a pole of the first term on the right-hand side. If we integrate G over a large circle, the second term will contribute nothing since it has no singularities in the finite part of the λ-plane. The integral of the first term will give the following expansion of the δ-function:

$$\delta(x - \xi) = 6x(1 - \xi) + 2\sum_{\lambda_n} \frac{\sin \sqrt{\lambda_n} x \sin \sqrt{\lambda_n}(1 - \xi)}{1 - \cos \sqrt{\lambda_n}}.$$

If we multiply this by $f(\xi)$ and integrate from 0 to 1, we shall get for $f(x)$ an expansion which is the same as that obtained previously in Example 3.

Consider now Example 4. The Green's function, as always, is the kernel of the integral operator that inverts the differential operator. Since the differential operator here is acting on the components of a two-dimensional vector space, it is really a two by two matrix differential operator; consequently, the Green's functions will have to be a two by two matrix G. Each column of G will be an element of our space, that is, its first component will be a function of x and its second component will be independent of x; consequently, we may define G as follows:

$$G = \begin{pmatrix} g_1(x, \xi, \lambda) & g_3(x, \lambda) \\ g_2(\xi, \lambda) & g_4(\lambda) \end{pmatrix}.$$

Then we have

$$(L - \lambda)G = \begin{pmatrix} -g_1'' - \lambda g_1 & -g_3'' - \lambda g_3 \\ g_1'(1, \xi, \lambda) - \lambda g_2(\xi, \lambda) & g_3'(1, \lambda) - \lambda g_4(\lambda) \end{pmatrix}.$$

Here primes denote differentiation with respect to x. G is the solution of the matrix equation

$$(L - \lambda)G = \begin{pmatrix} \delta(x - \xi) & 0 \\ 0 & 1 \end{pmatrix},$$

and G also satisfies the following boundary conditions:

$$g_1(0, \xi, \lambda) = g_3(0, \lambda) = 0,$$

$$g_1(1, \xi, \lambda) - g_2(\xi, \lambda) = g_3(1, \lambda) - g_4(\lambda) = 0.$$

These conditions imply that each column of G, considered as a function of x, belongs to the domain of the operator L.

We find that

$$g_1(x, \xi, \lambda) = \frac{\sqrt{\lambda} \sin \sqrt{\lambda}\xi}{\lambda(\cos \sqrt{\lambda} - \sqrt{\lambda} \sin \sqrt{\lambda})} \cdot [\cos \sqrt{\lambda}(1-x) - \sqrt{\lambda} \sin \sqrt{\lambda}(1-x)] H(x-\xi),$$

$$g_2(\xi, \lambda) = \frac{\sin \sqrt{\lambda}\xi}{\sqrt{\lambda}\,(\cos \sqrt{\lambda} - \sqrt{\lambda} \sin \sqrt{\lambda})}$$

$$g_3(x, \lambda) = \frac{\sin \sqrt{\lambda}x}{\sqrt{\lambda}\,(\cos \sqrt{\lambda} - \sqrt{\lambda} \sin \sqrt{\lambda})}$$

$$g_4(\lambda) = \frac{\sin \sqrt{\lambda}}{\sqrt{\lambda}\,(\cos \sqrt{\lambda} - \sqrt{\lambda} \sin \sqrt{\lambda})}.$$

It is clear that these functions have neither a pole nor a branch-point singularity at $\lambda = 0$, but they do have simple poles at the zeroes of the function $\cos \sqrt{\lambda} - \sqrt{\lambda} \sin \sqrt{\lambda}$. Evaluating the contour integral

$$\frac{1}{2\pi i} \oint G \, d\lambda$$

by means of residues, we obtain the following result:

$$\begin{pmatrix} \delta(x-\xi) & 0 \\ 0 & 1 \end{pmatrix} = \begin{pmatrix} 2\sum \dfrac{\sin \sqrt{\lambda_n}x \sin \sqrt{\lambda_n}\xi}{1 + \sin^2 \sqrt{\lambda_n}} & 2\sum \dfrac{\sin \sqrt{\lambda_n} \sin \sqrt{\lambda_n}x}{1 + \sin^2 \sqrt{\lambda_n}} \\ 2\sum \dfrac{\sin \sqrt{\lambda_n}\xi \sin \sqrt{\lambda_n}}{1 + \sin^2 \sqrt{\lambda_n}} & 2\sum \dfrac{\sin^2 \sqrt{\lambda_n}}{1 + \sin^2 \sqrt{\lambda_n}} \end{pmatrix}.$$

The result obtained agrees with those obtained previously. Note that in the evaluation we have replaced $\sqrt{\lambda_n}$ by $\cot \sqrt{\lambda_n}$ wherever it appears.

PROBLEM

4.12. Find the Green's matrix for the operator considered in Problem 4.9. Use complex integration to obtain the expansion theorem for two arbitrary functions $f(x)$ and $g(x)$.

Continuous Spectrum—Example

We shall consider one final example in order to illustrate the consequences of the presence of a branch cut in the integral. Suppose that L is again $-\dfrac{d^2}{dx^2}$ and that the domain D is the set of all twice-differentiable functions $u(x)$, $0 \leq x < \infty$, such that

$$u(0) = 0, \quad \int_0^\infty u^2 \, dx < \infty.$$

The Green's function G is now that solution of the equation

(4.34) $$-G'' - \lambda G = \delta(x - \xi)$$

which satisfies the conditions

$$G(0, \xi, \lambda) = 0, \quad \int_0^\infty G^2\,dx < \infty.$$

If λ is complex, these conditions will determine a unique Green's function. If λ is real, however, a solution of (4.33) satisfying the above conditions does not exist. This difficulty is due to the fact that infinity is a singular point of the differential equation.

We have seen in the preceding chapter that at such singular points an "outgoing wave" condition should be imposed; consequently, we find that

$$G = \frac{\sin\sqrt{\lambda}x}{\sqrt{\lambda}} e^{i\sqrt{\lambda}\xi} H(\xi - x) + e^{i\sqrt{\lambda}x}\frac{\sin\sqrt{\lambda}\xi}{\sqrt{\lambda}} H(x - \xi).$$

Note that since $\dfrac{\sin\sqrt{\lambda}x}{\sqrt{\lambda}}$ is an analytic function of λ without a branch point at $\lambda = 0$, it is immaterial how the square root of λ is defined; but since the function $e^{i\sqrt{\lambda}x}$ has a branch point at $\lambda = 0$, we must specify the sign of the square root. In order that G as written above should vanish for large real positive values of x, when λ is complex, the imaginary part of the square root of λ must be positive. This can be done if we assume $0 < \arg\lambda < 2\pi$ and then $0 < \arg\lambda^{1/2} < \pi$. We could just as well assume $-2\pi < \arg\lambda < 0$ and $-\pi < \arg\lambda^{1/2} < 0$ and then use $e^{-i\sqrt{\lambda}x}$ in the Green's function; or we might assume $-\pi < \arg\lambda < \pi$, and then G would contain $e^{+i\sqrt{\lambda}x}$ for $\arg\lambda > 0$, but it would contain $e^{-i\sqrt{\lambda}x}$ for $\arg\lambda < 0$. The simplest way is to assume $0 < \arg\lambda < 2\pi$, and we shall do so.

Now consider

$$\frac{1}{2\pi i}\oint G\,d\lambda.$$

Since G has a branch point at $\lambda = 0$, we introduce a branch cut in the complex λ-plane along the positive real axis and then take the contour as a large circle not crossing the branch cut (Fig. 4.1). We shall show later that the contour integral approaches $-\delta(x - \xi)$ as the radius of the circle approaches infinity. Since, by Cauchy's theorem, the integral over the circle is equal to the integral over the branch cut (Fig. 4.2) we have

$$-\delta(x - \xi) = \frac{1}{2\pi i}\int_C G\,d\lambda.$$

Put $\lambda = k^2$, $k > 0$, $d\lambda = 2k\,dk$; then on the upper side of the cut $\sqrt{\lambda} = k$ but on the lower side $\sqrt{\lambda} = -k$; consequently,

$$\delta(x-\xi)=\frac{1}{2\pi i}\left[2H(\xi-x)\left\{\int_0^\infty \sin kx\, e^{ik\xi}\,dk+\int_\infty^0 \sin kx\, e^{-ik\xi}\,dk\right\}\right.$$
$$\left.+\,2H(x-\xi)\left\{\int_0^\infty \sin k\xi\, e^{ikx}\,dk+\int_\infty^0 \sin k\xi\, e^{-ikx}\,dk\right\}\right]$$
$$=\frac{2}{\pi}\int_0^\infty \sin kx \sin k\xi\,dk.$$

This is a well-known result since it is another formulation of the Fourier sine transform theorem. We shall now investigate the behavior of the

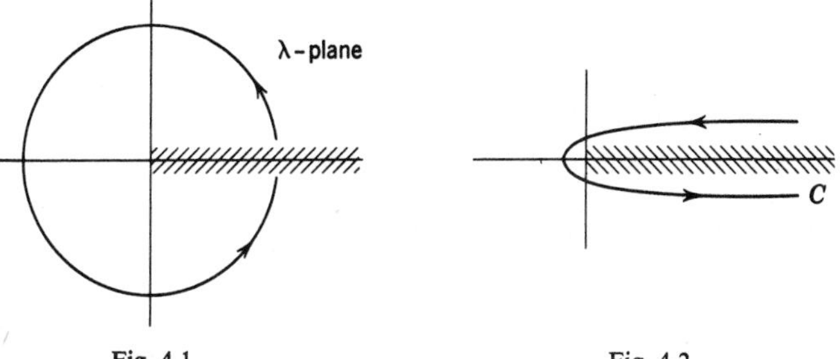

Fig. 4.1

Fig. 4.2

contour integral along the circle of radius R. Again, put $\lambda = k^2$, and the circle in the λ-plane becomes a semicircle of radius $R^{1/2}$ in the upper half of the k-plane. The integral is now

$$\frac{1}{2\pi i}\left[\int_{\curvearrowleft} 2\sin kx e^{ik\xi}\,dk\; H(\xi-x)+\int_{\curvearrowleft} 2\sin k\xi e^{ikx}\,dk\; H(x-\xi)\right]$$
$$=-\frac{1}{2\pi}\left[H(\xi-x)\int_{\curvearrowleft} e^{ik\xi}(e^{ikx}-e^{-ikx})\,dk\right.$$
$$\left.+\,H(x-\xi)\int_{\curvearrowleft} e^{ikx}(e^{ik\xi}-e^{-ik\xi})\,dk\right]$$
$$=-\frac{1}{2\pi}\left[\int_{\curvearrowleft} e^{ik(x+\xi)}\,dk-\int_{\curvearrowleft} e^{ik|x-\xi|}\,dk\right]$$
$$=-\frac{1}{2\pi}\left[-\int_{-\infty}^{\infty} e^{ik(x+\xi)}\,dk+\int_{-\infty}^{\infty} e^{ik(x-\xi)}\,dk\right].$$

The integrals here are obviously non-convergent. This is to be expected since the result we are looking for is not a function but a symbolic function $\delta(x-\xi)$; consequently, we must consider the integrals as defining a symbolic function. To obtain the symbolic function, let us consider the integral

$$I(x)=\int_{-\infty}^{\infty} e^{ikx}\,dk=\lim_{R\to\infty}\int_{-R}^{R} e^{ikx}\,dk$$

applied to a testing function $\phi(x)$ which is continuous, has a continuous derivative, and vanishes outside a finite interval. We define

$$\int_{-\infty}^{\infty} I(x)\phi(x)\,dx = \lim_{R\to\infty} \int_{-\infty}^{\infty} \phi(x)\,dx \int_{-R}^{R} e^{ikx}\,dk$$

$$= \lim 2 \int_{-\infty}^{\infty} \phi(x) \sin Rx \frac{dx}{x}.$$

In Problem 4.13 we shall show that the limit is $2\pi\phi(0)$; therefore $I(x) = 2\pi\delta(x)$. This result is so important that we state it for future reference:

$$\int_{-\infty}^{\infty} e^{ikx}\,dx = 2\pi\delta(x). \tag{4.35}$$

Applying this result to the evaluation of the contour integral, we find that the contour integral is $\delta(x+\xi) - \delta(x-\xi)$. This result implies that if $\phi(x)$ is a testing function as defined above, then

$$\lim_{R\to\infty} \frac{1}{2\pi i}\oint d\lambda \int_{-\infty}^{\infty} G(x,\xi,\lambda)\,\phi(\xi)\,d\xi = \int_{-\infty}^{\infty} [\delta(x+\xi) - \delta(x-\xi)]\phi(\xi)\,d\xi$$
$$= \phi(-x) - \phi(x).$$

We have shown above that the contour integral may be reduced to an integral over the k-axis; consequently, we have

$$\lim \frac{2}{\pi}\int_{0}^{R} \sin kx\,dk \int_{-\infty}^{\infty} \sin k\xi\,\phi(\xi)\,d\xi$$

$$= \lim \frac{1}{\pi}\int_{-R}^{R} \sin kx\,dk \int_{-\infty}^{\infty} \sin k\xi\,\phi(\xi)\,d\xi$$

$$= \phi(x) - \phi(-x).$$

Suppose that we restrict ourselves to testing functions which are odd, that is, $\phi(-x) = -\phi(x)$; then this formula gives

$$\lim \frac{2}{\pi}\int_{-R}^{R} \sin kx\,dk \int_{0}^{\infty} \sin k\xi\,\phi(\xi)\,d\xi = 2\phi(x),$$

when $\phi(x)$ is an odd function of x. Finally, since the integral over ξ is only from 0 to ∞, we may assume $\phi(\xi)$ to be defined only over 0 to ∞ and then extend it to an odd function by putting $\phi(-\xi) = -\phi(\xi)$. In this way we arrive at the following result:

$$\lim \frac{2}{\pi}\int_{0}^{R} \sin kx\,dk \int_{0}^{\infty} \sin k\xi\,\phi(\xi)\,d\xi = \phi(x)$$

if $\phi(x)$ is a continuous and continuously differentiable function over $0 \le x < \infty$, if $\phi(0) = 0$, and if $\phi(x)$ vanishes identically outside a finite interval.

This formula is known as the Fourier sine integral theorem. If we define $\psi(k)$, the Fourier sine transform of $\phi(x)$, by the formula

$$\int_0^\infty \sin k\xi\, \phi(\xi)\, d\xi = \psi(k),$$

we have the inverse formula

$$\frac{2}{\pi}\int_0^\infty \psi(k) \sin kx\, dk = \phi(x),$$

where both $\phi(x)$ and $\psi(k)$ are defined over the half-infinite interval from 0 to ∞.

It is worth noting that we could have obtained the Fourier sine transform theorem from the formula

$$\lim \frac{1}{2\pi i}\oint G\, d\lambda = \delta(x+\xi) - \delta(x-\xi)$$

by reasoning as follows: Since the domain of the operator L is restricted to functions $u(x)$ which are defined over the interval $0 \leq x < \infty$, and since the same is true for $u(\xi)$, it follows that $\delta(x+\xi)$ will contribute zero when applied to a testing function; consequently, we have

$$-\frac{1}{2\pi i}\oint G\, d\lambda = \frac{2}{\pi}\int_0^\infty \sin kx \sin k\xi\, dk = \delta(x-\xi)$$

for testing functions defined over $(0, \infty)$.

PROBLEMS

4.13. If $\phi(x)$ is continuous, has a continuous derivative, and vanishes outside a finite interval, the limit of $\int_{-\infty}^\infty \phi(x) \sin Rx \frac{dx}{x}$ as $R \to \infty$ is $\pi\phi(0)$. (*Hint.* $\int \phi(0) \frac{\sin Rx}{x}\, dx = \pi\phi(0)$. Suppose that $|\phi(x) - \phi(0)| < \varepsilon$ when $|x| < \eta$, and put $\int |\phi(x) - \phi(0)| \frac{\sin Rx}{x}\, dx = \int_{|x|<\eta} + \int_{|x|>\eta} = I_1 + I_2$. Integration by parts shows that I_2 goes to zero as $1/R$, no matter what η is. We have $|I_1| < 2\varepsilon R\eta$. Put $\varepsilon = R^{-2}$.)

4.14*a*. Use the Green's function for the operator $L = -\frac{d^2}{dx^2}$, with the conditions $u'(0) = 0$, $u(\infty)$ outgoing to obtain the following eigenfunction representation:

$$\frac{2}{\pi}\int_0^\infty \cos kx \cos k\xi\, dk = \delta(x-\xi).$$

b. Use the Green's function for the operator $L = -\frac{d^2}{dx^2}$, with the conditions $u(-\infty)$ and $u(+\infty)$ outgoing to obtain the following eigenfunction representation:

$$\frac{1}{2\pi}\int_{-\infty}^\infty e^{ikx}e^{-ik\xi}\, dk = \delta(x-\xi).$$

4.15. Prove (4.31) if G is given by (4.32).

(*Hint.* Put $\lambda = k^2$. Replace trigonometric functions by exponentials, and show that

$$G = \frac{\exp ik\,|x - \xi|}{2ik}[1 + \alpha(k)],$$

where $\alpha(k)$ is a function of k that vanishes exponentially with k if $0 < x < 1$ and $0 < \xi < 1$.)

Singularities of the Green's Function

The typical form of the Green's function of a differential equation is illustrated in (4.33). The second term of that formula is an analytic function of λ which has a singularity at $x = \xi$. The first term is a sum of solutions of the homogeneous equation divided by the conjunct of these solutions. That this is the behavior of the Green's function for a general differential operator may be seen from (3.61) in Chapter 3.

When the Green's function is integrated around a large circle in the λ-plane, the second term, being an analytic function, does not contribute anything. The first term is a linear combination of solutions of the homogeneous equation so constructed that the boundary conditions on the Green's function will be satisfied.

We shall now investigate the structure of the Green's function in the particular case where L is a second-order differential operator defined over the interval (0, 1). Let Lu be $-r^{-1}[(pu')' - qu]$, where we assume that p, q, and r are continuous functions and that $p > 0$, $q > 0$ in the closed interval (0, 1). Let $v_1(x, \lambda)$, $v_2(x, \lambda)$ be solutions of $Lu = \lambda u$ satisfying the respective boundary conditions:

$$v_1(0) = 1, \quad v_2(0) = 0,$$
$$p(0)v_1'(0) = 0, \quad p(0)v_2'(0) = 1;$$

consequently, the conjunct of v_1 and v_2 is unity. The function

$$g = p(\xi)[v_1(x)v_2(\xi) - v_2(x)v_1(\xi)]H(x - \xi)$$

is then a solution of

$$Lg - \lambda g = \delta(x - \xi).$$

To find the Green's function G we must add to g such a linear combination of v_1 and v_2 that the resulting function will satisfy some given boundary conditions such as $B_1(G) = B_2(G) = 0$. We assume B_1 and B_2 are independent of λ.

Put

$$G = g + \alpha v_1(x, \lambda) + \beta v_2(x, \lambda); \tag{4.36}$$

then

$$0 = B_1(G) = B_1(g) + \alpha B_1(v_1) + \beta B_1(v_2),$$
$$0 = B_2(G) = B_2(g) + \alpha B_2(v_1) + \beta B_2(v_2).$$

Put

$$\Delta = \begin{vmatrix} B_1(v_1) & B_1(v_2) \\ B_2(v_1) & B_2(v_2) \end{vmatrix},$$

and we find that

$$\alpha = \frac{B_2(g)B_1(v_2) - B_1(g)B_2(v_2)}{\Delta},$$

(4.37)

$$\beta = \frac{B_1(g)B_2(v_1) - B_2(g)B_1(v_1)}{\Delta}.$$

Note that α and β are functions of ξ and λ. Now clearly† $v_1(x, \lambda)$ and $v_2(x, \lambda)$ are entire functions of λ; consequently, g, $B_1(v_1)$, $B_1(v_2)$, $B_2(v_1)$, and $B_2(v_2)$ are also entire functions of λ. This shows that the only possible singularities of G are at the zeroes of Δ.

We shall assume, hereafter, that Δ is not identically zero because the case $\Delta \equiv 0$ is a very exceptional one. As an illustration of this, consider the equation

$$u'' + \lambda u = 0,$$

with the conditions

$$B_1(u) = u(0) - u(1) = 0,$$
$$B_2(u) = u'(0) + u'(1) = 0.$$

We have $v_1 = \cos \sqrt{\lambda} x$, $v_2 = \sin \sqrt{\lambda} x / \sqrt{\lambda}$, and then

$$\Delta = \begin{vmatrix} 1 - \cos \sqrt{\lambda} & -\sin \sqrt{\lambda}/\sqrt{\lambda} \\ -\sqrt{\lambda} \sin \sqrt{\lambda} & 1 + \cos \sqrt{\lambda} \end{vmatrix} \equiv 0.$$

It follows that every value of λ is an eigenvalue of this equation and that the eigenfunction is

$$u = \cos \sqrt{\lambda} x + \cos \sqrt{\lambda}(1 - x).$$

For this case there is no spectral representation possible.

Suppose that Δ is not identically zero. If $\Delta = 0$ for $\lambda = \lambda_0$, the columns of Δ must be linearly dependent; therefore, there exist scalars α_1, α_2, not both zero, such that

$$0 = \alpha_1 B_1(v_1) + \alpha_2 B_1(v_2) = B_1(\alpha_1 v_1 + \alpha_2 v_2),$$
$$0 = \alpha_1 B_2(v_1) + \alpha_2 B_2(v_2) = B_2(\alpha_1 v_1 + \alpha_2 v_2).$$

This shows that the function

$$u(x, \lambda_0) = \alpha_1 v_1(x, \lambda_0) + \alpha_2 v_2(x, \lambda_0)$$

† See Ince, *Ordinary Differential Equations*, Dover, 1944, pp. 72, 73.

satisfies the boundary conditions $B_1(u) = B_2(u) = 0$; consequently, $u(x, \lambda_0)$ is an eigenfunction of L corresponding to the eigenvalue λ_0. Conversely, suppose that λ_0 is an eigenvalue of L and $w(x, \lambda_0)$ is the corresponding eigenfunction. Since $v_1(x, \lambda_0)$ and $v_2(x, \lambda_0)$ are linearly independent solutions of $(L - \lambda_0)v = 0$, there must exist scalars β_1, β_2, not both zero, such that

$$w(x, \lambda_0) = \beta_1 v_1(x, \lambda_0) + \beta_2 v_2(x, \lambda_0).$$

We have

$$\begin{aligned} 0 &= B_1(w) = \beta_1 B_1(v_1) + \beta_2 B_1(v_2), \\ 0 &= B_2(w) = \beta_1 B_2(v_1) + \beta_2 B_2(v_2). \end{aligned}$$

This shows that the columns of Δ are linearly dependent and therefore $\Delta = 0$. We state these results as

Theorem 4.3. *The number λ_0 is an eigenvalue of L if, and only if, Δ vanishes for $\lambda = \lambda_0$.*

Let us introduce the functions

$$\begin{aligned} u_1(x, \lambda) &= B_2(v_2)v_1 - B_2(v_1)v_2, \\ u_2(x, \lambda) &= B_1(v_2)v_1 - B_2(v_1)v_2. \end{aligned}$$

Note that u_1 and u_2 are solutions of $Lu = \lambda u$ satisfying the conditions $B_2(u_1) = B_1(u_2) = 0$ and that u_1 and u_2 are linearly independent if, and only if, $\Delta \neq 0$. We may write

$$G = g - \gamma_1 u_1(x, \lambda) - \gamma_2 u_2(x, \lambda)$$

where

$$\gamma_1 = \frac{B_1(g)}{B_1(u_1)}, \gamma_2 = \frac{B_2(g)}{B_2(u_2)}.$$

Since $B_1(u_1) = - B_2(u_2) = \Delta$, we see that, if Δ vanishes for $\lambda = \lambda_0$, the Green's function has a pole at $\lambda = \lambda_0$.

The residue of the Green's function at this pole will give the eigenfunctions corresponding to λ_0. The number of eigenfunctions corresponding to λ_0 will depend on the structure of the Green's function and not on whether λ_0 is a simple or a multiple pole of the Green's function. In the next section we shall discuss an example in which every pole of the Green's function is simple, and yet there are two eigenfunctions corresponding to every eigenvalue except the lowest.

There is one important case in which there is only one eigenfunction corresponding to every eigenvalue; it is the case of a second-order differential operator with *unmixed* boundary conditions. This result is an immediate consequence of the fact that an unmixed boundary condition for a second-order differential equation determines the solution uniquely except for a multiplicative constant.

Suppose now that the Green's function has a multiple pole of order ν at $\lambda = \lambda_0$. From (4.36) and (4.37) we see that we may write

$$G = g + \frac{c_1(\xi, \lambda)v_1(x, \lambda) + c_2(\xi, \lambda)v_2(x, \lambda)}{\Delta},$$

where

$$c_1 = B_2(g)B_1(v_2) - B_1(g)B_2(v_2),$$
$$c_2 = B_1(g)B_2(v_1) - B_2(g)B_1(v_1).$$

Put $\Delta = (\lambda - \lambda_0)^\nu r(\lambda)$ where $r(\lambda_0) \neq 0$; then the residue of G at λ_0 will be

$$\frac{1}{(\nu - 1)!} \frac{\partial^{\nu-1}}{\partial \lambda^{\nu-1}} \left[\frac{c_1(\xi, \lambda)v_1(x, \lambda) + c_2(\xi, \lambda)v_2(x, \lambda)}{r(\lambda)} \right]_{\lambda=\lambda_0}$$

Differentiation will give terms containing the eigenfunctions $v_1(x, \lambda_0)$ and $v_2(x, \lambda_0)$ and also terms containing the derivatives of the eigenfunctions with respect to λ. These derivatives of the eigenfunctions will not be solutions of the equation $(L - \lambda)u = 0$ but of the equation $(L - \lambda)^\mu u = 0$. We shall give a fuller discussion of this possibility in the next section.

Multiple Eigenfunctions and Multiple Poles of the Green's Function

In this section we discuss two examples. The first is an example of a Green's function which has only simple poles but yet there are two eigenfunctions for every eigenvalue except the lowest. The second example is of a Green's function which has multiple poles.

For the first example, consider the equation $u'' + \lambda u = 0$, with the periodic boundary conditions $u(0) = u(1)$ and $u'(0) = u'(1)$. By the methods of the preceding section we find that the Green's function for this problem is

$$G = -\frac{\sin\sqrt{\lambda}(x - \xi)}{\sqrt{\lambda}} H(x - \xi) - \frac{\cos\sqrt{\lambda}x \cos\sqrt{\lambda}(\xi - 1/2)}{2\sqrt{\lambda}\sin(\sqrt{\lambda}/2)}$$
$$- \frac{\sin\sqrt{\lambda}x}{\sqrt{\lambda}} \frac{\sin\sqrt{\lambda}(\xi - 1/2)}{2\sin(\sqrt{\lambda}/2)}.$$

The poles of G are at the points where $\sin(\sqrt{\lambda}/2) = 0$, that is, at $\lambda_n = (2n\pi)^2, n = 0, 1, 2, \cdots$. Every pole is simple, but for every value of λ_n, except $\lambda = 0$, the eigenfunctions are $\cos nx$ and $\sin nx$.

From this example, it is clear how it would be possible for an eigenvalue of an mth-order differential equation to have *m-fold degeneracy*, that is, m independent eigenfunctions corresponding to the same eigenfunction. We shall see later that partial differential operators may have eigenvalues with arbitrarily high degeneracy.

For the second example, consider again the equation $u'' + \lambda u = 0$

but now with the boundary conditions $u(0) = 0$ and $u'(1) = cu(1)$. Here c is a complex constant equal to $\sqrt{\lambda_0} \cot \sqrt{\lambda_0}$, where λ_0 is that non-zero root of the equation $\sin 2\sqrt{\lambda_0} = 2\sqrt{\lambda_0}$ which has the smallest absolute value. The Green's function of this problem is

$$G = \frac{\sin \sqrt{\lambda} x}{\sqrt{\lambda}} \frac{\sqrt{\lambda} \cos \sqrt{\lambda}(1 - \xi) - c \sin \sqrt{\lambda}(1 - \xi)}{\sqrt{\lambda} \cos \sqrt{\lambda} - c \sin \sqrt{\lambda}} H(\xi - x)$$
$$+ \frac{\sin \sqrt{\lambda}\xi}{\sqrt{\lambda}} \frac{\sqrt{\lambda} \cos \sqrt{\lambda}(1 - x) - c \sin \sqrt{\lambda}(1 - x)}{\sqrt{\lambda} \cos \sqrt{\lambda} - c \sin \sqrt{\lambda}} H(x - \xi).$$

The singularities of G are poles at the zeroes of the function $\sqrt{\lambda} \cos \sqrt{\lambda} - c \sin \sqrt{\lambda}$. We shall show that λ_0 is a double pole at G whereas all the other poles are simple. The calculation will be simplified if, instead of considering

$$\frac{1}{2\pi i} \oint G \, d\lambda,$$

we put $\lambda = k^2$ and consider

$$\frac{1}{2\pi i} \int 2Gk \, dk.$$

Now G as a function of k is

$$\frac{\sin kx}{k} \frac{k \cos k(1 - \xi) - c \sin k(1 - \xi)}{k \cos k - c \sin k} H(\xi - x)$$

plus a similar term with x and ξ interchanged. If we expand the denominator in powers of $k - k_0$, we find that

$$\begin{aligned} k \cos k - c \sin k = {} & k_0 \cos k_0 - c \sin k_0 \\ & + (k - k_0)(\cos k_0 - k_0 \sin k_0 - c \cos k_0) \\ & + \frac{(k - k_0)^2}{2!} (c \sin k_0 - k_0 \cos k_0 - 2 \sin k_0) + \cdots. \end{aligned}$$

Since $c = k_0 \cot k_0$, where $\sin k_0 \cos k_0 = k_0$, the coefficients of the zeroth and first powers of $k - k_0$ vanish. We have then

$$k \cos k - c \sin k = -(k - k_0)^2 \sin k_0 + \cdots.$$

This shows that G has a double pole at $k = k_0$. A similar expansion shows that at every other zero of $k \cos k - c \sin k$ the Green's function has a simple pole.

The residue at the simple poles $k = k_n$, where $c = k_n \cot k_n$, is

$$\frac{2c \sin k_n x \sin k_n \xi}{c - \cos^2 k_n}.$$

At the double pole $k = k_0$, the residue is

$$\frac{2c}{k_0}[(\sin k_0x)(\xi \cos k_0\xi) + (x \cos k_0x)(\sin k_0\xi)].$$

We have finally the expansion

$$\delta(x - \xi) = 2c\sum_1^\infty \frac{\sin k_nx \sin k_n\xi}{c - \cos^2 k_n} + \frac{2c}{k_0}[\xi \cos k_0\xi \sin k_0x + x \cos k_0x \sin k_0\xi].$$

It is interesting to note that the functions $\sin k_nx$ $(n = 0, 1, 2, \cdots)$ are eigenfunctions but that the function $x \cos k_0x$ is not an eigenfunction. We may call it an eigenfunction of the second rank by analogy with the eigenvectors of second rank that were considered in Chapter 2 (page 68). In fact, we have

$$(L - \lambda_0)^2(x \cos k_0x) = 0$$

but

$$(L - \lambda_0)(x \cos k_0x) \neq 0.$$

The orthogonality properties and the expansion coefficients are obtained exactly as those obtained for eigenvectors of higher rank in Chapter 2.

PROBLEM

4.16. Find the expansion of a function $f(x)$ in terms of the eigenfunctions and generalized eigenfunctions of the second operator discussed in this section.

Perturbation of the Discrete Spectrum

As we have remarked before, often the solution of a differential equation cannot be expressed in terms of known functions. This fact makes the calculation of eigenvalues and eigenfunctions extremely difficult. In this section we present a method for approximating the eigenvalues and the eigenfunctions of a given differential operator by means of the eigenvalues and eigenfunctions of a different, simpler operator. This method is called a perturbation method because it is assumed that the difference between the given operator and the simpler operator is only a small perturbation of the latter.

Suppose that L is a self-adjoint operator, not necessarily a differential operator, and that we wish to find the eigenvalues and eigenfunctions of L. Suppose that L_0 is a self-adjoint operator having a complete set of normalized eigenfunctions $v_1, v_2, \cdots$ with the corresponding eigenvalues $\nu_1, \nu_2, \cdots$. Consider the eigenfunction equation for L, namely,

$$(L - \lambda)u = 0.$$

We shall show how to express λ and u in terms of the eigenvalues and the eigenfunctions of L_0. Put $L = L_0 + \Delta L$; then we may write the above equation as follows:

$$(L_0 - \lambda)u = -\Delta Lu. \tag{4.38}$$

This equation may be considered as a non-homogeneous equation involving the operator $L_0 - \lambda$ with $-\Delta Lu$ as the non-homogeneous term. The solution of (4.38) may then be written as follows:

$$u = -\frac{1}{L_0 - \lambda}\Delta Lu;$$

or, using the spectral representation of L_0, we find that

$$u = +\sum_1^\infty \frac{1}{\lambda - \nu_k}\langle v_k, \Delta Lu\rangle v_k. \tag{4.39}$$

This result, of course, is not a solution to the problem of finding the eigenfunctions of L since the right-hand side depends on the unknown function ΔLu. However, now we assume that ΔL is a small operator; more precisely, we assume that ΔL is a bounded operator with bound ε. We assume also that λ is close to the nth eigenvalue ν_n and that u is close to the nth eigenfunction v_n; therefore, we may put into (4.39)

$$\lambda = \nu_n + \alpha_1\varepsilon + \alpha_2\varepsilon^2 + \cdots,$$
$$u = v_n + w_1\varepsilon + w_2\varepsilon^2 + \cdots,$$

where $\alpha_1, \alpha_2, \cdots$ are unknown constants and $w_1, w_2, \cdots$ are unknown functions. Then, comparing the coefficients of ε, we obtain an infinite set of equations which may be solved for the unknown constants α and the unknown functions w. It can be shown† that the series so obtained converges if ε is small enough.

In many applications all that is needed is the term in the first power of ε, and this can be found quite easily. Since u is approximately equal to v_n, put u equal to v_n in the right-hand side of (4.39). We find that

$$u \sim \sum_1^\infty \frac{1}{\lambda - \nu_k}\langle v_k, \Delta Lv_n\rangle v_k.$$

Since u is equal to v_n approximately, the coefficient of v_n must be approximately one; therefore,

$$\lambda - \nu_n \sim \langle v_n, \Delta Lv_n\rangle. \tag{4.40}$$

Using this value of λ, we find that

$$u \sim v_n + \sum{}' \frac{1}{\lambda - \nu_k}\langle v_k, \Delta Lv_n\rangle v_k, \tag{4.41}$$

† Rellich, *Storungstheorie der Spektralzerlegung*, *Mathematischer Annalen*, Vol. 113, 1936, p. 600.

where the prime indicates that the sum is to be taken over all values of k except $k = n$ and where λ is given by (4.40). Equations (4.40) and (4.41) give the nth eigenvalue and nth eigenfunction of L correct to terms in the first power of ε.

As an example of this theory, consider the problem of finding the eigenvalues and eigenfunctions of the operator $L = -\dfrac{d^2}{dx^2}$, with the boundary conditions $u(0) = -\alpha u'(0)$, and $u(1) = 0$, where α is a given constant. This problem can be solved exactly very easily. The eigenvalues are the roots of the equation

$$\sin k = \alpha k \cos k.$$

However, for illustrative purposes we shall do this problem by considering L as a perturbation of the operator $L_0 = -\dfrac{d^2}{dx^2}$, with the boundary conditions $u(0) = 0$ and $u(1) = 0$. We can write the eigenfunction equation for L, namely,

$$(L - \lambda)u = 0$$

as follows:

$$(L_0 - \lambda)u = \alpha\delta'(x)u'(0). \tag{4.42}$$

Here we have used the extended definition of the operator L_0 as defined in Chapter 3, namely,

$$L_0 w = -w'' - w(0)\delta'(x) + w(1)\delta'(x-1),$$

where w is any function having piecewise continuous second derivatives.

Since the eigenfunctions of L_0 are $\sin k\pi x$ $(1 \leq k < \infty)$ and the eigenvalues are $(k\pi)^2$, the solution of (4.42) is

$$u(x) = +2\alpha u'(0) \sum_1^\infty \frac{k\pi \sin k\pi x}{\lambda - k^2\pi^2}.$$

We assume that the nth eigenfunction $u \sim \sin n\pi x$ and that $\lambda \sim n^2\pi^2$. Putting $u'(0) = n\pi$ in the above equation, we find that

$$\lambda - n^2\pi^2 \sim 2\alpha n^2\pi^2 \tag{4.43}$$

and

$$u(x) \sim \sin n\pi x + 2\alpha \sum{}' \frac{\pi^2 kn \sin k\pi x}{\lambda - k^2\pi^2},$$

where the prime indicates that the sum is over all values of k except $k = n$.

PROBLEMS

4.17. Show that the value of λ given by (4.43) is an approximate solution of $\sin k = \alpha k \cos k$ (put $\lambda = k^2$) if $n\pi\alpha$ is small compared to one.

4.18. Find the first eigenvalue and the first eigenfunction of $L = -\dfrac{d^2}{dx^2}$

$+ \dfrac{2}{(1+x)^2}$, with the boundary conditions $u(0) = u(1) = 0$, by perturbing the operator $L_0 = -\dfrac{d^2}{dx^2}$, with the boundary conditions $u(0) = u(1) = 0$. Compare the approximate result with the exact result. (*Hint.* The general solution of $Lu = k^2u$ is $u = a\left(\dfrac{\cos kx}{1+x} + k \sin kx\right) + b\left(\dfrac{\sin kx}{1+x} - k \cos kx\right)$.)

The Continuous Spectrum—Example

In order to acquire a better understanding of the part the continuous spectrum plays in the spectral representation of an operator, we shall consider the following example.

Let $L = -\dfrac{d^2}{dx^2}$ in the domain of functions $u(x)$ which have piecewise continuous second derivatives, which satisfy the condition

$$u'(0) = \alpha u(0),$$

and which are such that both $u(x)$ and $u''(x)$ are square integrable over the interval $(0, \infty)$. We shall obtain the spectral representation of this operator by the method we have used before, that is, by constructing the Green's function and integrating it around a large circle in the complex λ-plane.

The construction of the Green's function presents certain difficulties when the operators considered are defined over a semi-infinite or infinite interval. We know that the Green's function $G(x, \lambda, \xi)$ is a solution of the equation

$$G'' + \lambda G = -\delta(x - \xi), \tag{4.44}$$

which satisfies the boundary condition

$$G_x(0, \xi, \lambda) = \alpha G(0, \xi, \lambda). \tag{4.45}$$

However, these conditions are not enough to define G uniquely since to any solution of (4.44) satisfying (4.45) we may add an arbitrary multiple of $\cos \sqrt{\lambda}x + \alpha \sin \sqrt{\lambda}x/\sqrt{\lambda}$ and thus get another solution of (4.44) satisfying (4.45). To define G uniquely, some kind of boundary condition at infinity will be needed. One such condition is that, for λ-complex, G belong to $\mathcal{L}_2$ over the interval $(0, \infty)$. We shall show that whenever this condition can be satisfied, it is equivalent to the condition that G vanish at infinity.

From (4.44) we see that for $x > \xi$,

$$G = ae^{i\sqrt{\lambda}x} + be^{-i\sqrt{\lambda}x}.$$

If λ is complex or if λ is real and negative, one of the terms in this expression goes exponentially to ∞ and the other goes exponentially to zero. It is clear then that for these values of λ the Green's function belongs to $\mathcal{L}_2$ if, and only if, it goes exponentially to zero for large values of x. But what about real positive values of λ? In this case both exponentials are bounded at infinity and neither term, nor any linear combination, belongs to $\mathcal{L}_2$. We therefore apply the principle of analytic continuation and define the Green's function for real positive values of λ as the limit of the Green's function with complex λ as λ approaches the real axis.

Notice that this definition still does not specify the Green's function uniquely because λ may approach the positive real axis from above or from below, and these two different approaches will give different Green's functions. If we consider values of λ such that $\text{Im}\sqrt{\lambda} > 0$, the Green's function for real positive values of λ will behave like a multiple of $e^{i\sqrt{\lambda}x}$ for large values of x. On the other hand, if we consider values of λ such that $\text{Im}\sqrt{\lambda} < 0$, the Green's function for real values of λ will behave like a multiple of $e^{-i\sqrt{\lambda}x}$ for large values of x. Since we shall always assume a time factor in the form $e^{-i\omega t}$, the Green's function defined by analytic continuation from the Green's function defined for $\text{Im}\sqrt{\lambda} > 0$ will behave like "outgoing waves" for large values of x. Henceforth, the Green's function will mean that Green's function which behaves like outgoing waves at infinity. In a later chapter we shall see why this is a natural requirement.

We shall use the principle of analytic continuation in more general cases. We formulate it as the following

Rule. *Suppose that L is a differential operator which is in the limit-point case† at infinity; then we define the Green's function G for L by implicitly requiring that G vanish for large values of x if λ is complex and Im $\sqrt{\lambda} > 0$. For real values of λ, G is defined as the limit of G for complex values of λ as λ approaches the real axis.*

Let us now return to the solution of (4.44) and (4.45) for complex values of λ. We assume that $0 < \arg\lambda < 2\pi$ and choose that branch of the square root for which

$$0 < \arg\lambda^{1/2} < \pi. \tag{4.46}$$

The term $e^{i\sqrt{\lambda}x}$ will vanish exponentially as x approaches infinity; consequently,

$$G = \left(\cos\sqrt{\lambda}x + \frac{\alpha}{\sqrt{\lambda}}\sin\sqrt{\lambda}x\right)\frac{e^{i\sqrt{\lambda}\xi}}{i\sqrt{\lambda}-\alpha}H(\xi - x) +$$
$$\left(\cos\sqrt{\lambda}\xi + \frac{\alpha}{\sqrt{\lambda}}\sin\sqrt{\lambda}\xi\right)\frac{e^{i\sqrt{\lambda}x}}{i\sqrt{\lambda}-\alpha}H(x-\xi).$$

† A differential operator L is said to be in the *limit-point* case at infinity if the differential equation without boundary conditions, represented by $Lu = 0$, has at least one solution which is not of integrable square in an interval containing infinity.

Consider the integral of G over a large circle of radius R in the complex λ-plane. Just as in Problem 4.15, we can show that

$$\frac{1}{2\pi i}\oint G\,d\lambda = \frac{1}{2\pi i}\oint \frac{e^{i\sqrt{\lambda}|x-\xi|}}{i\sqrt{\lambda}-\alpha}\,d\lambda$$

plus terms of higher order in $1/R$. Put $\lambda = k^2$, and we get

$$\frac{1}{2\pi i}\oint G\,d\lambda = \frac{1}{2\pi i}\int \frac{e^{ik|x-\xi|}}{ik-\alpha}\,2k\,dk$$

plus terms which go to zero as $R \to \infty$.

We now use Cauchy's theorem to deform this circle into a contour around the singularities of G. Note that G has a branch-point singularity at $\lambda = 0$ and a possible pole at $\sqrt{\lambda} = -i\alpha$. Because of the branch-point singularity the value of G at A in Fig. 4.3 is not the same as the value

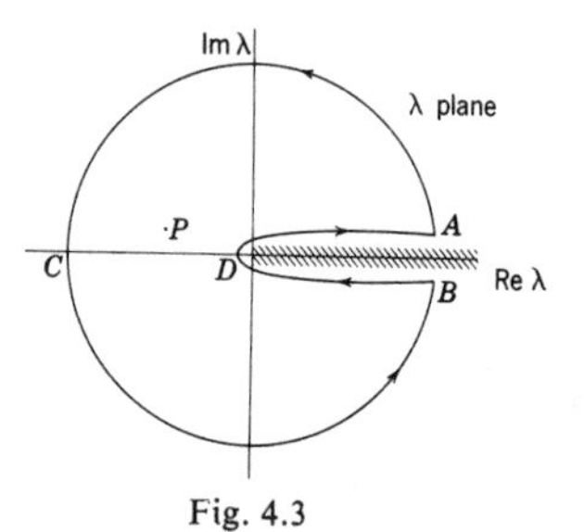

Fig. 4.3

of G at B; consequently, Cauchy's theorem does not apply to the circle ACB since G is not a single-valued function on this contour. However, on the curve $ACBDA$ (here the real axis has been taken as a branch cut) the function G is single-valued, and therefore by Cauchy's theorem

$$\frac{1}{2\pi i}\int_{ACBDA} G\,d\lambda$$

equals the sum of the residues of G inside the circle. The only possible singularity of G inside the circle is at the point P, where $\sqrt{\lambda} = -i\alpha$. Since by (4.46) the imaginary part of $\sqrt{\lambda}$ is positive, P will be inside the circle if, and only if, the real part of α is negative. Suppose that P is inside the circle; then the residue at P is

$$-2\alpha e^{\alpha(x+\xi)}[H(\xi - x) + H(x - \xi)] = -2\alpha e^{\alpha(x+\xi)}.$$

We have

$$-2\alpha e^{\alpha(x+\xi)} = \frac{1}{2\pi i}\int_{ACBDA} G\,d\lambda = \frac{1}{2\pi i}\int_{ACB} + \frac{1}{2\pi i}\int_{BDA};$$

and, therefore, since we assume that the first integral on the right-hand side gives a δ-function, we get

$$(4.47) \qquad -\delta(x-\xi) = \frac{1}{2\pi i}\int_{ACB} = -2\alpha e^{\alpha(x+\xi)} - \frac{1}{2\pi i}\int_{BDA} G\, d\lambda.$$

The last integral may be simplified by putting $\lambda = k^2$. Note that on the upper side of the cut $\sqrt{\lambda} = +k$, but on the lower side of the cut $\sqrt{\lambda} = -k$. We have

$$\begin{aligned}\frac{1}{2\pi i}\int_{BDA} G\, d\lambda = {} & \frac{1}{2\pi i}\int_0^\infty \left(\cos kx + \frac{\alpha}{k}\sin kx\right)\frac{e^{ik\xi}}{ik-\alpha}\, 2k\, dk\, H(\xi - x) \\ & + \frac{1}{2\pi i}\int_0^\infty \left(\cos k\xi + \frac{\alpha}{k}\sin k\xi\right)\frac{e^{ikx}}{ik-\alpha}\, 2k\, dk\, H(x-\xi) \\ & + \frac{1}{2\pi i}\int_\infty^0 \left(\cos kx + \frac{\alpha}{k}\sin kx\right)\frac{e^{-ik\xi}}{-ik-\alpha}\, 2k\, dk\, H(\xi - x) \\ & + \frac{1}{2\pi i}\int_\infty^0 \left(\cos k\xi + \frac{\alpha}{k}\sin k\xi\right)\frac{e^{-ikx}}{-ik-\alpha}\, 2k\, dk\, H(x-\xi).\end{aligned}$$

Combining the first and third integrals and also the second and fourth integrals, we get

$$\frac{1}{2\pi i}\int_{BDA} G\, d\lambda = -\frac{2}{\pi}\int_0^\infty \left(\cos kx + \frac{\alpha}{k}\sin kx\right)\left(\cos k\xi + \frac{\alpha}{k}\sin k\xi\right)\frac{k^2\, dk}{k^2+\alpha^2}.$$

From (4.47) we finally obtain the following spectral representation of L:

$$(4.48) \qquad \delta(x-\xi) = -2\alpha e^{\alpha(x+\xi)} + \frac{2}{\pi}\int_0^\infty \left(\cos kx + \frac{\alpha}{k}\sin kx\right)\left(\cos k\xi + \frac{\alpha}{k}\sin k\xi\right)\frac{k^2\, dk}{k^2+\alpha^2}.$$

Note that the first term is missing if the real part of α is positive.

A Direct Approach to the Continuous Spectrum

The previous sections have shown that the method of integrating the Green's function in the complex λ-plane gives the spectral representation for a differential operator whether the spectrum is discrete, continuous, or a mixture of both. However, if the operator has a spectrum that is purely discrete, we may avoid the use of the Green's function since it is comparatively easy to find all the eigenvalues and the eigenfunctions. If we assume these make up the entire spectrum, we have obtained the spectral representation.

We shall show that there exists a direct method for finding the values of λ in the continuous spectrum and the corresponding functions in the spectral representation. This method depends on the concept of the *approximate spectrum* of an operator L. The number λ is in the approximate spectrum of L if, given any $\varepsilon > 0$, there exists an element u_ε in the domain of L such that

$$\langle (L - \lambda)u_\varepsilon, (L - \lambda)u_\varepsilon \rangle < \varepsilon \langle u_\varepsilon, u_\varepsilon \rangle.$$

It can be shown (see page 127) that the approximate spectrum is contained in the spectrum of L. If L is self-adjoint, the approximate spectrum coincides with the spectrum (see Problem 2.53). In any case, if λ is an eigenvalue and u the corresponding eigenfunction, then $(L - \lambda)u$ is identically zero; consequently, all eigenvalues of λ are in the approximate spectrum.

Suppose that λ belongs to the continuous spectrum of L, that is, the inverse of $L - \lambda$ is an unbounded operator; then we can show that λ will belong to the approximate spectrum. Because the inverse of $L - \lambda$ is unbounded, there exists in the domain of the inverse operator $(L - \lambda)^{-1}$ a sequence of functions $v_n(x)$ having the property that

$$\langle (L - \lambda)^{-1}v_n, (L - \lambda)^{-1}v_n \rangle > n \langle v_n, v_n \rangle.$$

Put $u_n = (L - \lambda)^{-1}v_n$, and then we have

$$\langle (L - \lambda)u_n, (L - \lambda)u_n \rangle < \frac{1}{n} \langle u_n, u_n \rangle;$$

consequently, λ belongs to the approximate spectrum. If the functions $u_n(x)$ converge to a limit $u(x)$, we shall say that $u(x)$ is an *improper eigenfunction* corresponding to the *improper eigenvalue* λ. Since λ is in the continuous spectrum, the function $u(x)$ cannot belong to the domain of L, for otherwise it would be an eigenfunction and λ would be in the discrete spectrum.

If the differential operator L is such that it has no residual spectrum, we conclude that the following theorem holds:

Theorem 4.4. *Any value of λ in the approximate spectrum which is not an eigenvalue belongs to the continuous spectrum, and conversely any value of λ which belongs to the continuous spectrum is in the approximate spectrum.*

This theorem enables us to find the continuous spectrum directly. For example, consider the operator whose Green's function was defined in (4.44) and (4.45). We shall show that every real positive value of λ is in the approximate spectrum.

Consider the function†

$$\begin{aligned} u_n(x) &= \cos \sqrt{\lambda}x + \frac{\alpha}{\sqrt{\lambda}} \sin \sqrt{\lambda}x, && 0 \le x \le C \\ &= \frac{\alpha^2}{4}(x - B)^2, && C \le x \le B \\ &= 0, && B \le x \end{aligned}$$

† We assume $\alpha < 0$. A similar method can be used for any value of α, real or complex.

where

$$\sqrt{\lambda}C = 2n\pi$$

$$B = C - \frac{2}{\alpha}.$$

This function is continuous and has a continuous derivative everywhere from 0 to ∞. It satisfies the boundary condition $u'(0) = \alpha u(0)$ and it belongs to $\mathcal{L}_2$; therefore, $u_n(x)$ belongs to the domain of L. Now

$$\begin{aligned} (L - \lambda)u_n &= 0, & 0 \leq x \leq C \\ &= [-2 - \lambda(x - B)^2]\frac{\alpha^2}{4}, & C \leq x \leq B \\ &= 0, & B \leq x. \end{aligned}$$

We see that

$$\begin{aligned} \langle (L - \lambda)u_n, (L - \lambda)u_n \rangle &= \int_C^B [2 + \lambda(x - B)^2]^2 \frac{\alpha^2}{4}\, dx \\ &= \frac{\alpha^2}{4}\int_{C-B}^{0} (2 + \lambda y^2)^2\, dy < K \end{aligned}$$

where K is a constant independent of n. Since

$$\begin{aligned} \langle u_n, u_n \rangle &\geq \int_0^C \left(\cos \sqrt{\lambda}x + \frac{\alpha}{\sqrt{\lambda}} \sin \sqrt{\lambda}x\right)^2 dx \\ &= \frac{1}{2}\int_0^C \left(1 + \frac{\alpha^2}{\lambda}\right) dx + \frac{1}{2}\left(1 - \frac{\alpha^2}{\lambda}\right)\int_0^C \cos 2\sqrt{\lambda}x\, dx \\ &\quad + \frac{\alpha}{\sqrt{\lambda}}\int_0^C \sin 2\sqrt{\lambda}x\, dx \\ &= \frac{C}{2}\left(1 + \frac{\alpha^2}{\lambda}\right), \end{aligned}$$

we have

$$\frac{\langle (L - \lambda)u_n, (L - \lambda)u_n \rangle}{\langle u_n, u_n \rangle} < \frac{K}{\dfrac{C}{2}\left(1 + \dfrac{\alpha^2}{\lambda}\right)}$$

As n increases, C increases, and the right-hand side of this inequality approaches zero; consequently, λ is in the approximate spectrum.

Notice that what we have done is to use the fact that $u = \cos\sqrt{\lambda}x + \frac{\alpha}{\sqrt{\lambda}} \sin \sqrt{\lambda}x$ is a solution of the equation

$$(L - \lambda)u = 0.$$

Since u does not belong to $\mathcal{L}_2$ over the infinite interval, it does not belong to the domain of L and is not an eigenfunction. We, therefore, consider the function which is equal to u over the finite interval $(0, C)$ and equal to zero elsewhere. Since this function would be discontinuous and have a discontinuous derivative at $x = C$, we insert another function to make a smooth transition from u to zero. We shall call the resulting function u_n. Now $L - \lambda$ applied to this cut-off function u_n will give zero except in the transition region, but the integral square of u_n approaches infinity as C approaches infinity. If the function $(L - \lambda)u_n$ in the transition region is bounded for all C, we find that the ratio

$$\frac{\langle (L - \lambda)u_n, (L - \lambda)u_n \rangle}{\langle u_n, u_n \rangle}$$

approaches zero as C approaches infinity, and consequently λ is in the approximate spectrum. Note that the boundedness of the function $(L - \lambda)u_n$ in the transition region is due to the fact that $u(x)$ oscillates infinitely often and is bounded as x goes to infinity.

This same method can be applied to more general differential operators to obtain the continuous spectrum as well as its multiplicity. For example, consider the operator $L = -\dfrac{d^2}{dx^2}$ acting on the space of all piecewise twice-differentiable functions which are square integrable from $-\infty$ to ∞. We can show that every real positive value of λ is an improper eigenvalue of multiplicity two, that is, there exist two improper eigenfunctions, $\cos \sqrt{\lambda} x$ and $\sin \sqrt{\lambda} x$, corresponding to the value λ. The method is similar to that used above. Consider the function $u_n = \cos \sqrt{\lambda} x$, when x is between $-2n\pi/\sqrt{\lambda}$ and $2n\pi/\sqrt{\lambda}$. Add a transition function at both ends to smooth it off to zero. Then we see that the ratio

$$\frac{\langle (L - \lambda)u_n, (L - \lambda)u_n \rangle}{\langle u_n, u_n \rangle}$$

approaches zero as n approaches infinity. This shows that $\cos \sqrt{\lambda} x$ is an improper eigenfunction for the improper eigenvalue λ. A similar device shows that $\sin \sqrt{\lambda} x$ is also an improper eigenfunction for the same value of λ.

We state the following general result whose proof may be found in Titchmarsh *Eigenfunction Expansions:*

Theorem 4.5. *Consider the equation*

$$u'' + (\lambda + q)u = 0 \tag{4.49}$$

over a semi-infinite or infinite interval. If

$$\int |q| \, dx < \infty,$$

every real positive value of λ is an improper eigenvalue and the solutions of

(4.49) *which satisfy the boundary condition at the limit-circle*† *boundary point, if any, are improper eigenfunctions.*

Now that we have a method by which the values of λ in the continuous spectrum together with the corresponding eigenfunctions can be obtained, it is necessary to know how to normalize them so that the spectral representation can be obtained. For example, we have already shown that $\cos kx + \frac{\alpha}{k}\sin kx$, where $k = \sqrt{\lambda}$, are the improper eigenfunctions needed in the expansion (4.48). But the question of how to obtain the normalization factor $k^2(k^2 + \alpha^2)^{-1}$ still remains. For self-adjoint operators with a discrete spectrum, we normalize the eigenfunction v_n so that the integral square of v_n is one. Then we have

$$\langle v_n, v_m\rangle = \delta_{nm},$$

where $\delta_{nm} = 0$ if $n \neq m$ and $= 1$ if $n = m$.

For the continuous spectrum we shall make a similar normalization. Suppose that $u(x, \lambda)$ is an improper eigenfunction corresponding to the improper eigenvalue λ; then we normalize $u(x, \lambda)$ so that

$$\langle u(x, \lambda), u(x, \lambda')\rangle = \delta(\lambda - \lambda'). \tag{4.50}$$

Of course the scalar product does not exist in the ordinary sense since $u(x, \lambda)$ does not belong to $\mathcal{L}_2$. However, the scalar product does exist as a distribution in λ and λ' and must be evaluated as such. To show that (4.50) is necessary, we start with the representation

$$\delta(x - \xi) = \Sigma v_n(x)v_n(\xi) + \int u(x, \lambda)u(\xi, \lambda)\, d\lambda, \tag{4.51}$$

where $v_n(x)$ are the eigenfunctions and $u(x, \lambda)$ the improper eigenfunctions. If we take the scalar product of (4.51), with $u(x, \lambda')$ we get

$$u(\xi, \lambda') = \Sigma v_n(\xi)\langle v_n, u(\lambda)\rangle + \int d\lambda\, u(\xi, \lambda)\langle u(\lambda), u(\lambda')\rangle.$$

For this result to be valid, (4.50) and also $\langle v_n, u(\lambda)\rangle = 0$ must hold. We thus obtain

Theorem 4.6. *In the spectral representation of a self-adjoint operator the improper eigenfunctions are normalized according to* (4.50).

In practice it is not easy to evaluate the divergent integrals occurring in (4.50) and to show that they are δ-functions. Instead of showing how this may be done, we present in the next section a method which automatically obtains the improper eigenfunctions with the proper normalization.

† A *limit-circle* boundary point for a differential operator L is a point at which all solutions of the formal differential equation with no boundary conditions, represented by $Lu = 0$, are square integrable in some interval which has the boundary point as an end point.

The Continuous Spectrum by Perturbation

For some operators it is possible to obtain the improper eigenfunctions by means of perturbation methods. For example, suppose that we know the complete spectral representation of an operator L; then if M is a "small" operator we expect that the spectral representation for $L + M$ differs slightly from that for L. We shall show how the spectral representation for $L + M$ can be obtained by perturbation methods from the spectral representation for L.

For example, we may consider $L = -\dfrac{d^2}{dx^2}$ and $M = q(x)$ acting on the space of twice-differentiable functions which are square integrable over 0 to ∞ and which satisfy some specified boundary condition at $x = 0$. We assume also that $q(x)$ is absolutely integrable from 0 to ∞. The eigenfunctions and improper eigenfunctions of $L + M$ are solutions of the equation

$$u'' + (\lambda - q)u = 0.$$

We shall write this equation as follows:

$$u'' + \lambda u = qu, \tag{4.52}$$

and then, using the Green's function for the equation

$$u'' + \lambda u = 0,$$

we may transform (4.52) into an integral equation. Before formulating this integral equation, we shall consider exactly how the Green's function inverts the differential operator.

The Green's function $G(x, \xi, \lambda)$ is that solution of the equation

$$G'' + \lambda G = -\delta(x - \xi)$$

which satisfies the specified boundary condition at $x = 0$ and an "outgoing wave" condition at $x = \infty$. This outgoing wave condition is formulated as follows: If $\operatorname{Im} \sqrt{\lambda} > 0$, G should vanish exponentially at $x = \infty$; if λ is real, $G(x, \xi, \lambda)$ should be the limit of $G(x, \xi, \lambda + i\varepsilon)$ as ε approaches zero through positive values.

Consider now the solution of the non-homogeneous equation

$$w'' + \lambda w = f(x), \tag{4.53}$$

where $f(x)$ is absolutely integrable from 0 to ∞. If we put

$$w = -\int_0^\infty G(x, \xi, \lambda) f(\xi)\, d\xi, \tag{4.54}$$

w will be the solution of (4.53) which satisfies the specified boundary condition at $x = 0$ and also satisfies an outgoing wave condition at infinity.

There is a partial converse of this result. Suppose that $u(x)$ is any function with piecewise continuous second derivatives which satisfies the specified boundary condition at $x = 0$ and an outgoing wave condition at infinity, and suppose that u and u' are absolutely integrable from 0 to ∞; then consider the following integral:

$$F(x) = -\int_0^\infty G(x, \xi, \lambda)[u''(\xi) + \lambda u(\xi)]\, d\xi.$$

From (4.54) by putting $w = F, f = u'' + \lambda u$ we see that

$$F'' + \lambda F = u'' + \lambda u$$

and therefore $F = u$, unless the equation

$$v'' + \lambda v = 0$$

with the specified boundary condition at zero and the outgoing wave condition at ∞ has a non-trivial solution. It follows that $F(x) = u(x)$.

Let us return to the problem of finding the eigenfunctions and the improper eigenfunctions of (4.52). Suppose first that λ is an eigenvalue for (4.52) and that $u(x)$ is the corresponding eigenfunction; then $u(x)$ and $u''(x)$ are square integrable from 0 to ∞. It can be shown that $u(x)$ and $u''(x)$ are also absolutely integrable from 0 to ∞; therefore,

$$-\int_0^\infty G(x, \xi, \lambda) qu\, d\xi = -\int_0^\infty G(u'' + \lambda u)\, d\xi = u(x)$$

by the use of the partial converse proved above. This means that any eigenfunction of (4.52) will be a solution of the integral equation which results from replacing $f(\xi)$ in (4.54) by $q(\xi)u(\xi)$. The integral equation is:

$$u(x) = -\int_0^\infty G(x, \xi, \lambda) q(\xi) u(\xi)\, d\xi.$$

Suppose, however, that λ is an improper eigenvalue and $u(x)$ the corresponding improper eigenfunction; then u and u' will not be absolutely integrable, $F(x)$ will not equal $u(x)$, and we do not obtain the above integral equation for u.

In Appendix II of Chapter 3 we showed that at infinity $u(x)$ behaves like a linear combination of $e^{i\sqrt{\lambda}}$ and $e^{-i\sqrt{\lambda}x}$. Let us consider complex values of λ such that Im $\sqrt{\lambda} > 0$; then $e^{i\sqrt{\lambda}x}$ will go to zero, but $e^{-i\sqrt{\lambda}x}$ will become infinitely large as x approaches infinity. Since the improper eigenfunction $u(x)$ has an undetermined constant factor, we may choose the constant factor in such a way that, for large values of x, $u(x)$ behaves like $e^{-i\sqrt{\lambda}x}$ plus a constant multiple of $e^{i\sqrt{\lambda}x}$. This means that, if Im $\sqrt{\lambda} > 0$, the difference $u(x) - e^{-i\sqrt{\lambda}x}$ will go to zero exponentially fast as x approaches infinity. To emphasize that λ is complex, we shall replace λ by $\lambda + i\varepsilon$, where now λ will be real and ε will be a small positive number. Let $u_1(x)$ be that solution of the equation

$$u_1'' + (\lambda + i\varepsilon)u_1 = 0$$

which satisfies the specified boundary condition at $x = 0$ and is so normalized that $u_1(x) - \exp(-i\sqrt{\lambda + i\varepsilon}\, x)$ vanishes for large values of x. Put $u = u_1 + v$ in the left-hand side of (4.52), and we get

$$v'' + (\lambda + i\varepsilon)v = qu,$$

where $u_1 + v$ satisfies the specified boundary condition at $x = 0$ and v vanishes at infinity. Since the function $u(x)$ is bounded and, by hypothesis, $q(x)$ is absolutely integrable, qu is absolutely integrable. Since $v(x)$ satisfies the outgoing wave condition at infinity, we may use (4.54) to get the following equation:

$$v(x) = -\int_0^\infty G(x, \xi, \lambda + i\varepsilon)q(\xi)u(\xi)\, d\xi.$$

Finally, replacing v by $u - u_1$, and letting $\varepsilon \to 0$, we obtain the desired integral equation for $u(x)$, namely,

$$u(x) = \lim_{\varepsilon \to 0} \left[u_1 - \int_0^\infty G(x, \xi, \lambda + i\varepsilon)qu\, d\xi\right] \tag{4.55}$$

$$= u_1(x) - \int_0^\infty G(x, \xi, \lambda)qu\, d\xi,$$

where G is the "outgoing wave" Green's function.

We shall derive a similar equation in the general case of a differential operator L and a pertubation operator M. Suppose that we wish to obtain the eigenfunctions and the improper eigenfunctions of $L + M$. These functions will be solutions of the equation

$$(L + M - \lambda)u = 0.$$

In analogy with (4.52), we write this equation as follows:

$$(L - \lambda)u = -Mu.$$

If u is an eigenfunction of $L + M$, we may invert this equation just as before to obtain the integral equation

$$u = -(L - \lambda)^{-1}Mu.$$

If u is an improper eigenfunction of $L + M$, we consider the corresponding improper eigenfunction u_1 which is a solution of the equation

$$(L - \lambda)u_1 = 0.$$

We assume that M is a small perturbation in the sense that $v = u - u_1$ is a function which satisfies the outgoing wave condition at infinity. Replacing u by $u_1 + v$ in the eigenfunction equation for u, we get

$$(L - \lambda)v = -Mu$$

or

$$v = -\lim_{\varepsilon \to 0} (L - \lambda - i\varepsilon)^{-1}Mu.$$

Since $v = u - u_1$, we find that

$$u = u_1 - \lim_{\varepsilon \to 0} (L - \lambda - i\varepsilon)^{-1} M u,$$

an equation similar to (4.55). Henceforth, for convenience in writing, we shall omit the limit and simply write

$$u = u_1 - (L - \lambda - i\varepsilon)^{-1} M u, \tag{4.56}$$

but the second term on the right-hand side is still to be understood as the limit of the operator as $\varepsilon \to 0$.

We shall now show how to normalize the improper eigenfunctions of $L + M$. Suppose that $\phi(\lambda)$ is an improper eigenfunction of L corresponding to the improper eigenvalue λ; therefore,

$$(L - \lambda)\phi(\lambda) = 0.$$

Suppose also that $\phi^*(\lambda)$ is an improper eigenfunction of the adjoint operator L^* and that it is so normalized that

$$\langle \phi(\lambda), \phi^*(\lambda') \rangle = \delta(\lambda - \lambda'). \tag{4.57}$$

Let $\psi(\lambda)$ be that improper eigenfunction of $L + M$ which was obtained by perturbing $\phi(\lambda)$ just as in (4.56) u was obtained by perturbing u_1; then $\psi(\lambda)$ is the solution of the following integral equation:

$$\psi(\lambda) = \phi(\lambda) - (L - \lambda - i\varepsilon)^{-1} M \psi(\lambda). \tag{4.58}$$

A slight modification of the previous argument will show that the function $\psi^*(\lambda)$ which is the solution of the following integral equation

$$\psi^*(\lambda) = \phi^*(\lambda) - \lim_{\varepsilon \to 0} (L^* - \lambda + i\varepsilon)^{-1} M^* \psi^*(\lambda), \tag{4.59}$$

where M^* is the operator adjoint to M, is also an improper eigenfunction of $L^* + M^*$ corresponding to the improper eigenvalue λ. For convenience in writing, we shall omit the limit symbol in (4.59).

Theorem 4.7†. *The improper eigenfunctions $\psi(\lambda)$ and $\psi^*(\lambda)$ satisfy the following relation:*

$$\langle \psi(\lambda), \psi^*(\lambda') \rangle = \delta(\lambda - \lambda').$$

The proof of this result will be found in the appendix to this chapter.

Normalization of the Continuous Spectrum—Examples

Theorem 4.7 can be used to obtain the normalization of the functions $\cos kx + \dfrac{\alpha}{k} \sin kx$ which are the improper eigenfunctions for the

† This result is contained implicitly in a paper by Moeller, *Konigle Danske Videnskabernes Relsked*, Copenhagen, 1945.

previously considered operator $-\dfrac{d^2}{dx^2}$ for $0 \leq x < \infty$, with the boundary condition $u'(0) = \alpha u(0)$. We shall denote this operator by $L + M$, and we shall denote the operator $-\dfrac{d^2}{dx^2}$ for $0 \leq x < \infty$ with the boundary condition $u'(0) = 0$ by L. The spectral representation for L is known from Problem 4.14. It is given by the formula

$$\delta(x - \xi) = \frac{2}{\pi}\int_0^\infty \cos kx \cos k\xi \, dk,$$

where $\lambda = k^2$. Replacing x by k, ξ by k', and k by x, we get

$$\delta(k - k') = \frac{2}{\pi}\int_0^\infty \cos kx \cos k'x \, dx.$$

This shows that $(2/\pi)^{1/2} \cos kx$ are the normalized improper eigenfunctions.

According to Theorem 4.7, the normalized improper eigenfunctions of $L + M$ will be obtained if we find functions $\psi(x, \lambda)$ which are solutions of

$$(L + M)\psi = \lambda\psi \tag{4.60}$$

and which differ from $(2/\pi)^{1/2} \cos kx$ at infinity by a function which vanishes for $\operatorname{Im} k > 0$. We shall need also a function $\psi^*(x, \lambda)$ which is a solution of

$$(L + M)\psi^* = \lambda\psi^*$$

and which differs from $(2/\pi)^{1/2} \cos kx$ at infinity by a function which vanishes for $\operatorname{Im} k < 0$.

To find $\psi(x, \lambda)$ we note that (4.60) is $\psi'' + k^2\psi = 0$ with the boundary condition $\psi'(0, k^2) = \alpha\psi(0, k^2)$. Since e^{ikx} vanishes for large x if $\operatorname{Im} k > 0$, we put

$$\psi = (2/\pi)^{1/2}[\cos kx + \gamma e^{ikx}],$$

where γ is a constant to be determined so that ψ satisfies the specified boundary condition. We find that

$$\gamma = \alpha(ik - \alpha)^{-1}$$

and, therefore, if e^{ikx} is replaced by $\cos kx + i \sin kx$, we get

$$\psi = (2/\pi)^{1/2} ik(ik - \alpha)^{-1}\left[\cos kx + \frac{\alpha}{k}\sin kx\right].$$

Similarly, since L is self-adjoint, to find ψ^* we assume that

$$\psi^* = (2/\pi)^{1/2}\,[\cos kx + \gamma' e^{-ikx}],$$

where γ' is a constant to be determined so that ψ^* satisfies the specified boundary condition. Since the boundary condition and the differential equation are real, we recognize that ψ and ψ^*, γ and γ' must be complex conjugates. (The reader can verify this by finding γ' to satisfy the boundary condition.) We have

$$\psi^* = (2/\pi)^{1/2} ik(\alpha + ik)^{-1} \left[\cos kx + \frac{\alpha}{k} \sin kx\right].$$

From Theorem 4.7 we see that

$$\frac{2}{\pi}\int_0^\infty \frac{kk'}{(\alpha + ik')(\alpha - ik)} \left(\cos kx + \frac{\alpha}{k} \sin kx\right)\left(\cos k'x + \frac{\alpha}{k'} \sin k'x\right) dx = \delta(k - k').$$

Since the δ-function is zero unless $k = k'$, this equation may be rewritten as follows:

$$\frac{2}{\pi}\int_0^\infty \frac{k^2}{\alpha^2 + k^2}\left(\cos kx + \frac{\alpha}{k} \sin kx\right)\left(\cos k'x + \frac{\alpha}{k'} \sin k'x\right) dx = \delta(k - k');$$

consequently, the normalized improper eigenfunctions are

$$(2/\pi)^{1/2} k(\alpha^2 + k^2)^{-1/2} \left(\cos kx + \frac{\alpha}{k} \sin kx\right).$$

This result agrees with (4.48).

As another example of this method we shall obtain the spectral representation for the operator $L + M = -\dfrac{d^2}{dx^2} + \alpha\delta(x)$ acting on the domain of functions $u(x)$ which have piecewise continuous second derivatives for $-\infty < x < 0$ and for $0 < x < \infty$ and which are such that u and u'' are of integrable square from $-\infty$ to ∞. As we mentioned in Chapter 3, the solutions of

$$-u'' + \alpha\delta(x)u = \lambda u \tag{4.61}$$

are functions which are solutions of $-u'' = \lambda u$ for $x \neq 0$, and are continuous at $x = 0$, but are such that the first derivative has a jump at $x = 0$ equal to $\alpha u(0)$.

We can show that $L + M$ is self-adjoint and real if α is real. We shall assume that α is real; then λ will be real. If α is positive, $L + M$ will be positive-definite and λ can only be positive. If α is negative, $\lambda = \alpha/2$ is an eigenvalue and $u(x) = (2|\alpha|)^{-1} \exp(\alpha|x|/2)$ is a normalized eigenfunction.

To normalize the improper eigenfunctions we shall put $L = -\dfrac{d^2}{dx^2}$ over the domain of functions $u(x)$ which have piecewise continuous second derivatives and for which u and u'' are of integrable square from $-\infty$ to ∞. Every positive real value is an improper eigenvalue of L. To any real positive value of λ there are two corresponding improper eigenfunctions $\cos \sqrt{\lambda}x$ and $\sin \sqrt{\lambda}x$. If we put $\lambda = k^2$, we may let one improper eigenfunction correspond to a positive value of k and the other correspond to a negative value of k. In this way we would obtain the following spectral representation:

$$\frac{1}{\pi}\int_0^\infty \cos kx \cos k\xi \, dk + \frac{1}{\pi}\int_{-\infty}^0 \sin kx \sin k\xi \, dk = \delta(x - \xi).$$

Since the integrands are even functions of k, this result can also be written as follows:

$$\frac{1}{2\pi}\int_{-\infty}^\infty \cos kx \cos k\xi \, dk + \frac{1}{2\pi}\int_{-\infty}^\infty \sin kx \sin k\xi \, dk = \delta(x - \xi).$$

A particularly simple form is obtained for the spectral representation if we take $\cos kx + i \sin kx = e^{ikx}$ as one improper eigenfunction and $\cos kx - i \sin kx = e^{-ikx}$ as the other improper eigenfunction corresponding to $\lambda = k^2$. In that case, we have (see Problem 4.15)

$$\frac{1}{2\pi}\int_{-\infty}^\infty e^{ik(x-\xi)} \, dk = \delta(x - \xi).$$

The different normalizations for the improper eigenfunctions of this operator are a consequence of the fact that there are two improper eigenfunctions corresponding to each improper eigenvalue $\lambda = k^2$. We can find many more just by introducing different sets of basis vectors in the two-dimensional space of improper eigenfunctions.

To obtain the normalization for the improper eigenfunctions of $L + M$, we shall apply Theorem 4.7 with $\phi(\lambda) = e^{ikx}$ and $\phi^*(\lambda) = e^{-ikx}$. Then $\psi(\lambda)$ is the limit as ε approaches zero of the solution of

$$-\psi'' + \alpha\delta(x)\psi = (k + i\varepsilon)^2\psi$$

such that the difference between it and $e^{i(k+i\varepsilon)x}$ vanishes for $x = \pm\infty$. This means that $\psi - e^{i(k+i\varepsilon)x}$ must behave like some multiple of $e^{-i(k+i\varepsilon)x}$ for $x = -\infty$ and like some multiple of $e^{i(k+i\varepsilon)x}$ for $x = \infty$. It is clear that once we have this knowledge of the behavior of ψ we may take $\varepsilon = 0$ and put

$$\begin{aligned} \psi &= e^{ikx} + Re^{-ikx}, \quad x < 0 \\ &= e^{ikx} + Se^{ikx} = Te^{ikx}, \quad x > 0, \end{aligned}$$

where R and T are constants to be determined so that the discontinuity conditions at $x = 0$ are satisfied. These conditions imply

$$1 + R = T,$$

$$ikT - ik(1 - R) = \alpha(1 + R).$$

We find that

$$R = \alpha(2ik - \alpha)^{-1}, \quad T = 2ik(2ik - \alpha)^{-1};$$

consequently,

$$\begin{aligned} \psi &= e^{ikx} + \frac{\alpha e^{-ikx}}{2ik - \alpha}, \quad x < 0 \\ &= \frac{2ike^{ikx}}{2ik - \alpha}, \qquad x > 0. \end{aligned}$$

From the fact that $\phi^*(\lambda) = e^{-ikx}$, it follows that $\psi^*(\lambda)$ is the limit as ε approaches zero of the solution of

$$-\psi^{*''} + \alpha\delta(x)\psi^* = (k - i\varepsilon)^2\psi^*$$

such that the difference between it and $e^{-i(k-i\varepsilon)x}$ vanishes for $x = +\infty$. This means that $\psi - e^{-i(k-i\varepsilon)x}$ behaves like some multiple of $e^{i(k-i\varepsilon)x}$ for $x = -\infty$, and like some multiple of $e^{-i(k-i\varepsilon)x}$ for $x = +\infty$. Just as before, we may take $\varepsilon = 0$ and put

$$\begin{aligned} \psi^* &= e^{-ikx} + R^*e^{ikx}, \quad x < 0 \\ &= T^*e^{-ikx}, \qquad x > 0, \end{aligned}$$

where R^* and T^* are to be determined so that the discontinuity conditions are satisfied. We find that

$$R^* = -\alpha(2ik + \alpha)^{-1}, \quad T^* = 2ik(2ik + \alpha)^{-1};$$

consequently,

$$\begin{aligned} \psi^* &= e^{-ikx} - \frac{\alpha e^{ikx}}{2ik + \alpha}, \quad x < 0 \\ &= \frac{2ike^{-ikx}}{2ik + \alpha}, \qquad x > 0. \end{aligned}$$

From Theorem 4.7 we conclude that

$$\frac{1}{2\pi}\int_{-\infty}^{\infty} \psi(x, k)\psi^*(x, k')\, dx = \delta(k - k').$$

PROBLEM

4.19. Find the spectral representation for the operator $L = -\dfrac{d^2}{dx^2} + q(x)$ over $(0, \infty)$, with the boundary condition $u(0) = 0$, where $q(x) = 0$, for $x > 1$, and $q(x) = a$, for $0 \leq x < 1$. Is there any difference in the representations if a is negative instead of positive?

Normalization of the Continuous Spectrum and Scattering

For operators such as $-\dfrac{d^2}{dx^2} + q(x)$, where $q(x)$ vanishes for large values of x, there exists an intimate connection between the normalization of the continuous spectrum and the theory of the scattering of plane waves coming in from infinity. We shall first show the connection for the previously considered operator $-\dfrac{d^2}{dx^2}$ with the boundary condition $u'(0) = \alpha u(0)$. In the preceding section we showed that the improper eigenfunctions would be normalized if we start with a function

$$\psi = \left(\frac{2}{\pi}\right)^{1/2} [\cos kx + \gamma e^{ikx}]$$

which satisfies the equation

$$\psi'' + k^2\psi = 0$$

and the boundary condition $\psi'(0) = \alpha\psi(0)$. However, instead of perturbing $\cos kx$ to get the function ψ, we can start with a *plane wave incident from infinity*† $(2\pi)^{-1/2}e^{-ikx}$ (remember that we always assume a time factor $e^{-i\omega t}$) and perturb this to get the normalized improper eigenfunction. Put

$$\psi = (2\pi)^{-1/2}[e^{-ikx} + Se^{ikx}],$$

where S is a constant to be determined so that $\psi'(0) = \alpha\psi(0)$. We have

$$ik(S - 1) = \alpha(1 + S);$$

therefore,

$$S = \frac{ik + \alpha}{ik - \alpha},$$

and then

$$\psi = (2\pi)^{-1/2} \left[e^{-ikx} + \frac{ik + \alpha}{ik - \alpha} e^{ikx}\right].$$

It is easy to show that this is the same result as that we found before. Note also that the value of S could have been written down immediately by applying (3.73) in Chapter 3. The quantity S is the reflection coefficient for a plane wave coming in from ∞ and being reflected by an impedance discontinuity at $x = 0$. There, the impedance of the solution is $\psi'(0)/\psi(0) = \alpha$ whereas the incoming wave has the impedance $-ik$. The quantity z_m in (3.73) is the ratio of impedances, and in this case $z_m = -\alpha(ik)^{-1}$; consequently by (3.73)

$$S = r_m = \frac{1 + \alpha(ik)^{-1}}{1 - \alpha(ik)^{-1}} = \frac{ik + \alpha}{ik - \alpha},$$

which is in agreement with the previously derived result.

† Note that from Problem 4.14 we have the following normalization for plane waves:

$$(2\pi)^{-1}\int_{-\infty}^{\infty} e^{ik(x-x')}\, dk = \delta(x - x').$$

This approach to the normalization of the continuous spectrum is not simpler than the approach of the previous sections, but it is very useful because of the physical insight it gives into the formulas. We may state the following

Rule. *The improper eigenfunctions $\psi(x, k)$ of the real operator $-\dfrac{d^2}{dx^2}+$ $q(x)$, with some real boundary condition at $x = 0$ and where $q(x)$ vanishes for large values of x, will be properly normalized if they are obtained as the result of the scattering of the incoming plane wave $(2\pi)^{-1/2}e^{-ikx}$; that is, $\psi(x, k)$ is the solution of*

$$-\psi'' + q^2\psi = k^2\psi$$

satisfying the boundary condition at $x = 0$ and behaving at infinity like $(2\pi)^{-1/2}[e^{-ikx} + Se^{ikx}]$, where S is a constant. We shall have the relation

$$\int_0^\infty \psi(x, k)\overline{\psi(x, k')}\, dx = \delta(k - k'),$$

where $\bar{\psi}$ denotes the complex conjugate function to ψ.

The proof of this rule follows easily from Theorem 4.7. Note that if the operator is not real, the function $\overline{\psi(x, k')}$ must be replaced by $\psi^*(x, k')$, the function considered in Theorem 4.7.

A similar rule will hold for the real operator $-\dfrac{d^2}{dx}+q(x)$ over $(-\infty, \infty)$ if $q(x)$ vanishes outside some finite interval. Because there are two improper eigenfunctions for every real positive value of $\lambda = k^2$, we must consider plane waves incident from $-\infty$ as well as from $+\infty$.† The distinction between these waves will depend upon the sign of k. The function e^{ikx} for k positive will indicate a wave going from left to right and will be called a wave incident from $-\infty$, but the same function for k negative will indicate a wave going from right to left and will be called a wave incident from $+\infty$. We state the following

Rule. *The improper eigenfunctions $\psi(x, k)$ of the real operator $-\dfrac{d^2}{dx^2}+q(x)$ over $(-\infty, \infty)$, where $q(x)$ vanishes outside a finite interval, will be properly normalized if they are obtained as a result of the scattering of incoming waves $(2\pi)^{-1/2}e^{ikx}$; that is, $\psi(x, k)$ is the solution of*

$$-\psi'' + q\psi = k^2\psi$$

such that at $\pm\infty$, ψ behaves like

$$(2\pi)^{-1/2}[e^{ikx} + S_\pm \exp(i|kx|)]$$

† A proof of this will be found in Titchmarsh, *Eigenfunction Expansions Associated with Second Order Differential Equations*, Clarendon Press, Oxford, 1946.

where S_+ and S_- are suitable constants.

We shall have

$$\int_{-\infty}^{\infty} \psi(x, k)\overline{\psi(x, k')}\, dx = \delta(k - k'),$$

where $\overline{\psi(x, k')}$ is the complex conjugate of $\psi(x, k)$.

The proof of this will follow from Theorem 4.7 if we use a Green's function defined for $\lambda = (|k| + i\varepsilon)^2$. We shall illustrate this rule for the previously considered operator in which $q(x) = \alpha\delta(x)$. Put

$$\begin{aligned} \psi &= (2\pi)^{-1/2}[e^{ikx} + S_+ \exp(i|k|x)], \quad x > 0 \\ &= (2\pi)^{-1/2}[e^{ikx} + S_- \exp(-i|k|x)], \quad x < 0. \end{aligned}$$

From the discontinuity conditions satisfied by ψ we get

$$1 + S_+ = 1 + S_-$$

and

$$ik + i|k|S_+ - ik + i|k|S_- = \alpha(1 + S_+).$$

These equations imply

$$S_+ = S_- = \frac{\alpha}{2i|k| - \alpha};$$

consequently,

$$\begin{aligned} \psi &= (2\pi)^{-1/2}\left[e^{ikx} + \frac{\alpha}{2i|k| - \alpha}\exp(i|k|x)\right], \quad x > 0 \\ &= (2\pi)^{-1/2}\left[e^{ikx} + \frac{\alpha}{2i|k| - \alpha}\exp(-i|k|x)\right], \quad x < 0. \end{aligned} \tag{4.62}$$

It is easy to see that for $k > 0$ these functions are incident waves from $-\infty$ plus reflected and transmitted waves, but for $k < 0$ they are incident waves from $+\infty$ plus reflected and transmitted waves. The spectral representation will be given by the following formula:

$$(2|\alpha|)^{-1}\exp\left[\frac{\alpha}{2}(|x| + |\xi|)\right] + \int_{-\infty}^{\infty} \psi(x, k)\overline{\psi(\xi, k)}\, dk = \delta(x - \xi),$$

where the bar indicates the complex conjugate function. Note that the first term on the left-hand side is missing if α is positive.

PROBLEMS

4.20. Obtain (4.62) by using (3.73) of Chapter 3. (*Hint.* At $x = 0$ the impedance jumps by α.)

4.21. Find the spectral representation for the operator in Problem 4.19 by the rule of this section.

4.22. By using the rule of this section, find the spectral representation for the operator $-\dfrac{d^2}{dx^2} + q(x)$ over $(-\infty, \infty)$, where $q(x) = 0$ for $|x| > 1$ and where $q(x) = a$ for $-1 < x < 1$.

Summary of Spectral Representations

For reference we shall tabulate the spectral representation for some operators which occur frequently in applications.

First, we consider $L = -\dfrac{d^2}{dx^2}$ over (0, 1). In the first column we specify the boundary conditions; in the second column we give the spectral representation of $\delta(x - \xi)$.

Boundary conditions are	Then $\delta(x - \xi) =$
(*a*) $u(0) = u(1) = 0;$	$2\sum_1^\infty \sin n\pi x \sin n\pi\xi;$
(*b*) $u'(0) = u'(1) = 0;$	$1 + 2\sum_1^\infty \cos n\pi x \cos n\pi\xi;$
(*c*) $u(0) = u'(1) = 0;$	$2\sum_1^\infty \sin\left(n + \frac{1}{2}\right)\pi x \sin\left(n + \frac{1}{2}\right)\pi\xi;$
(*d*) $u(0) = u(1),\ u'(0) = u'(1);$	$1 + 2\sum_1^\infty (\cos 2n\pi x \cos 2n\pi\xi + \sin 2n\pi x \sin 2n\pi\xi).$

These results can be extended to the case where $L = -\dfrac{d^2}{dx^2}$ over the interval (a, b) by making the change of variable $y = a + (b - a)x$, $\eta = a + (b - a)\xi$ and using the fact that

$$\delta\left(\frac{y - \eta}{b - a}\right) = (b - a)\delta(y - \eta).$$

For example, if $a = 0$, $b = \pi$; then formula (*a*) becomes this:

If $L = -\dfrac{d^2}{dx^2}$ over $(0, \pi)$, with the conditions $u(0) = u(\pi) = 0$, then

$$\delta(x - \xi) = \frac{2}{\pi}\sum_1^\infty \sin nx \sin n\xi.$$

Next, we consider the case where $L = -\dfrac{d^2}{dx^2}$ over $(0, \infty)$.

Boundary conditions are	Then $\delta(x-\xi) =$
(a) $u(0) = 0$;	$\dfrac{2}{\pi}\int_0^\infty \sin kx \sin k\xi \, dk$;
(b) $u'(0) = 0$;	$\dfrac{2}{\pi}\int_0^\infty \cos kx \cos k\xi \, dk$;
(c) $u'(0) = \alpha u(0)$;	$-2\alpha e^{\alpha(x+\xi)} + \dfrac{2}{\pi}\int_0^\infty \left(\cos kx + \dfrac{\alpha}{k}\sin kx\right)\left(\cos k\xi + \dfrac{\alpha}{k}\sin k\xi\right)\dfrac{k^2\, dk}{k^2+\alpha^2}$; the first term is missing if $Re\ \alpha \geq 0$.
(d) $u(0) = u'(0) = 0$;	$\dfrac{1}{2\pi i}\int_{a-i\infty}^{a+i\infty} e^{p(x-\xi)}\, dp$, where $a > 0$.

Finally, if $L = -\dfrac{d^2}{dx^2}$ over $(-\infty, \infty)$, then

$$\delta(x-\xi) = \frac{1}{2\pi}\int_{-\infty}^{\infty} e^{ik(x-\xi)}\, dk.$$

APPENDIX

PROOF OF THEOREM 4.7

We shall prove

Theorem 4.7. *The improper eigenfunctions* $\psi(\lambda)$ *and* $\psi^*(\lambda')$ *satisfy the following relation:*

$$\langle \psi(\lambda), \psi^*(\lambda') \rangle = \delta(\lambda - \lambda').$$

From (4.58) and (4.59) we have

$$\text{(4A.1)} \quad \langle \psi(\lambda), \psi^*(\lambda') \rangle = \langle \phi(\lambda), \phi^*(\lambda') \rangle - \langle L_1^{-1} M\psi(\lambda), \phi^*(\lambda') \rangle - \langle \phi(\lambda), L_2^{*-1} M^* \psi^*(\lambda') \rangle + \langle L_1^{-1} M\psi(\lambda), L_2^{*-1} M^* \psi^*(\lambda') \rangle$$

where, for convenience, we have put

$$L_1 = L - \lambda - i\varepsilon,$$
$$L_2 = L - \lambda' + i\varepsilon.$$

Using the fact that $\phi^*(\lambda')$ is an improper eigenfunction of L^*, we get

$$\text{(4A.2)} \quad \begin{aligned} \langle L_1^{-1} M\psi(\lambda), \phi^*(\lambda') \rangle &= \langle M\psi(\lambda), L_1^{*-1} \phi^*(\lambda') \rangle \\ &= (\lambda' - \lambda - i\varepsilon)^{-1} \langle M\psi(\lambda), \phi^*(\lambda') \rangle, \end{aligned}$$

and, similarly,

$$\text{(4A.3)} \quad \langle \phi(\lambda), L_2^{*-1} M^* \psi^*(\lambda') \rangle = (\lambda - \lambda' + i\varepsilon)^{-1} \langle \phi(\lambda), M^* \psi^*(\lambda') \rangle.$$

Using (4.58) and (4.59) again, we see that

$$\text{(4A.4)} \quad \langle M\psi(\lambda), \phi^*(\lambda') \rangle = \langle M\psi(\lambda), \psi^*(\lambda') \rangle + \langle M\psi(\lambda), L_2^{*-1} M^* \psi^*(\lambda') \rangle$$

and

$$\text{(4A.5)} \quad \langle \phi(\lambda), M^* \psi^*(\lambda') \rangle = \langle \psi(\lambda), M^* \psi^*(\lambda') \rangle + \langle L_1^{-1} M\psi(\lambda), M^* \psi^*(\lambda') \rangle.$$

The sum of the second and third terms on the right-hand side of (4A.1) is the negative sum of the left-hand sides of (4A.2) and (4A.3). Using (4A.4) and (4A.5) to transform the right-hand sides of (4A.2) and (4A.3), we find that the sum of (4A.2) and (4A.3) is

$$\text{(4A.6)} \quad (\lambda' - \lambda - i\varepsilon)^{-1} \langle M\psi(\lambda), (L_2^{*-1} - L_1^{*-1}) M^* \psi^*(\lambda') \rangle.$$

But

$$L_2^{*-1} - L_1^{*-1} = (\lambda' - \lambda - 2i\varepsilon) L_1^{*-1} L_2^{*-1};$$

consequently, (4A.6) becomes

$$\frac{\lambda' - \lambda - 2i\varepsilon}{\lambda' - \lambda - i\varepsilon} \langle L_1^{-1} M\psi(\lambda), L_2^{*-1} M^* \psi^*(\lambda') \rangle.$$

Using the negative of this result for the sum of the second and third terms on the right-hand side of (4A.1), we find that (4A.1) becomes

$$\langle\psi(\lambda), \psi^*(\lambda')\rangle = \langle\phi(\lambda), \phi^*(\lambda')\rangle + \frac{i\varepsilon}{\lambda' - \lambda - i\varepsilon}\langle L_1^{-1}M\psi(\lambda), L_2^{*-1}M^*\psi^*(\lambda')\rangle.$$

It is clear that the limit of the last term is zero as $\varepsilon \to 0$; consequently, from (4.57), we get the desired result,

$$\langle\psi(\lambda), \psi^*(\lambda')\rangle = \delta(\lambda - \lambda').$$

5

PARTIAL DIFFERENTIAL EQUATIONS

Introduction

In this chapter we shall discuss some methods that can be used to solve explicitly the most frequently occurring equations in mathematical physics: the potential equation, the heat equation, and the wave equation. These equations have in common the fact that they contain a particular differential operator called the *Laplacian.* If $x_1, x_2, \cdots, x_n$ are the coordinates in an n-dimensional Euclidean space, the Laplacian of a function $u(x_1, x_2, \cdots, x_n)$ is

$$\Delta u = \frac{\partial^2 u}{\partial x_1^2} + \frac{\partial^2 u}{\partial x_2^2} + \cdots + \frac{\partial^2 u}{\partial x_n^2}.$$

The potential equation is

$$\Delta u = 0,$$

with appropriate boundary conditions on u. To define the heat equation and the wave equation, we use n space coordinates $x_1, x_2, \cdots, x_n$ and a time coordinate t. Let $u(x_1, x_2, \cdots, x_n, t)$ be a twice differentiable function of these $n + 1$ coordinates; then the heat equation is

$$\Delta u = \frac{\partial u}{\partial t},$$

with suitable conditions on u, and the wave equation is

$$\Delta u = \frac{\partial^2 u}{\partial t^2}$$

with suitable conditions on u.

Before discussing the conditions† suitable for a solution of the potential, heat, or wave equations, we shall define the linear vector space of functions and the scalar product in that space. Let R be any region in the n-dimensional space defined by the coordinates $x_1, x_2, \cdots, x_n$ or in the $(n + 1)$-dimensional space defined by the coordinates $x_1, \cdots, x_n, t$, and let dr denote the Euclidean volume element in either space. The linear

† A full treatment of this topic will be found in Courant-Hilbert, *Methods of Mathematical Physics*, Vol. II, reprint, Interscience, New York, 1943.

vector space we shall use is the space of all functions u of either n or $n + 1$ coordinates such that u is of integrable square over R, that is,

$$\int_R u^2 \, dr < \infty.$$

The scalar product of two functions u, v is then defined as follows:

$$\langle u, v \rangle = \int_R uv \, dr.$$

We shall consider linear differential operators such as the Laplacian Δ, the time derivative $\frac{\partial}{\partial t}$, and combinations of these two defined on manifolds in this vector space. For simplicity in writing, we shall denote the time derivative by the symbol P. The domain of the operator will be the set of all functions u in the vector space such that $\frac{\partial^2 u}{\partial x_i \, \partial x_j}$ $(i, j = 1, 2, \cdots n)$ are piecewise continuous and of integrable square and such that on the boundary B of R there exists a linear homogeneous relation involving u and its normal derivative. The domain of the operator P will be the set of all functions u in the vector space such that $\frac{\partial u}{\partial t}$ is piecewise continuous and of integrable square and such that $u = 0$ for $t = 0$. The domain of P^2 will be the set of all functions u such that u and Pu belong to the domain of P; consequently, u is in the domain of P^2 if $\frac{\partial^2 u}{\partial t^2}$ is piecewise continuous and of integrable square and if $u = \frac{\partial u}{\partial t} = 0$ for $t = 0$.

From Green's formula† in n-dimensions we have

$$\int_R (u \, \Delta v - v \, \Delta u) \, dr = \int_B \left(u \frac{\partial v}{\partial n} - v \frac{\partial u}{\partial n} \right) dS. \tag{5.1}$$

Here B is the boundary of the region R, dS is the surface element on the boundary, and $\frac{\partial}{\partial n}$ is the derivative normal to the boundary. Another one of Green's formulas that will be useful is this:

$$\int_R u \, \Delta u \, dr = \int_B u \frac{\partial u}{\partial n} \, dS - \int_R \left[\left(\frac{\partial u}{\partial x_1} \right)^2 + \cdots + \left(\frac{\partial u}{\partial x_n} \right)^2 \right] dr. \tag{5.2}$$

Formula (5.1) implies that Δ is a formally self-adjoint operator. Formula (5.2) shows that the operator $-\Delta$ is positive-definite if the conditions on the domain are such that the integral over B vanishes.

Note that since the adjoint of P is $-P$, the adjoint of the heat operator

† A proof of this result will be found in Kellogg, *Foundations of Potential Theory*, Springer, Berlin, 1929.

$\Delta - P$ is $\Delta + P$, and the wave equation operator $\Delta - P^2$ is formally self-adjoint.

Green's Functions for Partial Differential Operators

We shall find that many of the methods we have used in the study of ordinary differential equations can be extended to partial differential equations. As in Chapter 3, we shall find that the inverse of a partial differential operator M will be an integral operator. The kernel of the integral operator will be called the *Green's function* of the operator. The Green's function G is the actual or symbolic function which is the solution of the equation $MG = \delta$-function in the n or $(n + 1)$-dimensional space.

To explain the meaning of such a δ-function, consider the case of three-dimensional space.

Let the three-dimensional δ-function $\delta(x, y, z; \xi, \eta, \zeta)$ be a symbolic function defined by the following equation†:

$$\iiint \phi(\xi, \eta, \zeta)\delta(x, y, z; \xi, \eta, \zeta)\, d\xi\, d\eta\, d\zeta = \phi(x, y, z) \tag{5.3}$$

for all continuous functions $\phi(x, y, z)$ which vanish outside a finite region of xyz-space. If $\delta(x - \xi)$, $\delta(y - \eta)$, and $\delta(z - \zeta)$ are the one-dimensional δ-functions defined in Chapter 3, we have

$$\iiint \phi(\xi, \eta, \zeta)\delta(x - \xi)\delta(y - \eta)\delta(z - \xi)\, d\xi\, d\eta\, d\zeta = \phi(x, y, z). \tag{5.4}$$

Comparing (5.3) and (5.4) we see that

$$\delta(x, y, z; \xi, \eta, \zeta) = \delta(x - \xi)\delta(y - \eta)\delta(z - \zeta).$$

A similar argument may be used in n or $(n + 1)$-dimensional space; therefore, we conclude that *the n-dimensional δ-function is the product of n one-dimensional δ-functions.*

To illustrate these ideas we shall consider the case where

$$Mu = -\left(\frac{\partial^2 u}{\partial x^2} + \frac{\partial^2 u}{\partial y^2} + \frac{\partial^2 u}{\partial z^2}\right)$$

and the domain of M is the set of functions u with piecewise continuous second derivatives and such that

$$\iiint u^2\, dx\, dy\, dz < \infty, \quad \iiint (Mu)^2\, dx\, dy\, dz < \infty.$$

Just as in Chapter 3, we may extend the definition of M to functions w not in the domain of M. Since M is self-adjoint, we put

$$\iiint \phi M w\, dx\, dy\, dz = \iiint w M\phi\, dx\, dy\, dz,$$

where ϕ is any function in the domain of M. This equation defines the perhaps symbolic function Mw.

† The limits on all integrals will be from $-\infty$ to ∞, unless otherwise noted.

The Green's function for the operator M will be the perhaps symbolic function G, which is the solution of the equation

$$(5.5) \qquad MG = -G_{xx} - G_{yy} - G_{zz} = \delta(x-\xi)\delta(y-\eta)\delta(z-\zeta).$$

If $f(x, y, z)$ is any continuous function which vanishes outside a finite region, it is easy to show that the integral

$$(5.6) \qquad u(x, y, z) = \iiint G f(\xi, \eta, \zeta)\, d\xi\, d\eta\, d\zeta$$

is a solution of the equation

$$Mu = f.$$

This result shows that the integral operator whose kernel is G, that is, the integral operator in (5.6), is the inverse to M.

To solve (5.5), we use some special properties of the operator M. First, we note that if we put $x' = x - \xi$, $y' = y - \eta$, $z' = z - \zeta$, then (5.5) becomes

$$(5.7) \qquad MG = -G_{x'x'} - G_{y'y'} - G_{z'z'} = \delta(x')\delta(y')\delta(z').$$

Next, we introduce spherical polar coordinates as follows:

$$z' = r\cos\theta,$$
$$x' = r\sin\theta\cos\psi,$$
$$y' = r\sin\theta\sin\psi.$$

Note that $0 \le r < \infty$, $0 \le \theta \le \pi$, and $0 \le \psi \le 2\pi$. It is easy to show† that in these coordinates

$$(5.8) \qquad -MG = \frac{1}{r^2}\frac{\partial}{\partial r}\left(r^2\frac{\partial G}{\partial r}\right) + \frac{1}{r^2\sin\theta}\frac{\partial}{\partial\theta}\left(\sin\theta\,\frac{\partial G}{\partial\theta}\right) + \frac{1}{r^2\sin^2\theta}\frac{\partial^2 G}{\partial\psi^2},$$

but what happens to the three-dimensional δ-function? By definition, we have

$$\iiint \phi(x', y', z')\delta(x')\delta(y')\delta(z')\, dx'\, dy'\, dz' = \phi(0, 0, 0).$$

If we introduce spherical polar coordinates into this integral, it becomes

$$(5.9) \qquad \int_0^\infty\int_0^\pi\int_0^{2\pi} \phi(r, \theta, \psi)\delta(x')\delta(y')\delta(z')r^2\sin\theta\, dr\, d\theta\, d\psi = \phi|_{r=0}.$$

In spherical polar coordinates the origin $r = 0$ is a singular point of the coordinate system because the origin has every value of θ and ψ as a coordinate. Since $\phi(r, \theta, \psi)$ is continuous and single-valued at the origin, its value there is independent of the value of θ and ψ. For example, if $\phi(x, y, z) = 2 + x^2 + yz$, then $\phi(r, \theta, \psi) = 2 + r^2\sin^2\theta\cos^2\psi + r^2\sin\theta$

† See Courant, *Differential and Integral Calculus*, Vol. II, Interscience, New York 1936.

$\cos\theta \sin\psi$, and the value of ϕ for $r = 0$ is 2, no matter what the values of θ and ψ are. Consequently,

$$\int_0^{\pi}\int_0^{2\pi}\int_0^{\infty} \phi(r, \theta, \psi)\frac{\delta(r)}{r^2} r^2 \sin\theta \, d\theta \, d\psi \, dr$$

$$= \int_0^{\pi}\int_0^{2\pi} \phi\big|_{r=0} \sin\theta \, d\theta \, d\psi = 4\pi\phi\big|_{r=0}.$$

Comparing this result with (5.9), we see that

$$\delta(x')\delta(y')\delta(z') = \frac{\delta(r)}{4\pi r^2}. \tag{5.10}$$

Formula (5.10) is the expression in spherical polar coordinates for the three-dimensional δ-function located at the origin. Later, we shall obtain an expression in spherical polar coordinates for the three-dimensional δ-function located at an arbitrary point.

With the help of (5.8) and (5.10), equation (5.7) may be written in spherical polar coordinates as follows:

$$MG = \frac{\delta(r)}{4\pi r^2},$$

where the left-hand side is defined by (5.8). Since the right-hand side is independent of θ and ψ, the function G will not depend on these coordinates, and therefore the equation becomes

$$NG = \frac{1}{r^2}\frac{\partial}{\partial r}\left(r^2 \frac{\partial G}{\partial r}\right) = -\frac{\delta(r)}{4\pi r^2}$$

or, simply,

$$r^2 NG = -\frac{\delta(r)}{4\pi}. \tag{5.11}$$

The operator N will be formally self-adjoint if we use the following scalar product:

$$\langle \phi, \psi \rangle = \int_0^{\infty} \phi(r)\psi(r) r^2 \, dr.$$

If we recall the definition of the domain of M, we see that the domain of the self-adjoint operator N is the set of all functions ϕ, defined for $0 \leq r < \infty$, which have piecewise continuous second derivatives and are such that

$$\int_0^{\infty} \phi(r)^2 r^2 \, dr < \infty$$

and

$$\int_0^{\infty} (N\phi)^2 r^2 \, dr < \infty.$$

Let us consider the extended definition of this operator N. We have

$$\int_0^\infty G(N\phi)r^2\,dr = \int_0^\infty G\frac{1}{r^2}\frac{\partial}{\partial r}\left(r^2\frac{\partial\phi}{\partial r}\right)r^2\,dr$$

$$= r^2(G\phi_r - G_r\phi)\Big|_0^\infty + \int_0^\infty \phi\frac{1}{r^2}\frac{\partial}{\partial r}\left(r^2\frac{\partial G}{\partial r}\right)r^2\,dr;$$

therefore, since by the extended definition of N

$$\int_0^\infty G(N\phi)r^2\,dr = \int_0^\infty \phi(NG)r^2\,dr,$$

we see that

$$\int_0^\infty \phi(NG)r^2dr = \int_0^\infty \phi\frac{1}{r^2}\frac{\partial}{\partial r}\left(r^2\frac{\partial G}{\partial r}\right)r^2dr + g_0\phi(0) - g_1\phi'(0),$$

where

$$g_1 = \lim_{r\to 0}\,(r^2G) \quad\text{and}\quad g_0 = \lim_{r\to 0}\,(r^2G_r).$$

Consequently,

$$r^2NG = \frac{\partial}{\partial r}\left(r^2\frac{\partial G}{\partial r}\right) + g_1\delta'(r) + g_0\delta(r),$$

where the differentiation on the right-hand side is to be understood in the ordinary, not in the symbolic, sense.

These considerations show that, if G is to satisfy (5.11), we must have

$$\frac{\partial}{\partial r}\left(r^2\frac{\partial G}{\partial r}\right) = 0,$$

$g_1 = 0$ and $g_0 = -(4\pi)^{-1}$. The solutions of the homogeneous equation

$$\frac{\partial}{\partial r}\left(r^2\frac{\partial u}{\partial r}\right) = 0$$

are given by $u = ar^{-1} + b$, where a and b are arbitrary constants. From the fact that $g_0 = -(4\pi)^{-1}$, we see that†

$$G = (4\pi r)^{-1}.$$

In rectangular coordinates we have

$$G = (4\pi)^{-1}(x'^2 + y'^2 + z'^2)^{-1/2}$$

or

$$G = (4\pi)^{-1}[(x-\xi)^2 + (y-\eta)^2 + (z-\zeta)^2]^{-1/2}.$$

Putting this result into (5.6), we obtain the well-known result that

$$u(x, y, z) = \frac{1}{4\pi}\iiint\frac{f(\xi, \eta, \zeta)}{[(x-\xi)^2 + (y-\eta)^2 + (z-\zeta)^2]^{1/2}}\,d\xi\,d\eta\,d\zeta \tag{5.12}$$

is a solution of

$$u_{xx} + u_{yy} + u_{zz} = -f(x, y, z). \tag{5.13}$$

† We have implicitly assumed that G goes to zero as r goes to infinity.

Equation (5.13) is called Poisson's equation. It is the equation for the potential u produced by a charge distribution with volume density $f(x, y, z)$. The formula for G shows that it is the potential produced by a point charge located at $x = \xi$, $y = \eta$, $z = \zeta$. Formula (5.12) now has an interesting physical interpretation. It shows that the potential may be considered as if it were produced by point charges at every point (ξ, η, ζ) with strength $f(\xi, \eta, \zeta)$.

PROBLEMS

5.1. Find the Green's function for the operator $M = -\dfrac{\partial^2}{\partial x^2} - \dfrac{\partial^2}{\partial y^2} - \dfrac{\partial^2}{\partial z^2} + k^2$ over the entire xyz space. (*Hint.* Proceed as in the text by transforming the source point to the origin and introducing spherical polar coordinates.)

5.2. Show that the Green's function for the n-dimensional Laplacian ($n \geq 3$) over the entire n-dimensional space is

$$G = [(n-2)\omega_n]^{-1}\left[\sum_1^n (x_k - \xi_k)^2\right]^{-(n-2)/2}$$

where ω_n is the surface area of the unit sphere $\sum_1^n x_k^2 = 1$. (*Hint.* Transform the source point to the origin and introduce spherical polar coordinates. If $r^2 = \sum_1^n x_k'^2$, the equation for G reduces to $\dfrac{\partial}{\partial r}\left(r^{n-1}\dfrac{\partial G}{\partial r}\right) = -\omega_n^{-1}\delta(r)$.)

5.3. Show that the Green's function for the two-dimensional Laplacian over the entire two-dimensional space is $G = -(4\pi)^{-1}\log[(x-\xi)^2 + (y-\eta)^2]$.

Separation of Variables

The method which was used in the preceding section to find the Green's function is very special and depends strongly on the high degree of symmetry of the problem. This symmetry enabled us to reduce the partial differential equation in rectangular coordinates to an ordinary differential equation in polar coordinates. In many applications such a high degree of symmetry does not exist, and we must use other methods to solve the partial differential equation. In this chapter we shall discuss the method of *separation of variables*, one of the few methods available for finding explicit solutions to partial differential equations.

The method may be easily understood by noting how it is used in the following problem.

Find a function $u(x, y)$ defined over the rectangle $0 \leq x \leq a$, $0 \leq y \leq b$ such that

$$u_{xx} + u_{yy} = -\delta(x - \xi)\delta(y - \eta) \tag{5.14}$$

and such that

(5.15) $$u(0, y) = u(a, y) = u(x, 0) = u(x, b) = 0.$$

The classical approach to the method of separation of variables begins by considering the homogeneous equation

(5.16) $$w_{xx} + w_{yy} = 0$$

and by trying to find solutions of this equation which are the product of a function of x by a function of y. We put $w(x, y) = X(x)Y(y)$ and substitute this into (5.16); then we get

$$X''Y + XY'' = 0$$

or

$$\frac{X''}{X} = -\frac{Y''}{Y}.$$

The left-hand side of this equation is a function of x alone; the right-hand side is a function of y alone. The only way in which a function of x can be equal to a function of y is for both functions to be constants. We call this constant k^2; then we have

$$X'' = k^2X,$$
$$Y'' = -k^2Y.$$

This shows that functions of the form $w = e^{kx} \sin ky$ are solutions of (5.16). We now must try to find linear combinations of these functions which are solutions of (5.14) and which satisfy (5.15).

Instead of continuing with this classical approach, we shall discuss another approach which is more in keeping with the spirit of our discussion of operators. What is the operator that appears in (5.14) and (5.15)? It is the operator M such that

$$Mu = u_{xx} + u_{yy}$$

and such that its domain is the set of square-integrable functions $u(x, y)$ which have square-integrable second derivatives over the rectangle $0 \leq x \leq a$, $0 \leq y \leq b$ and which satisfy (5.15).

The distinguishing property of M which will enable us to find an explicit solution of the problem is this: *The operator M is the sum of two commutative operators N_1 and N_2.* The operator N_1 is defined as

$$N_1u = u_{xx}$$

for all square-integrable functions $u(x, y)$ which have square-integrable second derivatives in the rectangle and satisfy the conditions

$$u(0, y) = u(a, y) = 0.$$

Similarly, the operator N_2 is defined by the equation

$$N_2u = u_{yy}$$

for all square-integrable functions $u(x, y)$ which have square-integrable second derivatives in the rectangle and satisfy the conditions

$$u(x, 0) = u(x, b) = 0.$$

Note that the domain of M is made up of all functions $u(x, y)$ which are simultaneously in the domain of N_1 and N_2, and that for these functions $N_1N_2u = N_2N_1u$.

We shall show in the next section that if N_1 and N_2 commute, we may treat N_2 as a constant in the process of finding the inverse of $N_1 + N_2$. Let us see how this applies to (5.14). Write it as follows:

$$u_{xx} + N_2u = -\delta(x - \xi)\delta(y - \eta), \tag{5.17}$$

where $u(0, y) = u(a, y) = 0$ and where we consider N_2 as a constant. This means that (5.17) will be considered an ordinary differential equation for functions of x, which may depend also on a parameter y. Using the techniques of Chapter 3 to solve (5.17), we find that its solution is

$$\begin{aligned} u &= (k \sin ka)^{-1} \sin kx \sin k(a - \xi)\, \delta(y - \eta), \quad 0 \leq x < \xi \\ &= (k \sin ka)^{-1} \sin k(a - x) \sin k\xi\, \delta(y - \eta), \quad a \geq x > \xi, \end{aligned} \tag{5.18}$$

where $k = \sqrt{N_2}$. This result, however, must still be interpreted because it contains a complicated function of the operator N_2. Note that u has been written in such a form that the function of the operator comes before the function it acts on, namely, $\delta(y - \eta)$.

We have seen in Chapters 2 and 4 that the effect of a function of an operator acting on a vector can be determined easily if the vector is expressed in terms of the eigenvectors of the operator. In the case we are considering, N_2 is an ordinary differential operator acting on functions of y, which may depend also on a parameter x. The eigenvalues and eigenfunctions of N_2 are easily obtained. From Chapter 4 we find that

$$\delta(y - \eta) = \frac{2}{b}\sum_{1}^{\infty} \sin\left(\frac{n\pi y}{b}\right) \sin\left(\frac{n\pi\eta}{b}\right),$$

which shows that the eigenfunctions are $\sin n\pi y/b$ and the eigenvalues are $-(n\pi/b)^2$; then, if $\phi(N_2)$ is an analytic function of the operator N_2, we have

$$\phi(N_2)\, \delta(y - \eta) = \frac{2}{b}\sum_{1}^{\infty}\phi\left(-\frac{n^2\pi^2}{b^2}\right) \sin\frac{n\pi y}{b} \sin\frac{n\pi\eta}{b}.$$

Before applying this result to interpret the expression we obtained for u, we note that, since $k = \sqrt{N_2}$, the eigenvalues of k will be purely imaginary, equal to $in\pi/b$; and consequently the sin terms will become sinh terms.

We find that (5.18) becomes

$$(5.19)\qquad u = \frac{2}{b}\sum_{1}^{\infty}\frac{\sinh(n\pi x/b)\sinh\{n\pi(a-\xi)/b\}\sin(n\pi y/b)\sin(n\pi\eta/b)}{(n\pi/b)\sinh(n\pi a/b)},$$

$$0 \leq x < \xi$$

$$= \frac{2}{b}\sum_{1}^{\infty}\frac{\sinh(n\pi\xi/b)\sinh\{n\pi(a-x)/b\}\sin(n\pi y/b)\sin(n\pi\eta/b)}{(n\pi/b)\sinh(n\pi a/b)},$$

$$\xi \leq x \leq a.$$

These formulas give the solution to (5.14). They are also formulas for the Green's function of the two-dimensional Laplacian satisfying the conditions (5.15); consequently, they may be used to solve the equation

$$(5.20)\qquad w_{xx} + w_{yy} = -f(x, y),$$

where w satisfies the same conditions as u, namely,

$$w(0, y) = w(a, y) = w(x, 0) = w(x, b) = 0.$$

If we multiply (5.14) by $f(\xi, \eta)$ and integrate over the rectangle, we can see that

$$w = \int_0^a \int_0^b f(\xi, \eta)u(x, y; \xi, \eta)\, d\xi\, d\eta$$

with u given by (5.19) will be the desired solution of (5.20) satisfying the given boundary conditions.

The equation (5.20) could have been also solved without using the Green's function by writing it as follows:

$$w_{xx} + N_2 w = -f(x, y),$$

where $w(0, y) = w(a, y) = 0$. This equation is to be considered an ordinary differential equation for w, a function of x which incidentally depends also on y. By the techniques of Chapter 3 we find that

$$w(x, y) = (k \sin ka)^{-1}\Big[\sin k(a-x)\int_0^x \sin k\xi\, f(\xi, y)\, d\xi$$

$$+ \sin kx\int_x^a \sin k(a-\xi)\, f(\xi, y)\, d\xi\Big],$$

where $k = \sqrt{N_2}$. Note that again the function of the operator N_2 has been written before the function $f(\xi, y)$ it acts on.

Using the methods of Chapter 4, we may obtain the spectral representation of N_2, and from this representation we get the following expansion of f in terms of the eigenfunctions of N_2:

$$f(\xi, y) = \frac{2}{b}\sum_{1}^{\infty}\sin\left(\frac{n\pi y}{b}\right)\int_0^b f(\xi, \eta)\sin\left(\frac{n\pi\eta}{b}\right) d\eta.$$

We can now interpret the meaning of the function of the operator N_2 acting on $f(\xi, y)$. We find that

$$w(x, y) =$$

$$\frac{2}{b}\sum_1^\infty \frac{\sinh\{n\pi(a-x)/b\}\sin(n\pi y/b)\int_0^b\int_0^x \sinh(n\pi\xi/b)\sin(n\pi\eta/b)f(\xi,\eta)\,d\xi\,d\eta}{(n\pi/b)\sinh(n\pi a/b)}$$

$$+\frac{2}{b}\sum_1^\infty \frac{\sinh(n\pi x/b)\sin(n\pi y/b)\int_0^b\int_x^a \sinh\{n\pi(a-\xi)/b\}\sin(n\pi\eta/b)f(\xi,\eta)\,d\xi\,d\eta}{(n\pi/b)\sinh(n\pi a/b)}.$$

This result is the same as that we would find by using the Green's function.

PROBLEMS

5.4. Find the Green's function for the two-dimensional Laplacian with the conditions (5.15) by treating N_1 instead of N_2 as a constant.

5.5. Find the Green's function for the two-dimensional Laplacian with the following sets of conditions:

(*a*) $u(x, 0) = u(x, b) = u(0, y) = u_x(a, y) = 0, \quad 0 \le x \le a, \quad 0 \le y \le b.$

(*b*) $u(0, y) = \alpha u_x(0, y), \quad u(x, 0) = \beta u_y(x, 0), \quad u(a, y) = u(x, b) = 0,$
$0 \le x \le a, \quad 0 \le y \le b,$ α and β are fixed constants.

(*c*) $u(0, y) = u(x, 0) = u(x, b) = 0, \quad 0 \le x < \infty, \quad 0 \le y \le b.$

5.6. Solve $u_{xx} + u_{yy} + k^2u = -\delta(x - \xi)\delta(y - \eta)$, with the following sets of conditions:

(*a*) $u(x, 0) = 0, \quad -\infty < x < \infty, \quad 0 \le y < \infty.$

(*b*) $u(x, 0) = u_x(0, y) = 0, \quad 0 \le x < \infty, \quad 0 \le y < \infty.$

(*c*) $u_y(x, 0) = \alpha u(x, 0),$ α a fixed constant, $\quad -\infty < x < \infty, \quad 0 \le y < \infty.$

(*Hint.* Use the outgoing wave condition to define the operator, or, equivalently, assume that $\operatorname{Im} k > 0$ and assume that u is bounded at infinity.)

The Inverse of the Sum of Two Commutative Operators

We shall now justify the statement made in the previous section, namely, that if N_1 and N_2 commute, N_2 may be treated as a constant in the process of finding the inverse of $N_1 + N_2$. For the sake of simplicity, we shall treat precisely the case we discussed in the previous section. We consider the space $\mathcal{S}$ of real-valued functions $f(x, y)$ which are square integrable over a rectangle whose sides are parallel to the x and y axes. We assume that N_1 is a self-adjoint differential operator with respect to x and with boundary conditions acting on x alone and that N_2 is a self-adjoint differential operator with respect to y with boundary conditions acting on y alone. Let $\lambda_1, \lambda_2, \cdots$ be the eigenvalues and $v_1(y), v_2(y) \cdots$ be the corresponding normalized eigenfunctions of N_2. We assume that the inverse of $N_1 + \lambda_k$ is a bounded operator for every eigenvalue λ_k and that the bound is independent of λ_k. We assume finally that the eigenfunctions of N_2 are complete in the following sense:

For every function $f(x, y)$ in $\mathcal{S}$ the expansion

$$\Sigma \alpha_k(x) v_k(y),$$

where

$$\alpha_k(x) = \int f(x, y) v_k(y)\, dy,$$

converges to $f(x, y)$ in the sense of the norm in $\mathcal{S}$. We shall prove

Theorem 5.1.† *If N_1 and N_2 are the operators described above and if $f(x, y)$ is in $\mathcal{S}$, then*

(5.21)

$$(N_1 + N_2)^{-1} f = (N_1 + N_2)^{-1} \sum_1^\infty \alpha_k(x) v_k(y) = \sum_1^\infty (N_1 + \lambda_k)^{-1} \alpha_k(x) v_k(y).$$

Let C be the bound for all the operators $(N_1 + \lambda_k)^{-1}$; then for any function $g(x, y)$ in $\mathcal{S}$ we have

$$|(N_1 + \lambda_k)^{-1} g| < C|g|,$$

where $|g|$ denotes the norm of g in $\mathcal{S}$. Using this fact and the fact that the $v_k(y)$ are mutually orthogonal, we see that

$$\left|\sum_{k=j}^\infty (N_1 + \lambda_k)^{-1} \alpha_k(x) v_k(y)\right|^2 = \sum_{k=j}^\infty |(N_1 + \lambda_k)^{-1} \alpha_k(x)|^2$$

$$\leq C^2 \sum_{k=j}^\infty |\alpha_k(x)|^2 .$$

By assumption, the series $\sum_1^\infty |\alpha_k(x)|^2$ converges. Therefore, the sum $\sum_{k=j}^\infty |\alpha_k(x)|^2$ and also the sum

$$\left|\sum_{k=j}^\infty (N_1 + \lambda_k)^{-1} \alpha_k(x) v_k(y)\right|^2$$

converge to zero as j goes to infinity. This proves that the right-hand side of (5.21) converges in the sense of the norm.

To finish the proof, we must show that the right-hand side converges to the left-hand side. Because N_1 and N_2 commute, we have

$$(N_1 + N_2) \sum_1^m (N_1 + \lambda_k)^{-1} \alpha_k(x) v_k(y) = \sum_1^m (N_1 + \lambda_k)^{-1} (N_1 + N_2) \alpha_k(x) v_k(y).$$

† For a generalization of this theorem see *Proceedings of the Conference on Differential Equations*, 1956, University of Maryland Book Store, Maryland, pages 209–227.

Since $v_k(y)$ is an eigenfunction of N_2, the right-hand side of this result becomes

$$\sum_1^m (N_1 + \lambda_k)^{-1}(N_1 + \lambda_k)\alpha_k(x)v_k(y) = \sum_1^m \alpha_k(x)v_k(y).$$

By assumption, the sum $\sum_1^m \alpha_k(x)v_k(y)$ converges to $f(x, y)$; consequently, the sum

$$(N_1 + N_2)\sum_1^m (N_1 + \lambda_k)^{-1}\alpha_k(x)v_k(y)$$

converges to $f(x, y)$. This proves the theorem.

To see how Theorem 5.1 justifies the statement that N_2 may be considered as a constant in finding the inverse of $N_1 + N_2$, notice that in (5.21) each term is obtained by considering the inverse of N_1 plus some constant. Therefore, if we have a method for inverting N_1 plus an arbitrary constant λ, the inverse of $N_1 + N_2$ will be obtained from the inverse of $N_1 + \lambda$ by putting $\lambda = \lambda_1, \lambda_2, \cdots$, which are the eigenvalues of N_2, and combining these results according to (5.21). We may summarize this procedure as follows:

Rule. *If N_1 and N_2 commute, the inverse of $N_1 + N_2$ may be obtained by considering N_2 as a constant. The result will be a function of the operator N_2 and should be interpreted by using the spectral representation of N_2.*

Another way of looking at this rule may clarify the reason for its applicability. The essential reason is the fact that N_1 and N_2 commute. We have seen in Chapter 2 that, if two operators commute, the null space of one operator is an invariant manifold of the other; therefore, the null space of $N_2 - \lambda$ is an invariant manifold of N_1. If λ is not an eigenvalue of N_2, the null space of $N_2 - \lambda$ is just the zero vector; if λ is an eigenvalue, the null space of $N_2 - \lambda$ is the space of eigenvectors corresponding to that eigenvalue. Note that if v belongs to this null space, $N_2 v = \lambda v$ so that on this null space the effect of N_2 is just multiplication by a constant. If the eigenvectors of N_2 are complete, the whole space $\mathcal{S}$ will be the direct sum of these spaces of eigenvectors of N_2. This means that $\mathcal{S}$ is a direct sum of invariant manifolds of N_1, on each of which the effect of N_2 is multiplication by a constant λ; consequently, on each invariant manifold we may replace $N_1 + N_2$ by $N_1 + \lambda$.

A similar discussion may be applied to more general combinations of operators than the sum. For later use we state the following result:

Consider the space $\mathcal{S}$ defined before. Let $A_1, A_2, \cdots, A_n$ be self-adjoint *differential operators with respect to x and let $B_1, B_2, \cdots, B_n$ be self-adjoint differential operators with respect to y. Suppose that the operators $B_1, B_2, \cdots, B_n$ commute and have a complete set of normalized eigenvectors $v_1(y), v_2(y), \cdots$ such that*

$$B_i v_j(y) = \lambda_{ij} v_j, \quad (i = 1, 2, \cdots, n; j = 1, 2, \cdots).$$

Then, if $f(x, y)$ belongs to $\mathcal{S}$, we have

$$(5.22) \quad (A_1B_1 + \cdots + A_nB_n)^{-1}f = (A_1B_1 + \cdots + A_nB_n)^{-1}\Sigma\alpha_k(x)v_k(y)$$

$$= \sum_{k=1}^{\infty}(\lambda_{1k}A_1 + \cdots + \lambda_{nk}A_n)^{-1}\alpha_k(x)v_k(y).$$

For a proof of this result we refer to Problem 5.7. As an illustration of this result, consider the operator

$$Mu = r^{-1}(ru_r)_r + r^{-2}u_{\theta\theta} + p^2u,$$

where u is a real-valued square integrable function of r and θ, and where p is a constant. If we put

$$A_1u = r^{-1}(ru_r)_r, \quad A_2u = r^{-2}u, \quad A_3u = p^2u$$

and

$$B_1u = u, \quad B_2u = u_{\theta\theta}, \quad B_3u = u,$$

then

$$Mu = (A_1B_1 + A_2B_2 + A_3B_3)u.$$

From the above result, we see that

$$M^{-1}f(r, \theta) = \sum_{1}^{\infty}\left(r^{-1}\frac{\partial}{\partial r}r\frac{\partial}{\partial r} + \lambda_k r^{-2} + p^2\right)^{-1}\alpha_k(r)v_k(\theta)$$

if $\lambda_1, \lambda_2, \cdots$ are the eigenvalues of $\dfrac{\partial^2}{\partial\theta^2}$ and $v_1(\theta), v_2(\theta), \cdots$, the corresponding eigenfunctions.

PROBLEM

5.7. Prove (5.22). (*Hint.* Use the method of Theorem 5.1.)

Alternative Representations

Let us return to the question we considered previously, namely, to find a function $u(x, y)$ such that

$$(5.14) \qquad u_{xx} + u_{yy} = -\delta(x - \xi)\delta(y - \eta)$$

and such that

$$(5.15) \qquad u(0, y) = u(a, y) = u(x, 0) = u(x, b) = 0.$$

We solved this before by putting

$$N_1u = u_{xx}, \quad N_2u = u_{yy}$$

and considering N_2 as a constant while we found the inverse of $N_1 + N_2$. However, this problem could be solved just as readily by considering N_1 as a constant while we find the inverse of $N_2 + N_1$. The details of this solution are as follows.

Write (5.14) as

$$u_{yy} + N_1 u = -\delta(x-\xi)\,\delta(y-\eta),$$

where $u(x, 0) = u(x, b) = 0$. The solution of this equation is

$$\begin{aligned} u &= (p \sin pb)^{-1} \sin py \sin p(b-\eta)\,\delta(x-\xi), \quad 0 \le y \le \eta \\ &= (p \sin pb)^{-1} \sin p(b-y) \sin p\eta\,\delta(x-\xi), \quad \eta \le y \le b, \end{aligned}$$

where $p = \sqrt{N_1}$. Again, the solution has been written so that the function of the operator N_1 comes before the function $\delta(x-\xi)$ that it acts upon. From the spectral representation of N_1, we find that

$$\delta(x-\xi) = (2/a)\sum_{1}^{\infty} \sin\frac{m\pi x}{a} \sin\frac{m\pi\xi}{a}.$$

Using this, we obtain the following formula:

(5.23)

$$u(x, y) = \frac{2}{a}\sum_{1}^{\infty} \frac{\sinh(m\pi y/a)\sinh\{m\pi(b-\eta)/a\}\sin(m\pi x/a)\sin(m\pi\xi/a)}{(m\pi/a)\sinh(m\pi b/a)}$$

$$0 \le y \le \eta$$

$$= \frac{2}{a}\sum_{1}^{\infty} \frac{\sinh(m\pi\eta/a)\sinh\{m\pi(b-y)/a\}\sin(m\pi x/a)\sin(m\pi\xi/a)}{(m\pi/a)\sinh(m\pi b/a)}$$

$$\eta \le y \le b.$$

Note that, because of the symmetry of the problem, (5.23) could have been obtained from (5.19) by interchanging a and b, x and y, ξ and η.

The fact that there exist two different representations, (5.19) and (5.23), for the Green's function defined by (5.14) and (5.15) proves useful in applications. The series in (5.19) and (5.23) converge for all values of x and y except $x = \xi$, $y = \eta$, but for applications it is important that the series converge rapidly. In (5.19) $u(x, y)$ is expressed as a Fourier series in y but in (5.23) it is expressed as a Fourier series in x. The rapidity of convergence of these series will depend on the order of magnitude of the coefficients for large values of n.

Consider the coefficient in (5.23). We put

$$\begin{aligned} c_n &= \frac{\sinh\{n\pi(b-y)/a\}\sinh(n\pi\eta/a)}{\sinh(n\pi b/a)} \\ &= \frac{[e^{n\pi(b-y)/a} - e^{-n\pi(b-y)/a}][e^{n\pi\eta/a} - e^{-n\pi\eta/a}]}{2[e^{n\pi b/a} - e^{-n\pi b/a}]}. \end{aligned}$$

For large values of n, the negative exponentials can be neglected compared to the positive exponentials. Then, approximately,

$$c_n = \frac{e^{n\pi(b-y)/a}e^{n\pi\eta/a}}{2e^{n\pi b/a}} = \frac{e^{n\pi(\eta-y)/a}}{2}.$$

Note that c_n appears as part of the coefficient in the first expression in (5.23), where $y > \eta$. We see then that c_n approaches zero exponentially with n more rapidly as the difference between η and y increases. In the same way, it can be shown that a similar part of the coefficient in the first expression in (5.23) is approximately

$$\tfrac{1}{2}e^{n\pi(y-\eta)/a}$$

so that it, too, approaches zero exponentially with n since $\eta > y$.

Similarly, the corresponding coefficients in (5.19) are approximately equal to

$$\tfrac{1}{2}e^{-n\pi(x-\xi)/b}$$

and, therefore, also converge to zero exponentially with n, the more rapidly as the difference between x and ξ increases. We may conclude then that to find the value of the Green's function for points at which x differs markedly from ξ, the representation (5.19) should be used, but for points at which y differs markedly from η, the representation (5.23) should be used.

We shall find that the conclusions in this problem are typical of the general situation where the operator can be separated into the sum of two commutative operators. *There will exist two distinct representations of the Green's function, one using x-eigenfunctions, the other using y-eigenfunctions. For points whose x-coordinate differs markedly from the x-coordinate of the singularity of the Green's function, the value of the Green's function can be found more quickly from the second representation. For points whose y-coordinate differs markedly from the y-coordinate of the singularity of the Green's function, the value of the Green's function should be found from the representation using x-eigenfunctions.*

PROBLEM

5.8. Find the alternative representations for the solutions in Problems 5.5 and 5.6.

Boundary Value Problems

We saw in Chapter 3 that, by extending the definition of the operator, equations with non-homogeneous boundary conditions may be written as non-homogeneous equations with homogeneous boundary conditions. We have shown previously that the Green's function may be used to solve a non-homogeneous equation with homogeneous boundary conditions.

Consequently, the Green's function may be used to solve boundary value problems.

An illustration will clarify this. Suppose that we wish to find a function $v(x, y)$ such that

$$v_{xx} + v_{yy} = 0$$

in some two-dimensional region R bounded by a closed curve C and such that $v = h(s)$, a known function, on C. Let us denote by Δ the operator such that

$$\Delta u = u_{xx} + u_{yy}$$

and such that $u = 0$ on C. We shall extend the definition of the operator Δ. If w is any function in the domain of Δ, the extended meaning of Δ will be defined by the following:

$$\iint_R w\,\Delta v\,dx\,dy = \iint v\,\Delta w\,dx\,dy = \iint w(v_{xx} + v_{yy})\,dx\,dy + \int_C \left(v\,\frac{\partial w}{\partial n} - w\,\frac{\partial v}{\partial n}\right) ds, \tag{5.24}$$

where $\dfrac{\partial}{\partial n}$ is the outward-pointing normal derivative to C and where s is the arc length along C. From the definitions of Δ and of v and from (5.24) we see that

$$\Delta v = v_{xx} + v_{yy} - h(s)\delta'_C = -\,h(s)\delta'_C, \tag{5.25}$$

where δ'_C is the symbolic function defined by the equation

$$\iint_R \delta'_C w(x, y)\,dx\,dy = -\int_C \frac{\partial w}{\partial n}\,ds.$$

Suppose that $g(x, y; \xi, \eta)$ is the Green's function for Δ, that is, suppose that

$$\Delta g = -\,\delta(x - \xi)\delta(y - \eta);$$

then the solution of (5.25) is

$$\begin{aligned} V &= \iint_R h(s)\delta'_C g(x, y; \xi, \eta)\,d\xi\,d\eta \\ &= -\int_C h(s)\,\frac{\partial g}{\partial n}\,(x, y; \xi, \eta)\,ds. \end{aligned} \tag{5.26}$$

It is clear that a similar method may be used in more complicated problems. However, it is also possible to solve these problems directly without the use of the Green's function. For example, consider the following problem:

Find the function $u(x, y)$ such that

$$u_{xx} + u_{yy} = 0 \tag{5.27}$$

over the rectangle $0 \leq x \leq a$, $0 \leq y \leq b$ and such that

(5.28) $$u(0, y) = u(a, y) = u(x, b) = 0 \quad \text{and} \quad u(x, 0) = h(x).$$

Again, put $N_1 u = u_{xx}$ and $N_2 u = u_{yy}$. Let us first consider N_1 as a constant; then (5.27) becomes

$$u_{yy} + N_1 u = 0,$$

where

$$u(x, 0) = h(x), \quad u(x, b) = 0.$$

The solution of this problem is easily seen to be

(5.29) $$u = (\sin pb)^{-1} \sin p(b - y)\, h(x),$$

where $p = \sqrt{N_1}$. Note that again the function of the operator N_1 is written before the function $h(x)$ it acts upon. From the spectral representation of N_1, we have

$$h(x) = \frac{2}{a} \sum_{1}^{\infty} \sin (n\pi x/a) \int_0^a h(\xi) \sin (n\pi\xi/a)\, d\xi.$$

Substituting this in (5.29), we obtain the following solution for (5.27):

(5.30) $$u = \frac{2}{a} \sum_{1}^{\infty} [\sinh (n\pi b/a)]^{-1} \sinh \{n\pi(b - y)/a\} \sin (n\pi x/a) \int_0^a h(\xi) \sin (n\pi\xi/a)\, d\xi.$$

This result could have been obtained by using (5.26) with the representation (5.23) of the Green's function.

We may obtain an alternative representation for u by treating N_2 as a constant. Using the extended definition of N_2, we may write (5.27) as follows:

$$u_{xx} + N_2 u = h(x)\delta'(y),$$

with the condition $u(0, y) = u(a, y) = 0$.

The solution of this equation is

(5.31) $$u = (k \sin ka)^{-1}\Big[\sin k(a - x) \int_0^x \sin k\xi\, h(\xi)\, d\xi + \sin kx \int_x^a \sin k(a - \xi)\, h(\xi)\, d\xi\Big]\delta'(y),$$

where $k = \sqrt{N_2}$.

Using the spectral representation of N_2, we have

$$\delta(y - \eta) = \frac{2}{b} \sum \sin (n\pi y/b) \sin (n\pi\eta/b).$$

Differentiating this with respect to η and putting $\eta = 0$, we get

$$\delta'(y) = -\,(2/b) \sum_{1}^{\infty} (n\pi/b) \sin (n\pi y/b).$$

Substituting this in (5.31), we obtain another representation for the solution of (5.27). It is as follows:

(5.32)

$$u(x, y) = -\frac{2}{b}\sum \frac{\sin(n\pi y/b)}{\sinh(n\pi a/b)}\Big[\sinh\{n\pi(a-x)/b\}\int_0^x \sinh(n\pi\xi/b)h(\xi)\,d\xi$$

$$+ \sinh(n\pi x/b)\int_x^a \sinh\{n\pi(a-\xi)/b\}h(\xi)\,d\xi\Big].$$

This also could have been obtained by using (5.26) with the representation (5.19) of the Green's function.

An Apparent Contradiction

Equation (5.32) seems to contradict the last boundary condition in (5.28), that is, $u(x, 0) = h(x)$ because in (5.32) $u(x, 0) = 0$. However, the contradiction is easily explained. Formula (5.32) is not a continuous function of y as y approaches zero. As y approaches zero, $u(x, y)$ will approach $h(x)$ but $u(x, 0) = 0$. To prove this statement, we must give a short discussion on Fourier series.

Consider

$$S = \sum_1^\infty c_n \sin ny \tag{5.33}$$

and suppose that, for large n,

$$c_n = \frac{a_1}{n} + \frac{a_2(n)}{n^2},$$

where a_1 is a constant and $a_2(n)$ is a bounded function of n. If $a_1 = 0$,

$$S = \sum \frac{a_2(n)}{n^2}\sin ny.$$

Since $a_2(n)$ and $\sin ny$ are bounded for all values of n and y, we have

$$|S| \le C\sum \frac{1}{n^2},$$

where C is some constant. This shows that the series converges uniformly for all values of y; therefore, S will be a continuous function of y for all values of y.

If $a_1 \neq 0$,

$$S = a_1\sum \frac{\sin ny}{n} + \sum \frac{a_2(n)}{n^2}\sin ny. \tag{5.34}$$

The second sum is a continuous function of y for all values of y. The first sum is easily shown to be

$$a_1\frac{\pi - y}{2}, \quad \pi > y > 0.$$

This shows that as y approaches zero, S in (5.34) approaches $\pi a_1/2$.

If, instead of (5.33), we have

$$S = \Sigma c_n \sin(n\pi y/b),$$

and a_1 is defined as before, S approaches $a_1\pi/2$ as y approaches zero. Comparing this series with that in (5.32), we see that

$$c_n = \frac{2}{b}\frac{1}{\sinh(n\pi a/b)}[\cdots]$$

where $\cdots$ indicates the two terms in brackets in (5.32). An integration by parts gives

$$\int_0^x \sinh(n\pi\xi/b)h(\xi)\,d\xi = \frac{h(\xi)\cosh(n\pi\xi/b)}{(n\pi/b)}\Big|_0^x - \int_0^x \frac{h'(\xi)\cosh(n\pi\xi/b)}{(n\pi/b)}\,d\xi$$

and

$$\int_x^a \sinh\{n\pi(a-\xi)/b\}h(\xi)\,d\xi$$
$$= -\frac{h(\xi)\cosh\{n\pi(a-\xi)/b\}}{(n\pi/b)}\Big|_x^a + \int_x^a \frac{h'(\xi)\cosh\{n\pi(a-\xi)/b\}}{(n\pi/b)}\,d\xi;$$

consequently we have

$$c_n = \frac{2}{b}\frac{h(x)}{(n\pi/b)}\left[\frac{\sinh\{n\pi(a-x)/b\}\cosh(n\pi x/b) + \cosh\{n\pi(a-x)/b\}\sinh(n\pi x/b)}{\sinh(n\pi a/b)}\right]$$

plus higher order terms in $1/n$. We find that

$$a_1 = \frac{2h(x)}{\pi};$$

therefore,

$$\lim_{y\to 0} u(x, y) = h(x).$$

PROBLEMS

5.9. Solve the following boundary value problems for the equation $u_{xx} + u_{yy} + k^2u = 0$, k a constant:

(*a*) $u(x, 0) = f(x)$, $u(x, \infty)$ outgoing, $0 \leq y < \infty$, $-\infty < x < \infty$.

(*b*) $u(x, 0) = 0$, $u_x(0, y) = g(y)$, $u(x, \infty)$ and $u(\infty, y)$ outgoing, $0 \leq x < \infty$, $0 \leq y < \infty$.

(*c*) $u(x, 0) = f_1(x)$, $u(x, b) = f_2(x)$, $u(0, y) = g_1(y)$, $u(a, y) = g_2(y)$, $0 < x \leq a$, $0 \leq y \leq b$.

Find two representations for each solution.

5.10. Solve the following boundary value problem:

$$u_{xx} + u_{yy} + u_{zz} = 0, \quad (0 \leq x \leq a, \quad 0 \leq y \leq b, \quad 0 \leq z \leq c),$$

where $u(x, y, 0) = f(x, y)$, $u(x, y, c) = u(0, y, z) = u(a, y, z) = u(x, 0, z) = u(x, b, z) = 0$. (*Hint.* The operators $\frac{\partial^2}{\partial x^2}$ and $\frac{\partial^2}{\partial y^2}$ may be treated as constants while the z-equation is being solved. Then the spectral representation for

$\frac{\partial^2}{\partial x^2}$ and $\frac{\partial^2}{\partial y^2}$ will interpret the solution.) Also, solve this problem by assuming that $\frac{\partial^2}{\partial y^2}$ and $\frac{\partial^2}{\partial z^2}$ are constant for the x-equation.

5.11. Solve the following boundary value problems for the equation $u_{rr} + \frac{1}{r}u_r + \frac{1}{r^2}u_{\theta\theta} = 0$;

(*a*) $u(r, 0) = u(r, 2\pi)$, $u_\theta(r, 0) = u_\theta(r, 2\pi)$, $u(0, \theta)$ regular, $u(a, \theta) = h(\theta)$, $0 \leq \theta \leq 2\pi$, $0 \leq r \leq a$.

(*b*) $u(r, 0) = u(r, \pi) = 0$, $u(a, \theta) = h(\theta)$, $u(\infty, \theta) = 0$, $0 \leq \theta \leq \pi$, $a \leq r < \infty$.

(*c*) $u(r, 0) = 0$, $u_\theta(r, \pi) = 0$, $u(0, \theta)$ regular, $u(a, \theta) = h(\theta)$, $0 \leq \theta \leq \pi$, $0 \leq r \leq a$.

Changing One Representation into Another

In the examples of separation of variables that have been discussed so far, we have seen that the solution of the problems considered could be represented in two different ways: either in terms of x-eigenfunctions or in terms of y-eigenfunctions. We have seen that one representation may be more useful than another because it converges more rapidly in a certain region.

In this section we shall obtain a contour integral representation from which the two representations can be derived. Or, alternatively, if we start with one eigenfunction representation, the contour integral representation can be obtained and then from this the other eigenfunction representation. This latter procedure will be used in some cases to obtain a spectral representation for an operator which is not of the type considered in the previous chapter.

For the sake of simplicity we shall consider only the case where the operator is a sum of two commutative operators N_1 and N_2. Just as in the hypothesis of Theorem 5.1 we shall assume that N_1 is a self-adjoint differential operator acting on x alone and N_2 is a self-adjoint differential operator acting on y alone. Let $\lambda_1, \lambda_2, \cdots$ be the eigenvalues and $v_1(y), v_2(y), \cdots$ be the corresponding normalized eigenfunctions of N_2. We assume that the inverse of $N_1 + \lambda_k$ is a bounded operator for every value of λ_k, real or complex, with a bound independent of λ_k. We assume finally that the eigenfunctions of N_2 are complete. We shall prove

Theorem 5.2. *If N_1 and N_2 are the operators described above and if $f(x, y)$ is in the space $\mathcal{S}$; then*

$$(N_1 + N_2)^{-1}f = \lim_{n\to\infty} \frac{1}{2\pi i}\oint_{C_n} (N_1 + \lambda)^{-1}(\lambda - N_2)^{-1}f\,d\lambda, \tag{5.35}$$

where C_n is any closed contour in the λ-plane which contains the points $\lambda = \lambda_1, \cdots, \lambda_n$, and does not contain any point of the spectrum of $-N_1$.

The expansion of f in terms of the eigenfunctions of N_2 is

$$f(x, y) = \Sigma \alpha_k(x) v_k(y),$$

where

$$\alpha_k(x) = \int f(x, y) v_k(y)\, dy.$$

From the properties of the eigenfunctions, we have

$$(\lambda - N_2)^{-1} f = \Sigma(\lambda - \lambda_k)^{-1} \alpha_k(x) v_k(y).$$

Substituting this formula in the right-hand side of (5.35), we get

$$\frac{1}{2\pi i}\oint_{C_n} (N_1 + \lambda)^{-1}(\lambda - N_2) f\, d\lambda = \frac{1}{2\pi i}\oint_{C_n} (N_1 + \lambda)^{-1} \Sigma(\lambda - \lambda_k)^{-1} \alpha_k(x) v_k(y)\, d\lambda.$$

Because of the definition of C_n, this integral reduces to a sum of integrals around the poles $\lambda_1, \lambda_2, \cdots, \lambda_n$ of the integrand, that is, the integral is equal to

$$\frac{1}{2\pi i}\sum_{k=1}^{n} \oint (N_1 + \lambda)^{-1}(\lambda - \lambda_k)^{-1} \alpha_k(x) v_k(y)\, d\lambda = \sum_k (N_1 + \lambda_k)^{-1} \alpha_k(x) v_k(y).$$

By Theorem 5.1 the limit of the right-hand side of this equation is $(N_1 + N_2)^{-1} f$. This proves the theorem.

Since the eigenvalues of N_2 have infinity as a limit point, the contours C_n will go to infinity. We may usually replace the contours C_n by a single contour C which starts and ends at infinity, encloses the spectrum of N_2, and does not contain any point in the spectrum of $-N_1$. In practice, such an integral can be obtained from the N_2-eigenfunction representation of the inverse by the well-known methods for writing an infinite sum as a contour integral. Often, by using Cauchy's theorem the contour may be shifted so that it does not contain the spectrum of N_2 but does enclose the spectrum of $-N_1$. In such a case the evaluation of the contour integral by residues will give an expansion in terms of the N_1-eigenfunctions. The example of the following section will illustrate this point.

Example—Changing One Representation into Another

We shall illustrate the method of the preceding section by showing how the formula (5.23) for the Green's function can be transformed into formula (5.19). First, however, we shall prove the formula

$$(5.36) \qquad \sum_{1}^{\infty} c_n \sin (n\pi x/a) = (2i)^{-1} \int_C \frac{c(k) \sin \{k\pi(a - x)/a\}\, dk}{\sin k\pi}.$$

Here C is a contour starting at $\infty - i\varepsilon$ in the complex k-plane, going in a negative direction around the positive real axis, and ending at $\infty + i\varepsilon$, while $c(k)$ is an analytic function of k in the neighborhood of the positive real axis and is such that $c(n) = c_n$, for $n = 1, 2, \cdots$

To prove (5.36), we notice that the only singularities of the integral are poles at the zeroes of $\sin k\pi$, that is, at the points $k = 1, 2, \cdots$. Evaluating the residues at these points, we find that the integral is

$$-\sum_1^\infty \frac{c_n \sin \{n\pi(a-x)/a\}}{\cos n\pi} = \sum_1^\infty c_n \sin (n\pi x/a),$$

which proves (5.36).

As an example of this formula, consider the case where $c_n = 1/n$, and $a = \pi$. We find that

$$\Sigma n^{-1} \sin nx = (2i)^{-1} \int \frac{\sin \{k(\pi - x)\}\, dk}{k \sin k\pi}.$$

The integral may be evaluated by the following manipulations: In the integral over the lower half of the contour C, replace k by $-k$, and we get

$$\Sigma n^{-1} \sin nx = (2i)^{-1} \int_{C'P'} + (2i)^{-1} \int_{PC_+}$$

as illustrated in Fig. 5.1. Since the integrand has no singularities in the

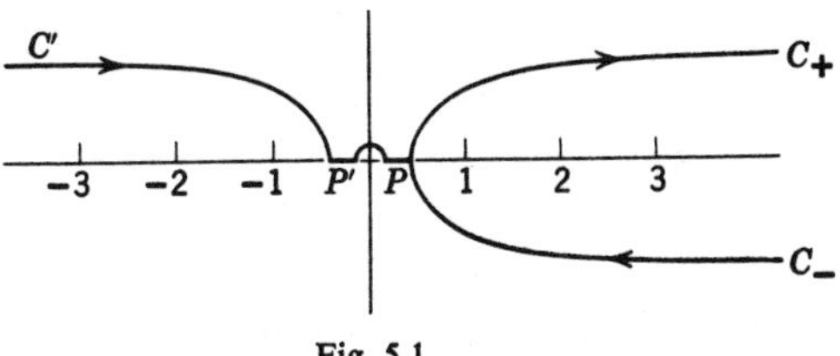

Fig. 5.1

upper half plane, then, by Cauchy's theorem, the integral taken over the contour $C'P'PC_+$ and the infinite semicircle must vanish. However, since

$$\left| \frac{\sin \{k(\pi - x)\}}{\sin k\pi} \right| < \exp \{-x \operatorname{Im}(k)\}$$

if $x \leq \pi$, the integral over the infinite semicircle also vanishes. This shows that

$$(2i)^{-1}\left[\int_{C'P'} + \int_{PC_+}\right] = -(2i)^{-1} \int_{P'P}.$$

This last integral may be reduced to an integral along the real axis from P' to P plus an integral along a small semicircle in the negative direction around the origin. Because the integrand is an odd function of k, the integral along the real axis vanishes and, after evaluating the residue at $k = 0$, we conclude that

$$\Sigma n^{-1} \sin nx = (\pi - x)/2, \quad 0 \leq x \leq \pi.$$

This result may be checked by expanding $(\pi - x)/2$ in a sine series.

Formula (5.36) is a typical case of a method which shows how an expansion in a series of eigenfunctions can be represented as a contour integral over the spectrum of the operator. We shall give the formula for the general case in Problem 5.14.

Let us now return to the discussion of (5.23). It will be a sine series of the same form as that in (5.36) if we put

$$c_n = 2a^{-1} \frac{\sinh \{n\pi(b - y)/a\} \sinh \{n\pi\eta/a\} \sin (n\pi\xi/a)}{(n\pi/a) \sinh (n\pi b/a)}, \qquad \eta < y.$$

Using (5.36), we get

(5.37) $\quad u(x, y)$

$$= (\pi i)^{-1} \int_C \frac{\sinh \{k\pi(b - y)/a\} \sinh (k\pi\eta/a) \sin (k\pi\xi/a) \sin \{k\pi(a - x)/a\}\, dk}{k \sinh (k\pi b/a) \sin k\pi},$$

$$\eta < y.$$

There is a similar formula for $\eta > y$, but we shall not consider it.

Again, just as before, we have

$$\int_C = \int_{C'P'} + \int_{PC_+}.$$

Since the integrand is an odd function of k and since it does not have any singularity at $k = 0$, we may take

$$\int_C = \int_{C'P'PC_+}.$$

By Cauchy's theorem, this integral equals the integral over the infinite semicircle plus the sum of the residues at the poles in the upper half-plane.

Using the following estimates,

$$\left| \frac{\sinh \{k\pi(b - y)/a\} \sinh (k\pi\eta/a)}{\sinh (k\pi b/a)} \right| < \exp \{(\eta - y)|\mathrm{Re}\,(k\pi/a)|\}, \quad 0 < \eta, y < b$$

and

$$\left| \frac{\sin (k\pi\xi/a) \sin \{k\pi(a - x)/a\}}{\sin k\pi} \right| < \exp \{(\xi - x)\,\mathrm{Im}\,(k\pi)/a\}, \quad 0 < \xi, x < a,$$

we conclude that the integrand, and therefore also the integral over an infinite semicircle, will vanish if $\eta < y$ and $\xi < x$.

The poles of the integrand are the zeroes of $\sinh (k\pi b/a)$, that is, the points

$$bk = ai, 2ai, 3ai, \cdots.$$

Evaluating the residues at these poles, we obtain finally the representation (5.19) for $u(x, y)$, namely:

$$u(x, y) = 2b^{-1} \sum \frac{\sinh(n\pi\xi/b) \sinh\{n\pi(a - x)/b\} \sin(n\pi\eta/b) \sin(n\pi y/b)}{(n\pi/b) \sinh(n\pi a/b)}.$$

Note that so far this formula has been proved only if $\eta < y$ and $\xi < x$. However, because of the symmetry of the result in y and η, it is clear that the result holds also for $\eta > y$. To obtain the formula when $\xi > x$, we should interchange x and ξ in (5.37).

It should be noted that (5.37) is merely Theorem 5.2 applied to this special case. Of course, the fact that there are two representations for $u(x, y)$ was known previously; consequently, the manipulations of this section were unnecessary. However, we shall see that in more complicated cases, where the spectral representation for one operator is unknown, the method presented here leads to useful results.

PROBLEMS

5.12. Use the method of this section to evaluate $\sum_1^\infty n^{-1} \sin nx$ for $\pi < x < 2\pi$.

(*Hint.* Use the integral

$$(2i)^{-1} \int_C \frac{\sin\{k(x - \pi)\}}{k \sin k\pi} dk.)$$

5.13. Evaluate $\sum_0^\infty \frac{\cos nx}{1 + n^2}$ by a method similar to that of this section.

5.14. Suppose that $u_n(x)$ are the eigenfunctions of a self-adjoint second-order differential operator L, with unmixed boundary conditions $B_1(u) = B_2(u) = 0$. Let $v(x, \lambda)$ be that solution of $(L - \lambda)u = 0$ which satisfies the condition $B_1(v) = 0$, and let $w(x, \lambda)$ be that solution of $(L - \lambda)u = 0$ which satisfies the condition $B_2(w) = 0$; then show that

$$\Sigma c_n u_n(x) = \int_C \frac{c(\lambda) w(x, \lambda) \alpha(\lambda)\, d\lambda}{J[v, w]},$$

where C is a contour enclosing the spectrum λ_n of L and $\alpha(\lambda)$ is a normalizing factor such that

$$\frac{w(x, \lambda)\alpha(\lambda)}{J[v, w]} \to \frac{u_n(x)}{\lambda - \lambda_n}$$

as λ approaches λ_n.

Spectral Representation for the Sum of Two Commutative Operators

The preceding sections have shown us how to find the inverse for an operator such as $N_1 + N_2$ which is the sum of two commutative operators. In this section we shall show how to obtain the spectral representation for the sum $N_1 + N_2$ if the spectral representation is known for each of the operators N_1 and N_2. When it is obtained, the spectral representation will enable us to extend the methods of Theorem 5.1 or of separation of variables to problems involving the sum of three commutative operators. Later, we shall discuss such a problem.

Just as before, we assume that N_1 is a self-adjoint differential operator acting on functions of x and that N_2 is a self-adjoint differential operator acting on functions of y. Let $\lambda_1, \lambda_2, \cdots$ be the eigenvalues and $v_1(y)$, $v_2(y), \cdots$ be the corresponding normalized eigenfunctions of N_2. Let $\mu_1, \mu_2, \cdots$ be the eigenvalues and $u_1(x), u_2(x), \cdots$ be the corresponding normalized eigenfunctions of N_1. We shall assume that the set of eigenfunctions of N_1 is complete in the linear vector space of real-valued functions $f(x)$, and the set of eigenfunctions of N_2 is complete in the linear vector space of real-valued functions $g(y)$ which are of integrable square over the interval $0 \leq y \leq b$. We shall now prove

Theorem 5.3. *The eigenvalues of $N_1 + N_2$, where N_1 and N_2 are the operators described above, are the numbers $\mu_j + \lambda_k$ $(j, k = 1, 2, \cdots)$. The corresponding eigenfunctions are the functions $u_j(x)v_k(y)$. These eigenfunctions are complete in the linear vector space of all real-valued functions $h(x, y)$ such that*

$$\int_0^a \int_0^b h(x, y)^2 \, dx \, dy < \infty. \tag{5.38}$$

The proof of the first two statements of this theorem follows from the fact that

$$(N_1 + N_2)u_j v_k = N_1 u_j v_k + N_2 u_j v_k = \mu_j u_j v_k + \lambda_k u_j v_k.$$

To prove the completeness, let $h(x, y)$ be any function satisfying (5.38); then a well-known theorem of analysis† tells us that

$$\int_0^a h(x, y)^2 \, dx < \infty$$

for almost all values of y. Since the eigenfunction of N_1 are complete, it follows that the series

$$\sum_j \beta_j(y) u_j(x),$$

where

$$\beta_j(y) = \int_0^a h(x, y) u_j(x) dx,$$

converges to $h(x, y)$ in the mean with respect to x for almost all values of y. Also, by the orthogonality of the eigenfunctions, we have that

$$\int_0^a h(x, y)^2 \, dx = \sum_j \beta_j(y)^2 \tag{5.39}$$

† Fubini's theorem. See Hobson, *Theory of Functions of a Real Variable*, p. 631. Cambridge University Press, London, 1926.

for almost all values of y. Comparing (5.38) with (5.39), we see that, for all j,

$$\int_0^b \beta_j(y)^2 \, dy < \infty.$$

Now, since the eigenfunctions of N_2 are complete, we may write

$$\beta_j(y) = \sum_k \gamma_{jk} v_k(y)$$

where

$$\gamma_{jk} = \int_0^b \beta_j(y) v_k(y) \, dy = \int_0^b \int_0^a h(x, y) u_j(x) v_k(y) \, dx \, dy.$$

These series converge in the mean with respect to y.

Consider the double series

$$\sum_j \sum_k \gamma_{jk} u_j(x) v_k(y).$$

We shall show that this series converges in the mean to $h(x, y)$, that is, we shall show that if

$$I_{mn} = \int_0^b \int_0^a \left[h(x, y) - \sum_{j=1}^m \sum_{k=1}^n \gamma_{jk} u_j(x) v_k(y) \right]^2 dx \, dy$$

then I_{mn} converges to zero as m and n approach infinity. From the orthogonality of the u_j and v_k, we obtain the relation

$$\int_0^b \int_0^a \left[\sum_1^m \sum_1^n \gamma_{jk} u_j(x) v_k(y) \right]^2 dx \, dy$$

$$= \int_0^b \int_0^a \left[\sum_j \sum_{j'} \sum_k \sum_{k'} \gamma_{jk} \gamma_{j'k'} u_j(x) u_{j'}(x) v_k(y) v_{k'}(y) \right] dx \, dy$$

$$= \sum \sum \gamma_{jk}^2.$$

Using this to evaluate I_{mn}, we see that

$$I_{mn} = \int_0^b \int_0^a \left[h(x, y)^2 - \sum_1^m \sum_1^n \gamma_{jk}^2 \right] dx \, dy.$$

From the fact that the set of eigenfunctions $v_k(y)$ are complete, we conclude that the series $\sum_{k=1}^n \gamma_{jk}^2$ converges to $\int_0^b \beta_j(y)^2 \, dy$. From the fact that the set of eigenfunctions $u_j(x)$ are complete, we conclude that the series $\sum_{j=1}^m \beta_j(y)^2$

converges to $\int_0^a h(x, y)^2\,dx$. These results imply that

$$\sum_{j=1}^{m}\sum_{k=1}^{n}\gamma_{jk}^2 \to \sum_1^m \int_0^b \beta_j(y)^2\,dy = \int_0^b \sum_1^m \beta_j(y)^2\,dy$$

$$\to \int_0^b \int_0^a h(x, y)^2\,dx\,dy.$$

This proves our theorem.

Theorem 5.3 may be formulated in easily remembered form. Since the eigenfunctions of both N_1 and N_2 are complete, we have

$$\delta(x - \xi) = \Sigma u_j(x)u_j(\xi), \tag{5.40}$$
$$\delta(y - \eta) = \Sigma v_k(y)v_k(\eta).$$

Multiplying these equations together, we have a result equivalent to that of Theorem 5.3, namely,

$$\delta(x - \xi)\delta(y - \eta) = \sum_j \sum_k u_j(x)v_k(y)u_j(\xi)v_k(\eta). \tag{5.41}$$

The justification for this formal multiplication of the series in (5.40) is given by the proof of Theorem 5.3 which shows that (5.41) is valid.

If $\phi(N_1, N_2)$ is an analytic function of the operators N_1 and N_2 described above, then, since $u_j(x)$ are eigenfunctions of N_1 and $v_k(y)$ are eigenfunctions of N_2, we have

$$\phi(N_1, N_2)\delta(x - \xi)\delta(y - \eta) = \sum_j \sum_k \phi(\mu_j, \lambda_k)u_j(x)v_k(y)u_j(\xi)v_k(\eta). \tag{5.42}$$

This result will be used in the next section to find the inverse of a sum of three commutative operators.

Example—The Inverse of a Sum of Three Commutative Operators

Consider the following problem:

Find a function $u(x, y, z)$ such that

$$u_{xx} + u_{yy} + u_{zz} + k^2u = -\delta(x - \xi)\delta(y - \eta)\delta(z - \zeta), \tag{5.43}$$

such that

$$u(0, y, z) = u(a, y, z) = u(x, 0, z) = u(x, b, z) = 0,$$

and such that $u(x, y, z)$ is outgoing for $z = \pm\infty$.

If we put $N_1 = -\dfrac{\partial^2}{\partial x^2}$, with the boundary conditions $u(0) = u(a) = 0$, and $N_2 = -\dfrac{\partial^2}{\partial y^2}$, with the boundary conditions $v(0) = v(b) = 0$, then (5.43) may be written as follows:

$$u_{zz} + (k^2 - N_1 - N_2)u = -\delta(x - \xi)\delta(y - \eta)\delta(z - \zeta),$$

where u should be outgoing for $z = \pm\infty$. Since the operator $N_1 + N_2$ commutes with the operator $\frac{\partial^2}{\partial z^2}$, we may treat $N_1 + N_2$ as a constant. In that case, the solution of the equation is

$$u(x, y, z) = -(2ip)^{-1}e^{ip|z-\zeta|}\delta(x - \xi)\delta(y - \eta),$$

where $p = (k^2 - N_1 - N_2)^{1/2}$. Using (5.42) to interpret this solution, we get

(5.44) $u(x, y, z)$

$$= (2i/ab)\sum_m\sum_n\left(k^2 - \frac{m^2\pi^2}{a^2} - \frac{n^2\pi^2}{b^2}\right)^{-1/2}\sin\frac{m\pi x}{a}\sin\frac{n\pi y}{b}\sin\frac{m\pi\xi}{a}\sin\frac{n\pi\eta}{b}$$

$$\cdot\exp\left[i\left(k^2 - \frac{m^2\pi^2}{a^2} - \frac{n^2\pi^2}{b^2}\right)^{1/2}|z - \zeta|\right].$$

Here we have used the fact that $(m^2\pi^2/a^2)$ are the eigenvalues and $(2/a)^{1/2}\sin(m\pi x/a)$ the normalized eigenfunctions of N_1 and also the corresponding result for N_2. Note that the outgoing wave condition requires that the square root be so chosen that it has a non-negative imaginary part.

This problem has some interesting physical aspects. Equation (5.43) is the equation for the acoustic pressure field produced by a point source at (ξ, η, ζ) which is oscillating in a cylinder of rectangular cross section which extends from $-\infty$ to ∞ along the z-axis. The source is oscillating at a frequency $kc/2\pi$, where c is the velocity of sound. Equation (5.43) is also the equation satisfied by one component of the electromagnetic field produced by a point source at (ξ, η, ζ) oscillating with a frequency $kc/2\pi$, where c is the velocity of propagation of electromagnetic waves. The solution (5.44) shows that in either case the field produced by the source is a sum of terms. We shall write an individual term of the solution as follows:

(5.45) $$A_{mn}\exp(i\beta_{mn}|z - \zeta|)\sin(m\pi x/a)\sin(n\pi y/b),$$

where A_{mn} is a quantity independent of x, y, z and where

(5.46) $$\beta_{mn}^2 + m^2\pi^2/a^2 + n^2\pi^2/b^2 = k^2.$$

Again, we require β_{mn} to be so chosen that it has a non-negative imaginary part. For any fixed value of m and n, we shall call the expression

$$T_{mn}(x, y) = \sin(m\pi x/a)\sin(n\pi y/b)$$

a *mode of vibration* of the cross-section of the cylinder. Note that it is an eigenfunction of the operator $N_1 + N_2$ corresponding to the eigenvalue $m^2\pi^2/a^2 + n^2\pi^2/b^2$. We see that if $w_{mn}(z)$ satisfies the equation

$$\frac{d^2w_{mn}}{dz^2} + \beta_{mn}^2 w_{mn} = 0,$$

where β_{mn} satisfies (5.46), then $T_{mn}(x, y)w_{mn}(z)$ is a solution of (5.43) with the right-hand side equal to zero. Of particular interest are the cases where

$$w_{mn}(z) = \exp(\pm i\beta_{mn}z).$$

Suppose that the exponent has the positive sign, then, if β_{mn} is real, $T_{mn}(x, y)w_{mn}(z)$ represents a wave traveling in the direction of positive z. If the exponent has the negative sign and if β_{mn} is real, then $T_{mn}(x, y)w_{mn}(z)$ represents a wave traveling in the direction of negative z. The modes for which β_{mn} is real will be called *propagating modes.* If β_{mn} is not real, the exponential term in z goes to zero or to infinity as z goes to infinity. The modes for which β_{mn} is not real will be called *non-propagating* or *attenuated modes* because in all physical situations the exponential will go to zero.

Let us return to the solution (5.44) and the individual modes (5.45). For propagating modes, that is, β_{mn} real, the term (5.45) represents a wave going in the direction of positive z when $z > \zeta$ and a wave going in the direction of negative z when $z < \zeta$. This means that the source at $z = \zeta$ has excited waves which travel away from the source. For non-propagating modes, that is, β_{mn} complex, the term (5.45) will go to zero as z goes to $\pm\infty$. Combining these results, we conclude that a source excites all the modes and that the propagating modes go away from the source to infinity, but the non-propagating modes become exponentially small as we go away from the source.

Suppose that, instead of (5.43), we consider the equation

$$u_{xx} + u_{yy} + u_{zz} + k^2u = f(x, y, z),$$

where $f(x, y, z) = 0$ for $|z| > z_0$ and where $u(x, y, z)$ satisfies the same boundary conditions as the solution of equation (5.43). The physical interpretation of this equation is that it is the equation for either the acoustic pressure field or one component of the electromagnetic field produced by a distribution of sources whose density is $f(x, y, z)$. It is easy to show that for $|z| > z_0$ the solution of this problem is a sum

$$\sum_{m,n} A_{mn}T_{mn}(x, y) \exp(i\beta_{mn}|z|),$$

where $T_{mn}(x, y)$ and β_{mn} are the quantities defined previously and A_{mn} are constants depending on $f(x, y, z)$. This type of solution exhibits the following physically reasonable result: a distribution of sources produces waves which propagate to infinity along the z-axis and also waves which are exponentially attenuated.

This example is typical of a general case. Suppose that we consider the equation

$$Nu + u_{zz} + k^2u = \delta(x - \xi)\,\delta(y - \eta)\,\delta(z - \zeta),$$

where N is a self-adjoint differential operator in x and y acting on functions $f(x, y)$, which are of integrable square over some region R in the x, y plane and satisfy some linear homogeneous conditions on the boundary of R. We suppose that $-\infty < z < \infty$ and that $u(x, y, z)$ satisfies outgoing wave conditions as z goes to $\pm\infty$. The modes will be the eigenfunctions $T_{mn}(x, y)$ of the operator. Put

$$NT_{mn} = \lambda_{mn} T_{mn};$$

then $T_{mn}(x, y) w_{mn}(z)$ will be a solution of the homogeneous equation

$$Nu + u_{zz} + k^2 u = 0$$

if

$$\frac{d^2 w_{mn}}{dz^2} + \beta_{mn}^2 w_{mn} = 0,$$

where

$$\beta_{mn}^2 = k^2 + \lambda_{mn}.$$

Again, the modes for which β_{mn} is real will be called propagating modes; those for which β_{mn} is complex will be called non-propagating modes. We can also show, just as before, that a point source will excite all modes and that the propagating modes will go to infinity while the non-propagating ones will be exponentially attenuated as we go to infinity. Similar results can be obtained in a wide class of problems.

PROBLEMS

5.15. What are the modes for the equation

$$u_{xx} + u_{yy} + u_{zz} + k^2 u = 0,$$

where $u_x(0, y, z) = u_x(a, y, z) = u_y(x, 0, z) = u_y(x, b, z) = 0$ and where $u(x, y, z)$ is an outgoing wave at infinity? Which modes propagate? Which do not propagate?

5.16. Solve

$$\begin{aligned} u_{xx} + u_{yy} + u_{zz} + k^2 u &= \sin(2\pi x/a), \quad |z| < z_0 \\ &= 0, \quad |z| > z_0, \end{aligned}$$

with the same boundary conditions as in Problem 5.15.

5.17. Use (5.46) to find the range of values of k for which either no, exactly one, or exactly two propagating modes exist in the equation (5.43).

The Spectral Representation for Partial Differential Operators

In the preceding chapter we showed that the spectral representation for an ordinary differential operator L could be obtained by integrating the Green's function for $L - \lambda$ around a large circle in the λ-plane. The poles of the Green's function produce the eigenfunctions of L, and the branch-cut integrals in the λ-plane produce the continuous spectrum.

The same procedure can be used to find the spectral representation for a partial differential operator M. The integral of $(M - \lambda)^{-1}$ over a large circle in the complex λ-plane can be reduced to a sum of residues and integrals over branch cuts. The residue terms will give the eigenfunctions of M; the branch-cut integrals will give the continuous spectrum.

As an illustration of this procedure, we shall obtain the spectral representation of the two-dimensional Laplacian Δ over the entire x, y plane. We find the Green's function by solving

$$(\Delta - \lambda)g = g_{xx} + g_{yy} - \lambda g = -\,\delta(x - x')\,\delta(y - y').$$

The solution of this equation may be found by considering $\dfrac{\partial^2}{\partial x^2}$ a constant in the ordinary differential equation for y. We find that

$$g = \frac{1}{4\pi}\int_{-\infty}^{\infty} \frac{e^{ip(x-x')}e^{-\sqrt{\lambda+p^2}|y-y'|}\,dp}{\sqrt{\lambda + p^2}};$$

just as in the preceding chapter, it can be shown that

$$\frac{1}{2\pi i}\oint g\,d\lambda = \delta(x - x')\,\delta(y - y').$$

The contour integral has no poles, only a branch cut extending from $\lambda = -\,p^2$ to $\lambda = -\,\infty$ along the negative λ-axis. We evaluate the contour integral by interchanging the λ-integral with the p-integral and putting $\lambda = -\,p^2 - q^2$. In this way we get†

$$\text{(5.47)}\qquad \delta(x - x')\,\delta(y - y') = \frac{1}{2\pi i}\oint g\,d\lambda$$
$$= \left(\frac{1}{2\pi}\right)^2 \int_{-\infty}^{\infty} e^{ip(x-x')}\,dp \int_{-\infty}^{\infty} e^{-iq(y-y')}\,dq.$$

This is the desired spectral representation. To see this, let $f(x, y)$ be a function of integrable square over $(-\,\infty, \infty)$. Multiply (5.47) by $f(x', y')$ and integrate over all values of x' and y'. We get

$$f(x, y) = \frac{1}{4\pi^2}\iint_{-\infty}^{\infty} e^{i(px-qy)}\,dp\,dq\,g(p, q),$$

where

$$g(p, q) = \iint_{-\infty}^{\infty} e^{-i(px-qy)}f(x, y)\,dx\,dy.$$

These are just the well-known formulas for the double Fourier transform of a function of two variables. Note that the improper eigenvalues of Δ corresponding to the improper eigenfunctions $e^{i(px-qy)}$ are $-\,p^2 - q^2$; consequently,

† The term $|y - y'|$ in the exponent can be replaced by $y - y'$ because the integral containing it is a δ-function, which is an even function of its argument.

$$\Delta f = -\frac{1}{4\pi^2}\iint_{-\infty}^{\infty} (p^2 + q^2)e^{i(px-qy)}g(p, q)\, dp\, dq,$$

and

$$\phi(\Delta)f = -\frac{1}{4\pi^2}\iint_{-\infty}^{\infty} \phi(-p^2 - q^2)e^{i(px-qy)}g(p, q)\, dp\, dq$$

if $\phi(t)$ is an analytic function of t which is of less than exponential growth as t goes to $\pm\infty$. The last condition is necessary in order to ensure that the integral have a meaning, if not as an ordinary function, at least as a symbolic function.

PROBLEM

5.18. Find the spectral representation of the three-dimensional Laplacian over the entire (x, y, z) space.

A Physical Interpretation of the Continuous Spectrum

The eigenfunctions ϕ of a differential operator M are solutions of the differential equation $M\phi = \lambda\phi$ and usually have an immediate physical significance in any problem in which they appear. However, the improper eigenfunctions ψ, even though they also satisfy a differential equation, do not have individual physical significance since the functions ψ do not belong to the vector space over which the operator is defined. Physically, this means, for example, that the physical state defined by the function ψ would be a state of infinite energy.

In this section we shall consider a particular example and we shall investigate the physical significance of the continuous spectrum. The example we shall consider is this:

Find a function $u(x, y)$ such that

$$u_{xx} + u_{yy} + k^2u = \delta(x)\,\delta(y - y') \tag{5.48}$$

over the half plane $-\infty < x < \infty$, $0 \leq y < \infty$, such that $u_y(x, 0) = -\alpha u(x, 0)$, and such that $u(x, y)$ is outgoing for $x = \pm\infty$. The function $u(x, y)$ could be interpreted physically as the acoustic pressure field produced by a source located at $(0, y')$. The source acts in the half plane $y \geq 0$ bounded by a semi-rigid diaphragm at $y = 0$. Note that $\alpha = 0$ would correspond to a rigid diaphragm.

Put $N_2u = -u_{yy}$; then we may write (5.48) as follows:

$$u_{xx} + (k^2 - N_2)u = \delta(x)\delta(y - y').$$

The solution of this equation is

$$u = \frac{e^{i\sqrt{k^2-N_2}|x|}}{2i\sqrt{k^2 - N_2}}\delta(y - y'). \tag{5.49}$$

From (4.48) we obtain the following spectral representation for N_2:

$$\delta(y-y') = 2\alpha e^{-\alpha(y+y')} + \frac{2}{\pi}\int_0^\infty \left((\cos py - \frac{\alpha}{p}\sin py\right)\left(\cos py' - \frac{\alpha}{p}\sin py'\right)\frac{p^2\,dp}{p^2+\alpha^2}.$$

Putting this into (5.49), we get

$$(5.50)\quad u(x,y) = \frac{-i\alpha e^{i\sqrt{k^2+\alpha^2}\,|x|}e^{-\alpha(y+y')}}{\sqrt{k^2+\alpha^2}}$$
$$-\frac{i}{\pi}\int_0^\infty \left(\cos py - \frac{\alpha}{p}\sin py\right)\left(\cos py' - \frac{\alpha}{p}\sin py'\right)\frac{e^{i\sqrt{k^2-p^2}|x|}}{\sqrt{k^2-p^2}}\,\frac{p^2\,dp}{p^2+\alpha^2}.$$

The first term on the right-hand side is a plane wave which travels along the x-axis and is exponentially damped in the direction of the y-axis. Such a wave is called a *surface wave*. Similar results hold in more general cases. Each eigenfunction produces a surface wave which is exponentially damped in the direction transverse to the direction in which the wave is moving.

To interpret the integral term in (5.50), we notice first that the integrand is an even function of p and consequently we can write the given integral as one-half the integral from $-\infty$ to ∞. Next, replace the trigonometric functions by exponentials. We have

$$\cos py - \frac{\alpha}{p}\sin py = \frac{1}{2}e^{ipy}\left(1+\frac{i\alpha}{p}\right) + \frac{1}{2}e^{-ipy}\left(1-\frac{i\alpha}{p}\right),$$

and then

$$\left(\cos py - \frac{\alpha}{p}\sin py\right)\left(\cos py' - \frac{\alpha}{p}\sin py'\right)\frac{p^2}{p^2+\alpha^2}$$
$$= e^{ip(y-y')} + e^{-ip(y-y')} + e^{ip(y+y')}\frac{p+i\alpha}{p-i\alpha} + e^{-ip(y+y')}\frac{p-i\alpha}{p+i\alpha}.$$

When this result is substituted into the integral in (5.50), it becomes

$$-\frac{i}{2\pi}\int_{-\infty}^\infty \left[e^{ip(y-y')} + e^{-ip(y-y')} + e^{-ip(y+y')}\frac{p+i\alpha}{p-i\alpha} + e^{-ip(y+y')}\frac{p-i\alpha}{p+i\alpha}\right]$$
$$\frac{e^{i\sqrt{k^2-p^2}|x|}}{\sqrt{k^2-p^2}}\,dp.$$

If we replace p by $-p$ in the second and fourth terms, we see that these terms are the same as the first and third, respectively; consequently, this integral can be expressed as the sum of I_1 and I_2, where

$$I_1 = -\frac{i}{\pi}\int_{-\infty}^\infty e^{ip(y-y')+i\sqrt{k^2-p^2}|x|}\,\frac{dp}{\sqrt{k^2-p^2}}$$

and

$$I_2 = -\frac{i}{\pi}\int_{-\infty}^{\infty} e^{ip(y+y')+i\sqrt{k^2-p^2}|x|}\frac{p-i\alpha}{p+i\alpha}\frac{dp}{\sqrt{k^2-p^2}}.$$

The integral I_1 can be evaluated in terms of Bessel functions. We shall not do this, but instead we shall obtain the asymptotic behavior of I_1, that is, its behavior as the distance between the observation point (x, y) and the source point $(0, y')$ goes to infinity. The method we shall use to obtain this asymptotic behavior is applicable in many other cases. The essence of the method is the introduction of the exponent as a new variable of integration and then the study of the new integrand in the neighborhood of its singularities. Let us see how this works for I_1. We put

(5.51) $$p(y-y') + \sqrt{k^2-p^2}|x| = \sigma;$$

then differentiation gives

(5.52) $$[\sqrt{k^2-p^2}(y-y') - p|x|]\frac{dp}{\sqrt{k^2-p^2}} = d\sigma.$$

But

$$[p(y-y') + \sqrt{k^2-p^2}|x|]^2 + [\sqrt{k^2-p^2}(y-y') - p|x|]^2 = k^2r^2,$$

where

$$r^2 = (y-y')^2 + x^2;$$

consequently, (5.52) becomes

$$\frac{dp}{\sqrt{k^2-p^2}} = \frac{d\sigma}{\sqrt{k^2r^2-\sigma^2}}.$$

When this result is substituted in I_1, we have

$$I_1 = -\frac{i}{\pi}\int_C e^{i\sigma}\frac{d\sigma}{\sqrt{k^2r^2-\sigma^2}},$$

where C is the contour indicated in Fig. 5.2. This curve C is obtained by

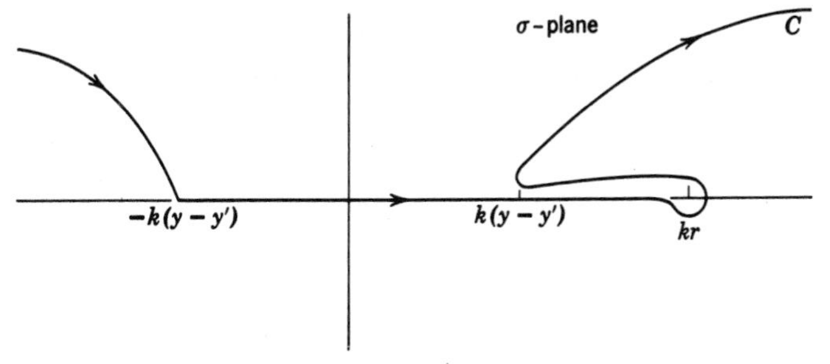

Fig. 5.2

letting p go from $-\infty$ to $+\infty$ in (5.51). Since the square root of $k^2 - p^2$ must have a non-negative imaginary part for the solution to be outgoing, we see that as p approaches $\pm\infty$,

$$\sqrt{k^2 - p^2} \sim +i|p|.$$

This shows that, for large $|p|$,

$$\sigma \sim p(y - y') + i|px|.$$

When $-k \leq p \leq k$, σ is real. Considering the derivative of σ with respect to p as given in (5.52), we see that for $p > -k$, σ increases as p increases to the value p_0, where

$$\sqrt{k^2 - p_0^2}\,(y - y') = p_0|x|.$$

When $p = p_0$, $\sigma = kr$. For $p > p_0$, the value of σ decreases until $p = k$ and $\sigma = k(y - y')$. This explains the bending back of the curve C. Note that there is a branch point in the σ-plane at $\sigma = kr$; consequently, we must distinguish between the values of the integrand on the part of C above kr from the values of the integrand on the part of C below kr.

The next step in the treatment of I_1 is to shift the contour C so that the exponential will go to zero as rapidly as possible. This is done by straightening the bend in the curve and then moving the left-hand and right-hand parts of C until they are parallel to the imaginary axis. In this way we obtain the contour C' of Fig. 5.3. Shifting the contour

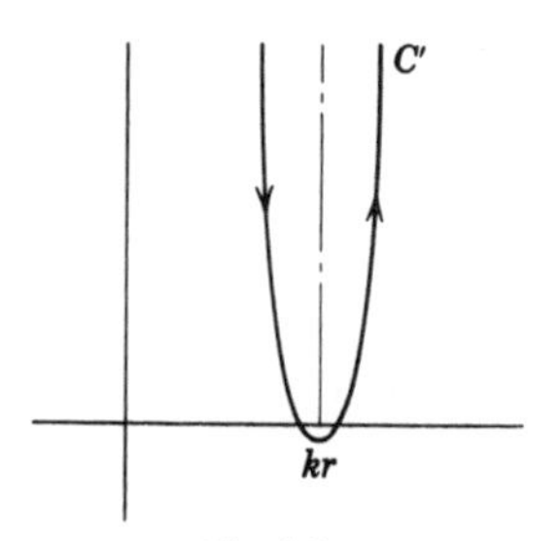

Fig. 5.3

C to C' can be justified by the facts that the contours can be connected by an arc of large radius and that the value of the integral on the arc can be shown to approach zero as the radius increases.

It should be noted that the left-hand and right-hand parts of C' can be brought as close together as desired as long as they are not made to coincide. The presence of the branch point of the integrand at $\sigma = kr$ indicates that the value of the integrand on the right-hand part of C differs from the value of the integrand at the neighboring point on the left-hand part of C by the factor minus one.

The final step in the method is to put

$$\sigma = kr + i\tau$$

in the integral for I_1. We get

$$I_1 = -\frac{i}{\pi}\int_{C'} \frac{e^{i\sigma}d\sigma}{\sqrt{k^2r^2 - \sigma^2}} = \frac{1}{\pi}e^{ikr}\Big[\int_{\infty L}^{0} + \int_{0R}^{\infty}\Big](-i\tau)^{-1/2}(2kr + i\tau)^{-1/2}e^{-\tau}\,d\tau.$$

Here the L and the R under the integrals indicate that the integrands should be evaluated on the left-hand and right-hand sides of C', respectively. Consider first the right-hand side of C'. The points on it correspond to values of $p > k$. From (5.52) and the equation beneath it, we see that

$$\sqrt{k^2 - p^2}(y - y') - p|x| = \sqrt{k^2r^2 - \sigma^2}.$$

For $p > k$, the imaginary part of the left-hand side is non-negative (this assumes $y > y'$); consequently, we must take that square root of $k^2r^2 - \sigma^2$ which has a non-negative imaginary part. This implies that

$$(k^2r^2 - \sigma^2)^{1/2} = (-i\tau)^{1/2}(2kr + i\tau)^{1/2} = e^{3\pi i/4}\tau^{1/2}(2kr + i\tau)^{1/2}$$

on the right-hand part of C'. On the left-hand part of C', the value of the square root will be just the negative of the above value. Using these results, we see that

$$I_1 = -\frac{2}{\pi}e^{i(kr+\pi/4)}\int_0^\infty \frac{\tau^{-1/2}e^{-\tau}\,d\tau}{(2kr + i\tau)^{1/2}}.$$

For large values of kr this integral approaches the integral

$$(5.53) \qquad -\frac{2}{\pi}e^{i(kr+\pi/4)}\int_0^\infty \frac{\tau^{-1/2}e^{-\tau}\,d\tau}{(2kr)^{1/2}} = -\left(\frac{2}{\pi kr}\right)^{1/2}e^{i(kr+\pi/4)}.$$

The error in using this value for I_1 is given by the difference

$$E = \frac{2}{\pi}e^{i(kr+\pi/4)}\int_0^\infty \tau^{-1/2}e^{-\tau}\,d\tau\left[\frac{1}{(2kr + i\tau)^{1/2}} - \frac{1}{(2kr)^{1/2}}\right].$$

We shall show in Problem 5.19 that

$$|e^{-i(kr+\pi/4)}E| < \left(\frac{2}{\pi kr}\right)^{1/2}(8kr)^{-1}.$$

This proves that, as kr approaches infinity, I_1 converges to the value given in (5.53).

In a similar manner the integral I_2 can be evaluated for large values of kr. We shall show in Problem 5.20 that

$$(5.54) \qquad I_2 \sim -\left(\frac{2}{\pi kr'}\right)^{1/2}e^{i(kr'+\pi/4)},$$

where

$$r'^2 = (y + y')^2 + x^2,$$

if the ratio $r\alpha/k$ is small.

Going back to $u(x, y)$, we see that

$$-u(x, y) = -i\alpha \frac{e^{i\sqrt{k^2+\alpha^2}|x|}e^{-\alpha(y+y')}}{\sqrt{k^2+\alpha^2}} - I_1 - I_2$$

$$\sim i\alpha \frac{e^{i\sqrt{k^2+\alpha^2}|x|}e^{-\alpha(y+y')}}{\sqrt{k^2+\alpha^2}} + \left(\frac{2}{\pi kr}\right)^{1/2} e^{i(kr+\pi/4)} + \left(\frac{2}{\pi kr'}\right)^{1/2} e^{i(kr'+\pi/4)}.$$

The last two terms have an immediate physical interpretation. The first of them is the field produced by a cylindrical source located at $(0, y')$; the second term is the field produced by a cylindrical source located at the image point $(0, -y')$. These terms came from the continuous part of the spectrum. We conclude then that *the discrete part of the spectrum produces surface waves while the continuous part of the spectrum produces cylindrical waves or space waves, depending on the number of dimensions.*

PROBLEMS

5.19. Prove the estimate for E given above. (*Hint.* The difference $(2kr + i\tau)^{-1/2} - (2kr)^{-1/2} = -i\tau(2kr)^{-1/2}(2kr + i\tau)^{-1/2}[(2kr)^{1/2} + (2kr + i\tau)^{1/2}]^{-1}$. This last expression is less in absolute value than $\tau(2kr)^{-3/2}/2$.)

5.20. Find the asymptotic behavior of I_2. (*Hint.* Put $\sigma = p(y + y') + \sqrt{k^2 - p^2}|x|$.)

5.21. Show that the integral $I_1 = g(x, y)$, where $g(x, y)$ is the solution of $g_{xx} + g_{yy} + k^2 g = \delta(x)\delta(y - y')$ over the whole x, y plane. (*Hint.* Solve for g by treating $\frac{\partial^2}{\partial x^2}$ as a constant.)

δ-Functions in Different Coordinate Systems

We have shown previously that the three-dimensional δ-function is equal to the product of three one-dimensional δ-functions. A similar result is valid for n-dimensional δ-functions. Let $\delta(x_1, x_2, \cdots, x_n)$ be the n-dimensional δ-function, that is, suppose that

$$\int \cdots \int \phi(x_1, \cdots, x_n)\, \delta(x_1, \cdots, x_n)\, dx_1 \cdots dx_n = \phi(0, 0, \cdots, 0).$$

Then, since

$$\int \cdots \int \phi(x_1, \cdots, x_n)\, \delta(x_1) \cdots \delta(x_n)\, dx_1 \cdots dx_n = \phi(0, \cdots, 0),$$

we conclude that

$$\delta(x_1, \cdots, x_n) = \delta(x_1) \cdots \delta(x_n). \tag{5.55}$$

The expressions for the δ-function become more complicated when we introduce curvilinear coordinates. Before, we obtained an expression in spherical polar coordinates for the δ-function at the origin in three-dimensional space. We wish to generalize this result. For simplicity, we shall discuss the special case of two-dimensional space. Suppose that we change from Cartesian coordinates x_1, x_2 to curvilinear coordinates ξ_1, ξ_2 by the formulas

$$x_1 = u(\xi_1, \xi_2),$$
$$x_2 = v(\xi_1, \xi_2).$$

Here we assume that u and v are continuously differentiable single-valued functions of their arguments. Suppose that the coordinates $\xi_1 = \beta_1$, $\xi_2 = \beta_2$ correspond to $x_1 = \alpha_1$, $x_2 = \alpha_2$. By a change of coordinates, the equation

$$\iint \delta(x_1 - \alpha_1)\delta(x_2 - \alpha_2)\phi(x_1, x_2)\, dx_1\, dx_2 = \phi(\alpha_1, \alpha_2)$$

becomes

$$\iint \delta[u_1(\xi_1, \xi_2) - \alpha_1]\delta[v(\xi_1, \xi_2) - \alpha_2]\phi(u, v)|J|\, d\xi_1\, d\xi_2 = \phi(\alpha_1, \alpha_2), \tag{5.56}$$

where J, the Jacobian of the transformation, is given by the formula

$$J = \frac{\partial u}{\partial \xi_1}\frac{\partial v}{\partial \xi_2} - \frac{\partial u}{\partial \xi_2}\frac{\partial v}{\partial \xi_1}.$$

Equation (5.56) shows that the symbolic function

$$\delta[u(\xi_1, \xi_2) - \alpha_1]\delta[v(\xi_1, \xi_2) - \alpha_2]|J|$$

assigns to any testing function the value of that testing function at the point where $u_1 = \alpha_1$, $v_2 = \alpha_2$, that is, at the point where $\xi_1 = \beta_1$, $\xi_2 = \beta_2$; consequently, we may write

$$\delta[u(\xi_1, \xi_2) - \alpha_1]\delta[v(\xi_1, \xi_2) - \alpha_2]|J| = \delta(\xi_1 - \beta_1)\delta(\xi_2 - \beta_2)$$

or, if $|J| \neq 0$, we have

$$\delta(x_1 - \alpha_1)\delta(x_2 - \alpha_2) = \frac{\delta(\xi_1 - \beta_1)\delta(\xi_2 - \beta_2)}{|J|}. \tag{5.57}$$

As an illustration of this theorem, consider the transformation from rectangular coordinates x, y to polar coordinates r, θ, where $x = r\cos\theta$, $y = r\sin\theta$. Since $J = r$, we have

$$\delta(x - x_0)\delta(y - y_0) = \frac{\delta(r - r_0)\delta(\theta - \theta_0)}{r} \tag{5.58}$$

if $x_0 = r_0 \sin\theta_0$, $y_0 = r_0 \cos\theta_0$.

What happens to (5.57) if $J = 0$ at $\xi_1 = \beta_1$, $\xi_2 = \beta_2$? Then the transformation from x_1, x_2 to ξ_1, ξ_2 breaks down and is no longer one-to-one.

For example, in the transformation from rectangular coordinates to polar coordinates, the Jacobian becomes zero at the origin $x = y = 0$, and we know that there the transformation is no longer one-to-one since there $r = 0$ but θ may have any value. This situation is typical of the general case. We shall call a coordinate such as θ, which is either many-valued or else has no determinate value at a singular point of the transformation, an *ignorable* coordinate at that point.

Suppose that $x_1 = \alpha_1$, $x_2 = \alpha_2$ is a singular point of the transformation, that ξ_2 is an ignorable coordinate there, and that the testing function $\phi(x_1, x_2)$ in x_1, x_2 space becomes the testing function $\phi(\xi_1, \xi_2)$ in ξ_1, ξ_2 space. When the point $x_1 = \alpha_1$, $x_2 = \alpha_2$ is a singular point, we have $\xi_1 = \beta_1$, ξ_2 indeterminate, and then the testing function $\phi(\xi_1, \xi_2)$ depends only on β_1. We denote its value by $\phi(\beta_1)$. (The case of the origin in polar coordinates illustrates this. The function $\phi(r, \theta)$ reduces at the origin to a function of r alone.) The symbolic function corresponding to $\delta(x_1 - \alpha_1)\,\delta(x_2 - \alpha_2)$ will consequently be a symbolic function only of ξ_1.

Formula (5.56) may be written as follows:

$$\phi(\alpha_1, \alpha_2) = \iint \delta(u - \alpha_1)\delta(v - \alpha_2)\phi(u, v)|J|\, d\xi_1\, d\xi_2 = \iint t(\xi_1)\phi(\xi_1, \xi_2)|J|\, d\xi_1\, d\xi_2 = \phi(\beta_1), \tag{5.59}$$

where $t(\xi_1)$ is some symbolic function of ξ_1. Put

$$J_1 = \int |J|\, d\xi_2;$$

then (5.59) will be satisfied if we put

$$t(\xi_1) = \delta(\xi_1 - \beta_1)/J_1;$$

consequently, in case $J = 0$ for $x_1 = \alpha_1$, $x_2 = \alpha_2$, we have

$$\delta(x_1 - \alpha_1)\delta(x_2 - \alpha_2) = \delta(\xi_1 - \beta_1)/|J_1|.$$

For example, in the transformation from rectangular to polar coordinates, θ is an ignorable coordinate at the origin and $\int_0^{2\pi} r\, d\theta = 2\pi r$; consequently,

$$\delta(x)\delta(y) = (2\pi r)^{-1}\delta(r).$$

Similar results hold for transformations in n-dimensional space. We state them in

Theorem 5.4. *Let $x_1, \cdots, x_n$ be n-dimensional rectangular coordinates and let $\xi_1, \cdots, \xi_n$ be any other coordinate system not necessarily orthogonal, with n-dimensional volume element $|J|\, d\xi_1 \cdots d\xi_n$. Suppose that the point P with coordinates $x_1 = \alpha_1, \cdots, x_n = \alpha_n$ has coordinates $\xi_1 = \beta_1, \cdots, \xi_n = \beta_n$ and that $J \neq 0$ at P; then*

$$\delta(x_1 - \alpha_1) \cdots \delta(x_n - \alpha_n) = |J|^{-1}\, \delta(\xi_1 - \beta_1) \cdots \delta(\xi_n - \beta_n).$$

If $J = 0$ at P, suppose that the equations $\xi_1 = \beta_1, \cdots, \xi_k = \beta_k$ define P and that therefore the coordinates $\xi_{k+1}, \cdots, \xi_n$ are ignorable. Put

$$J_k = \int \cdots \int J \, d\xi_{k+1} \cdots d\xi_n,$$

that is, J_k is the integral of the Jacobian over the ignorable coordinates; then

$$\delta(x_1 - \alpha_1) \cdots \delta(x_n - \alpha_n) = |J_k|^{-1} \delta(\xi_1 - \beta_1) \cdots \delta(\xi_k - \beta_k).$$

The proof of this theorem is similar to the proof in two dimensions. We shall illustrate it by considering the transformation from three-dimensional rectangular coordinates x, y, z to spherical coordinates r, θ, ψ. Here

$$x = r \sin \theta \cos \psi, \quad y = r \sin \theta \sin \psi, \quad z = r \cos \theta.$$

The Jacobian is $J = r^2 \sin \theta$. It vanishes for all points on the z-axis where ψ is an ignorable coordinate and also at the origin where both θ and ψ are ignorable coordinates. Let the point (x', y', z') in rectangular coordinates have spherical coordinates (r', θ', ψ'); then, if $r' \neq 0, \theta' \neq 0$, we have

$$\delta(x - x')\delta(y - y')\delta(z - z') = \frac{\delta(r - r')\delta(\theta - \theta')\delta(\psi - \psi')}{r^2 \sin \theta}.$$

If $x' = y' = 0$; then $\theta' = 0$, the angle ψ is ignorable, and we have

$$\delta(x)\delta(y)\delta(z - z') = \frac{\delta(r - r')\delta(\theta)}{2\pi r^2 \sin \theta}.$$

If $x' = y' = z' = 0$; then $r' = 0$, the angles θ and ψ are ignorable, and we have

$$\delta(x)\delta(y)\delta(z) = (4\pi r^2)^{-1} \delta(r).$$

PROBLEMS

5.22. Verify the above relations between δ-functions in rectangular and spherical coordinates by operating on both sides with testing functions.

5.23. Solve the problem $g_{xx} + g_{yy} = \delta(x - x')\delta(y - y')$ by introducing polar coordinates. (*Hint.* Use (5.58) to transform the δ-functions, and then consider $\dfrac{\partial^2}{\partial \theta^2}$ as a constant.)

5.24. Show that the solution of Problem 5.23 is just the expansion in a Fourier series of $(4\pi)^{-1} \log (r^2 + r'^2 - 2rr' \cos \psi)$, where $r^2 = x^2 + y^2$, $r'^2 = x'^2 + y'^2$, and where ψ is the angle between the line joining the origin to x, y and the line joining the origin to x', y'. (*Hint.* $\psi = \theta - \theta'$, where $x = r \cos \theta$, $x' = r' \cos \theta'$.)

Initial-Value Problems

The partial differential equations we have considered up to now were all of the elliptic type, and therefore the appropriate conditions on the solutions were boundary conditions. In this section and the next, we shall consider parabolic and hyperbolic type partial differential equations, and therefore the solutions will be required to satisfy certain initial conditions with respect to the time besides, possibly, boundary conditions. The methods we shall use to solve these equations will be the same as those we have used so far inasmuch as we shall assume one part of the operator constant and solve the remaining part as an ordinary differential equation. Again, by treating different parts of the operator as constant, we shall obtain different representations for the solution, and we shall find that each representation is useful in a certain region.

There will be, however, two main differences. First, we shall be using the operator $\dfrac{\partial}{\partial t}$ or $\dfrac{\partial^2}{\partial t^2}$, with the conditions $u(0) = 0$ or $u(0) = u'(0) = 0$, respectively. The spectral representation for this operator will be obtained from the following formula, which was given also at the end of Chapter 4:

$$\delta(t - t') = \frac{1}{2\pi i}\int_{a-i\infty}^{a+i\infty} e^{p(t-t')}\, dp, \tag{5.60}$$

where $a > 0$. This formula contains implicitly the *Laplace transform* theorem. For, let $f(t)$ be a function of integrable square over $(0, \infty)$; then, multiplying (5.60) by $f(t')\, dt'$ and integrating over $(0, \infty)$, we get

$$f(t) = \frac{1}{2\pi i}\int_{a-i\infty}^{a+i\infty} e^{pt} g(p)\, dp,$$

where

$$g(p) = \int_0^\infty f(t')e^{-pt'}\, dt'.$$

We call $g(p)$ the *Laplace transform* of $f(t)$ and $f(t)$ the *inverse Laplace transform* of $g(p)$. Note that, if $f(t)$ is in the domain of the operator $\dfrac{\partial}{\partial t}$, that is, if $f(0) = 0$ and if $f'(t)$ is of integrable square over $(0, \infty)$, then

$$f'(t) = \frac{1}{2\pi i}\int_{a-i\infty}^{a+i\infty} pe^{pt} g(p)\, dp. \tag{5.61}$$

This shows that (5.60) and (5.61) give the spectral representation for $\dfrac{\partial}{\partial t}$.

Consider the extended definition of $\dfrac{\partial}{\partial t}$. We shall denote the ordinary derivative of a differentiable function $f(t)$ by $f'(t)$. Then, to find the extended derivative $\dfrac{\partial f}{\partial t}$, we have, since the adjoint operator is $-\dfrac{\partial}{\partial t}$, acting on functions which vanish in a neighborhood of ∞, that

$$\int_0^\infty \frac{\partial f}{\partial t} g(t)\,dt = \int_0^\infty f\left(-\frac{\partial g}{\partial t}\right) dt = f(0)g(0) + \int_0^\infty gf'\,dt,$$

where $g(t)$ is any function in the domain of the adjoint operator $-\dfrac{\partial}{\partial t}$. This shows that

$$\frac{\partial f}{\partial t} = f(0)\delta(t) + f'(t).$$

Consequently, (5.61) may be generalized as follows:

$$\frac{\partial}{\partial t} f = f'(t) + f(0)\delta(t) = \frac{1}{2\pi i}\int_{a-i\infty}^{a+i\infty} pe^{pt} g(p)\,dp.$$

This result is easily verified when $f(t) = 1$ because then we have $\dfrac{\partial f}{\partial t} = \delta(t)$ and $g(p) = p^{-1}$.

The second difference from the previous theory is that the Green's function for the wave equation in more than one space variable will be a symbolic function instead of an ordinary function. This is in contrast to all our previous examples in which the Green's functions were ordinary functions. We shall discuss this fact in greater detail when we come to it.

Consider the following problem of the parabolic type:

Find $u(x, t)$ such that

$$u_{xx} = u_t, \quad -\infty < x < \infty, \quad 0 \leq t < \infty, \tag{5.62}$$

such that u satisfies the initial condition

$$u(x, 0) = \delta(x - x'), \tag{5.63}$$

and such that $u(x, t)$ is bounded for all t as x approaches $\pm\infty$.

We solve this first by putting $N_1 u = u_{xx}$ and treating N_1 as a constant. The solution of the equation

$$u_t = N_1 u,$$

with the condition $u(0) = \delta(x - x')$, is

$$u = e^{N_1 t}\delta(x - x'),$$

or, using the spectral representation of N_1, we get

$$u = \frac{1}{2\pi}\int_{-\infty}^{\infty} e^{-k^2 t} e^{ik(x-x')}\,dk.$$

This integral may be evaluated by completing the square in the exponential. We have

$$\begin{aligned} u &= \frac{1}{2\pi}\int_{-\infty}^{\infty} \exp\left[-t\left(k - \frac{i(x-x')}{2t}\right)^2\right] dk \exp\left[-(4t)^{-1}(x-x')^2\right] \\ &= (4\pi t)^{-1/2} \exp\left[-(4t)^{-1}(x-x')^2\right]. \end{aligned}$$

This last result can be used to solve the following initial-value problem for the heat equation:

Find $w(x, t)$ such that

$$w_{xx} = w_t, \quad -\infty < x < \infty, \quad 0 \leq t < \infty,$$

such that

(5.64) $$w(x, 0) = f(x),$$

and such that $w(x, t)$ is bounded as x approaches $\pm\infty$.

We notice that, if we multiply (5.63) by $f(x')$ and integrate from $-\infty$ to ∞, it becomes (5.64); therefore we get

$$\begin{aligned} w(x, t) &= \int_{-\infty}^{\infty} u(x, t) f(x')\, dx' \\ &= (4\pi t)^{-1/2} \int_{-\infty}^{\infty} f(x') \exp\left[-\frac{(x - x')^2}{4t}\right] dx', \end{aligned}$$

the desired solution.

For the sake of illustration, we shall solve (5.62) in another way by putting $P = \dfrac{\partial}{\partial t}$ and treating P as a constant. Since u is not zero for $t = 0$, the extended definition of $\dfrac{\partial}{\partial t}$ shows that (5.62) becomes

(5.65) $$Pu = u_{xx} + \delta(x - x')\delta(t).$$

The solution of this is

$$u = \frac{e^{-\sqrt{P}|x-x'|}}{2\sqrt{P}}\, \delta(t);$$

or, using the spectral representation of P, it becomes

$$u = \frac{1}{2\pi i} \int_{a-i\infty}^{a+i\infty} e^{pt} e^{-\sqrt{p}|x-x'|} \frac{dp}{2\sqrt{p}}.$$

Putting $p = -k^2$ and shifting the path of integration will reduce this integral to the previous form.

Finally, we may solve (5.62) by treating both N_1 and P as constants. This technique will be useful in a later problem. From (5.65) we have

$$(P - N_1)u = \delta(x - x')\delta(t);$$

therefore,

$$u = \frac{1}{P - N_1}\, \delta(x - x')\delta(t)$$

and, using the spectral representations of both P and N_1, we get

(5.66) $$u = \frac{1}{2\pi} \int_{-\infty}^{\infty} \frac{1}{2\pi i} \int_{a-i\infty}^{a+i\infty} \frac{1}{p + k^2} e^{pt} e^{ik(x-x')}\, dk\, dp.$$

If we carry out the integration with respect to either p or k, we would have a previously obtained result. Instead of doing this, we write

$$(p + k^2)^{-1} = \int_0^\infty e^{-\tau(p+k^2)}\, d\tau \tag{5.67}$$

and substitute it in (5.66). The reason for this substitution is that the resulting integral can be evaluated easily with respect to p, τ, and k. We shall see that a similar substitution will enable us to solve the n-dimensional heat equation (see Problem 5.27).

We wish to evaluate the integral for u, namely,

$$u = \frac{1}{4\pi^2 i}\int_{-\infty}^{\infty}\int_0^\infty \int_{-i\infty}^{i\infty} e^{p(t-\tau)}e^{-\tau k^2}e^{ik(x-x')}\, dk\, d\tau\, dp.$$

If we put $p = iq$, we find that the p-integral is

$$i\int_{-\infty}^{\infty} e^{iq(t-\tau)}\, dq = 2\pi i\delta(t-\tau).$$

Evaluating the τ-integral next, we get

$$u(x, t) = \frac{1}{2\pi}\int_{-\infty}^{\infty} \exp\,[ik(x - x') - k^2 t]\, dk,$$

and just as before we find

$$u(x, t) = (4\pi t)^{-1/2}\exp\,[-(4t)^{-1}(x - x')^2].$$

PROBLEMS

5.25. If $P = \dfrac{\partial}{\partial t}$, use the extended definition of the operator to show that $P^2 f = f''(t) + f(0)\delta'(t) + f'(0)\delta(t)$.

5.26. Solve $g_{xx} - g_t = \delta(x - x')\delta(t - t')$ if $g(x, 0) = 0$ and $g(x, t)$ is bounded as x approaches $\pm\infty$.

5.27. Show that the function $u(x_1, \cdots, x_n, t)$ which satisfies

$$\Delta u = u_t, \quad -\infty < x_1, \cdots, x_n < \infty, \quad 0 \le t < \infty$$

and the initial condition

$$u(x_1, \cdots, x_n, 0) = \delta(x_1 - x_1') \cdots \delta(x_n - x_n')$$

and which is such that $u(x_1, \cdots, x_n, t)$ is bounded for all t as any x_j $(j = 1, \cdots, n)$ approaches $\pm\infty$ is

$$u = (4\pi t)^{-n/2}\exp\,[-(4t)^{-1}|r - r'|^2].$$

Here, Δ is the n-dimensional Laplacian and r, r' are n-dimensional vectors with components $x_1, \cdots, x_n$ and $x_1', \cdots, x_n'$, respectively. (*Hint.* Consider both Δ and P as constants. Use the fact that the spectral representation of Δ is given by the n-dimensional Fourier transform. Replace $(\Delta + P)^{-1}$ by an integral as was done in the text with (5.67). Carry out the P and τ integrals, and the remaining integrals reduce to a product of integrals of the same type.)

The Green's Function for the Wave Equation

We shall apply the methods of the preceding section to find the Green's function for the wave equation in one space dimension, that is, to find the solution of the equation

$$g_{xx} - g_{tt} = -\delta(x - x')\delta(t - t') \tag{5.68}$$

such that

$$g(x, 0) = g_t(x, 0) = 0$$

and such that $g(x, t)$ is bounded as x approaches $\pm\infty$. Put $P = \dfrac{\partial}{\partial t}$; then (5.68) becomes

$$g_{xx} - P^2 g = -\delta(x - x')\delta(t - t').$$

The solution of this equation is

$$g = \frac{e^{-P|x-x'|}}{2P}\delta(t - t')$$

or

$$g = \frac{1}{2\pi i}\int_{-i\infty}^{i\infty} \exp\left[p(t - t') - p|x - x'|\right]\frac{dp}{2p} \tag{5.69}$$

if the spectral representation of P is used. Note that the contour of integration should be indented around the origin on the right side since the inverse Laplace transform should have the path of integration from $a - i\infty$ to $a + i\infty$, where $a > 0$.

The integral (5.69) may be evaluated by closing the contour with an infinite semicircle in the right or left half plane according as the coefficient of p in the exponent is negative or positive. We find that

$$g = \tfrac{1}{2} H[(t - t') - |x - x'|], \tag{5.70}$$

where $H(x)$ is the Heaviside unit function, equal to one if $x > 0$, equal to zero otherwise.

This expression (5.70) is called the *Riemann function* for the wave equation. The formula shows that $g = 0$ unless the point (x, t) is inside the "*characteristic cone*" defined by the inequality

$$t - t' > |x - x'|.$$

The angle of the "cone" depends upon the wave velocity; that is, if equation (5.68) were

$$g_{xx} - c^{-2}g_{tt} = -\delta(x - x')\,\delta(t - t'),$$

where c is the wave velocity, then the characteristic cone would be given by the inequality

$$c(t - t') > |x - x'|.$$

We shall now consider the wave equation in three space dimensions. We wish to find a function $g(x_1, x_2, x_3, t)$ satisfying the equation

(5.71) $$\Delta g - g_{tt} = -\delta(x_1 - x_1')\,\delta(x_2 - x_2')\,\delta(x_3 - x_3')\,\delta(t - t'),$$

satisfying the initial conditions

$$g(x_1, x_2, x_3, 0) = g_t(x_1, x_2, x_3, 0) = 0$$

and such that $g(x_1, x_2, x_3, t)$ is bounded for all t as either x_1, x_2, or x_3 approaches $\pm\infty$.

Consider Δ and $P = \dfrac{\partial}{\partial t}$ as constants; then

$$g = -(\Delta - P^2)^{-1}\,\delta(x_1 - x_1')\,\delta(x_2 - x_2')\,\delta(x_3 - x_3')\,\delta(t - t').$$

Since the spectral representation of Δ is given by the three-dimensional Fourier transform, we see that

(5.72)

$$g = (2\pi)^{-3}(2\pi i)^{-1}\iiint_{-\infty}^{\infty} dk_1\,dk_2\,dk_3 \int_{-i\infty}^{i\infty} dp\,\frac{\exp\,[p(t - t') - i\mathbf{k}\cdot(\mathbf{x} - \mathbf{x}')]}{p^2 + k^2}.$$

Here, $\mathbf{k}$, $\mathbf{x}$, and $\mathbf{x}'$ are vectors with components (k_1, k_2, k_3), (x_1, x_2, x_3), and (x_1', x_2', x_3'), respectively; also

$$k^2 = k_1^2 + k_2^2 + k_3^2.$$

The p-integral may be evaluated by closing the contour in the right or left half-plane according as the difference $t - t'$ is negative or positive. Since the p-integration is to the right of the imaginary axis, we find that for $t < t'$ (5.72) is zero, but for $t > t'$ it becomes

(5.73) $$g = (2\pi)^{-3}\iiint_{-\infty}^{\infty} dk_1\,dk_2\,dk_3\,(2ik)^{-1}\,[\exp\{ik(t - t') - i\mathbf{k}\cdot(\mathbf{x} - \mathbf{x}')\} - \exp\{-ik(t - t') - i\mathbf{k}\cdot(\mathbf{x} - \mathbf{x}')\}].$$

To finish the evaluation, we introduce spherical coordinates in the k-integrals. We take the polar axis along the direction of the vector $\mathbf{x} - \mathbf{x}'$; then

$$\mathbf{k}\cdot(\mathbf{x} - \mathbf{x}') = k|x - x'|\cos\theta$$

where θ is the polar angle. The volume element $dk_1\,dk_2\,dk_3$ becomes $k^2\,dk\sin\theta\,d\theta\,d\psi$, where ψ is the meridional angle. Equation (5.73) becomes

(5.74)

$$g = (2\pi)^{-3}(2i)^{-1}\int_0^{\infty} k\,dk \int_0^{\pi} \sin\theta\,d\theta \int_0^{2\pi} d\psi\,[\exp\{ik(t - t') - ik|\mathbf{x} - \mathbf{x}'|\cos\theta\} - \exp\{-ik(t - t') - ik|\mathbf{x} - \mathbf{x}'|\cos\theta\}]H(t - t').$$

The θ and ψ integrations can be performed immediately, and the above expression reduces to the following:

$$(5.75)\quad g = (2\pi)^{-2}(|x - x'|)^{-1}\int_0^\infty dk\,[\cos k\{(t - t') - |\mathbf{x} - \mathbf{x}'|\} - \cos k\{(t - t') + |\mathbf{x} - \mathbf{x}'|\}]H(t - t')$$

$$= \frac{1}{4\pi}\,\frac{\delta(t - t' - |\mathbf{x} - \mathbf{x}'|) - \delta(t - t' + |\mathbf{x} - \mathbf{x}'|)}{|\mathbf{x} - \mathbf{x}'|}\,H(t - t'),$$

a formula given by Dirac.

It should be noted that the Green's function given by (5.75) is really a symbolic function representating a distribution. In the other examples of Green's functions, it was always a piecewise analytic function. The difference between this example and the others is that now we are dealing with a hyperbolic differential equation. It can be shown that the Green's function for an elliptic equation is analytic whereas the Green's function for a hyperbolic equation is, in general, a symbolic function representing a distribution.†

PROBLEMS

5.28. Use (5.70) to find the function $u(x, t)$ such that $u_{xx} - u_{tt} = f(x, t)$, $-\infty < x < \infty$, $0 \leq t < \infty$ and such that $u(x, 0) = u_t(x, 0) = 0$.

5.29. Solve the following initial-value problem: $u_{xx} - u_{tt} = 0$, $-\infty < x < \infty$, $0 \leq t < \infty$; $u(x, 0) = a(x)$, $u_t(x, 0) = b(x)$. (*Hint.* Treat $P = \dfrac{\partial}{\partial t}$ as a constant and use Problem 5.25.)

5.30. Solve $\Delta u = u_{tt}$, $-\infty < x_1, x_2, x_3 < \infty$, $0 \leq t < \infty$, and $u(x_1, x_2, x_3, 0) = a(x_1, x_2, x_3)$, $u_t(x_1, x_2, x_3, 0) = b(x_1, x_2, x_3)$. (*Hint.* Treat Δ as a constant.)

† See L. Schwartz, *Théorie des distributions*, Vol. 2, *Actualitiés scientifiques et industrieles*, 1091 and 1122, Hermann & Cie, Paris, 1950, 1951.

BIBLIOGRAPHY

Banach, S., *Théorie des opérations linéaires*, Warszawa, 1932.

Bateman, H., *Partial Differential Equations of Mathematical Physics*, Dover, New York, 1944.

Birkhoff, G., and S. MacLane, *A Survey of Modern Algebra*, Macmillan, New York, 1948.

Courant, R., *Differential and Integral Calculus*, Interscience, New York, 1936.

Courant, R., and D. Hilbert, *Methoden der Mathematischen Physik*. Springer, Berlin, 1931, 2 vols. (Reprint, Interscience, New York, 1943.)

Dirac, P. A. M., *The Principles of Quantum Mechanics*, Clarendon Press, Oxford, 1947.

Halmos, P. R., *Finite Dimensional Vector Spaces*, Princeton University Press, Princeton, 1942.

Halmos, P. R., *Introduction to Hilbert Space*, Chelsea Publishing Co., New York, 1951.

Hille, E., *Functional Analysis and Semi-groups*, American Mathematical Society, New York, 1948.

Hobson, E. W., *Theory of Functions of a Real Variable*, Cambridge University Press, London, 1926.

Ince, E. L., *Ordinary Differential Equations*, Dover, New York, 1944.

Jeffreys, H., and B. S. Jeffreys, *Methods of Mathematical Physics*, Cambridge University Press, London, 1950.

Kellogg, *Foundations of Potential Theory*, Springer, Berlin, 1929.

Morse, P., *Vibration and Sound*, McGraw-Hill Book Co., New York, 1948.

Morse, P., and H. Feshbach, *Methods of Theoretical Physics*, McGraw-Hill Book Co., New York, 1953.

Murnaghan, F. D., *Theory of Group Representations*, John Hopkins Press, Baltimore, 1938.

Nagy, B. v. Sz., *Spektraldarstellung Linearer Transformationen des Hilbertschen Raumes*, Springer, Berlin, 1942.

Schiff, L., *Quantum Mechanics*, McGraw-Hill Book Co., New York, 1949.

Schwartz, L., *Théorie des distributions*, Tome I et II, *Actualitiés scientifiques et industrielles* 1091 and 1122, Hermann & Cie, Paris, 1950, 1951.

Stone, M. H., *Linear Transformations in Hilbert Space and Their Applications to Analysis*, American Mathematical Society, New York, 1932.

Stratton, J. A., *Electromagnetic Theory*, McGraw-Hill Book Co., New York, 1941.

Titchmarsh, E. C., *Theory of Functions*, Clarendon Press, Oxford, 1939.

Titchmarsh, E. C., *Eigenfunction Expansions Associated with Second Order Differential Equations*, Clarendon Press, Oxford, 1946.

Whittaker, E. T., and G. N. Watson, *A Course of Modern Analysis*, Cambridge University Press, London, 1943.

INDEX

A CATALOG OF SELECTED
DOVER BOOKS
IN SCIENCE AND MATHEMATICS

A CATALOG OF SELECTED

DOVER BOOKS

IN SCIENCE AND MATHEMATICS

SPECIAL FUNCTIONS, N.N. Lebedev. Translated by Richard Silverman. Famous Russian work treating more important special functions, with applications to specific problems of physics and engineering. 38 figures. 308pp. 5⅜ × 8½.
60624-4 Pa. $6.95

OBSERVATIONAL ASTRONOMY FOR AMATEURS, J.B. Sidgwick. Mine of useful data for observation of sun, moon, planets, asteroids, aurorae, meteors, comets, variables, binaries, etc. 39 illustrations 384pp. 5⅜ × 8¼. (Available in U.S. only)
24033-9 Pa. $5.95

INTEGRAL EQUATIONS, F.G. Tricomi. Authoritative, well-written treatment of extremely useful mathematical tool with wide applications. Volterra Equations, Fredholm Equations, much more. Advanced undergraduate to graduate level. Exercises. Bibliography. 238pp. 5⅜ × 8½.
64828-1 Pa. $6.95

CELESTIAL OBJECTS FOR COMMON TELESCOPES, T.W. Webb. Inestimable aid for locating and identifying nearly 4,000 celestial objects. 77 illustrations. 645pp. 5⅜ × 8½.
20917-2, 20918-0 Pa., Two-vol. set $12.00

MODERN NONLINEAR EQUATIONS, Thomas L. Saaty. Emphasizes practical solution of problems; covers seven types of equations. ". . . a welcome contribution to the existing literature. . . ."—*Math Reviews*. 490pp. 5⅜ × 8½.
64232-1 Pa. $9.95

FUNDAMENTALS OF ASTRODYNAMICS, Roger Bate et al. Modern approach developed by U.S. Air Force Academy. Designed as a first course. Problems, exercises. Numerous illustrations. 455pp. 5⅜ × 8½.
60061-0 Pa. $8.95

INTRODUCTION TO LINEAR ALGEBRA AND DIFFERENTIAL EQUATIONS, John W. Dettman. Excellent text covers complex numbers, determinants, orthonormal bases, Laplace transforms, much more. Exercises with solutions. Undergraduate level. 416pp. 5⅜ × 8½.
65191-6 Pa. $8.95

INCOMPRESSIBLE AERODYNAMICS, edited by Bryan Thwaites. Covers theoretical and experimental treatment of the uniform flow of air and viscous fluids past two-dimensional aerofoils and three-dimensional wings; many other topics. 654pp. 5⅜ × 8½.
65465-6 Pa. $14.95

INTRODUCTION TO DIFFERENCE EQUATIONS, Samuel Goldberg. Exceptionally clear exposition of important discipline with applications to sociology, psychology, economics. Many illustrative examples; over 250 problems. 260pp. 5⅜ × 8½.
65084-7 Pa. $6.95

LAMINAR BOUNDARY LAYERS, edited by L. Rosenhead. Engineering classic covers steady boundary layers in two- and three-dimensional flow, unsteady boundary layers, stability, observational techniques, much more. 708pp. 5⅜ × 8½.
65646-2 Pa. $15.95

LECTURES ON CLASSICAL DIFFERENTIAL GEOMETRY, Second Edition, Dirk J. Struik. Excellent brief introduction covers curves, theory of surfaces, fundamental equations, geometry on a surface, conformal mapping, other topics. Problems. 240pp. 5⅜ × 8½.
65609-8 Pa. $6.95